An Introduction to

GAUGE-HIGGS UNIFICATION

Extra Dimensions in Particle Physics

An Introduction to GAUGE-HIGGS UNIFICATION

Extra Dimensions in Particle Physics

Yutaka Hosotani
Osaka University, Japan

World Scientific

NEW JERSEY • LONDON • SINGAPORE • BEIJING • SHANGHAI • HONG KONG • TAIPEI • CHENNAI

Published by

World Scientific Publishing Co. Pte. Ltd.
5 Toh Tuck Link, Singapore 596224
USA office: 27 Warren Street, Suite 401-402, Hackensack, NJ 07601
UK office: 57 Shelton Street, Covent Garden, London WC2H 9HE

Library of Congress Control Number: 2024946691

British Library Cataloguing-in-Publication Data
A catalogue record for this book is available from the British Library.

AN INTRODUCTION TO GAUGE-HIGGS UNIFICATION
Extra Dimensions in Particle Physics

ISBN 978-981-98-0097-1 (hardcover)
ISBN 978-981-98-0098-8 (ebook for institutions)
ISBN 978-981-98-0099-5 (ebook for individuals)

For any available supplementary material, please visit
https://www.worldscientific.com/worldscibooks/10.1142/14045#t=suppl

Preface

In 2012 the Higgs boson (the Higgs particle) was discovered at the LHC (Large Hadron Collider) at CERN (European Organization for Nuclear Research) in Switzerland. The existence of the Higgs boson has been firmly established by numerous experiments since then. In the standard model of particle physics the Higgs boson had been the last piece to be found in addition to leptons, quarks and gauge bosons. The discovery of the Higgs boson, besides being the discovery of a new particle, has profound implications to particle physics.

It is often said in the news media that the Higgs boson explains the origin of masses in the universe. This is quite misleading. Most masses of matter which we see in our daily life are explained in terms of the mass of protons and neutrons. It is known that only 2% of the mass of protons and neutrons is due to the Higgs boson. The rest is generated by strong interactions. This fact had been pointed out by Y. Nambu before P. Higgs proposed his theory.

The importance of the discovery of the Higgs boson lies in a different place. The Higgs boson is absolutely necessary in the standard model in which electromagnetic and weak forces are unified. In order for two distinct forces to be unified in a single force, one needs a mechanism by which one force differentiates into two distinct forces at low energies. The Higgs boson plays an essential role in this process. In other words, the discovery of the Higgs boson has established the scenario of the unification of forces.

However, we still do not have good understanding of what the Higgs boson is. From the mathematical point of view, the process of a single force differentiating into two distinct forces is a process in which gauge symmetry partially breaks down spontaneously. The Higgs mechanism is a mechanism for spontaneous gauge symmetry breaking, and is employed in the standard model. The details of the mechanism are yet to be explored experimentally.

There is another mechanism for spontaneous gauge symmetry breaking. It is the Hosotani mechanism in which gauge symmetry in four-dimensional spacetime is spontaneously broken by dynamics of the fifth-dimensional component of gauge fields. The mechanism is the generalization of the Aharonov-Bohm effect in quantum mechanics to the fifth dimension. The Higgs boson appears as a four-dimensional fluctuation mode of Aharonov-Bohm phases. Gauge fields and the Higgs boson are unified, thus the scenario being termed gauge-Higgs unification. The interaction of the Higgs boson is controlled by the gauge principle. The finite mass of the Higgs boson is generated by quantum effects.

The Hosotani mechanism was discovered in 1983. At the time it was difficult to describe the chiral nature of the weak interactions of quarks and leptons, and no realistic models were constructed. Since then good progress has been made in field theory on orbifolds and the Randall-Sundrum warped space. In the 2010's realistic gauge-Higgs unification models have become available.

In this book the Hosotani mechanism and gauge-Higgs unification scenario are explained from the basics. The book is intended both for advanced undergraduate and graduate students, as well as for frontier researchers. Minimal knowledge of field theory is assumed. After reviewing the Higgs mechanism and the standard model in Chapter 2, we introduce the Hosotani mechanism in Chapter 3 which generalizes the Aharonov-Bohm effect to non-Abelian gauge theory. The effective potential is introduced in Chapter 4, with which we see how dynamical gauge symmetry breaking occurs by the Hosotani mechanism. Gauge symmetry breaking on the lattice is also examined.

In Chapter 5 general properties of field theory in higher dimensions are reviewed. In addition to the universal extra dimension model, we examine a few gauge-Higgs unification models on the flat $M^4 \times (S^1/Z_2)$ orbifold. Difficulties in having realistic gauge-Higgs unification models on flat orbifolds are recognized.

To overcome those difficulties one needs to consider field theory in the Randall-Sundrum (RS) warped space. In Chapter 6 the RS space is introduced as a solution to the Einstein equations, and general properties of field theory in the RS space are investigated. Minimal knowledge of general relativity is sufficient to handle field theory in the RS space. In the flat space trigonometric functions ($\sin z$ and $\cos z$) play the basic role. In the

RS space they are generalized to four kinds of basis functions expressed in terms of the Bessel functions. With these basis functions all manipulations in the RS space are simplified.

In Chapter 7 gauge-Higgs unification models in the RS space are introduced. $SO(5) \times U(1) \times SU(3)$ models inspired by grand unification scenario are described in detail. The mass spectrum of all fields and particles in the standard model is reproduced. The effective potential is evaluated and it is shown that the electroweak symmetry is dynamically broken by the Hosotani mechanism.

In Chapter 8 wave functions of all particles including Kaluza-Klein excited modes are determined, with which all of the gauge couplings and Higgs couplings are evaluated. It is shown that the resultant phenomenology at low energies is nearly the same as that in the standard model.

Many predictions follow from gauge-Higgs unification. In Chapter 9 we derive predictions for the W boson mass, Higgs self-couplings and Z' bosons, which await for confirmation in future experiments. Furthermore theory of gauge-Higgs unification has many remarkable properties from the theoretical point of view, which include special coupling sum rules and the phenomenon of anomaly flow. Gauge-Higgs grand unification has been attempted, too.

Theory of gauge-Higgs unification is far from complete. Physics of the Hosotani mechanism has not been well understood. I wish this book serves as a good useful guide for further exploration.

Yutaka Hosotani
July 2024

Contents

Preface v

1. Introduction 1

1.1 Gauge theory and forces 1
1.2 Unification of forces 2
1.3 Mechanisms of spontaneous symmetry breaking 4

2. Higgs Mechanism 7

2.1 Spontaneous symmetry breaking 7
2.2 Gauge invariance 9
2.3 Higgs mechanism 10
2.4 Higgs mechanism in non-Abelian gauge theory 11
2.5 Standard model 15
2.6 W boson, Z boson and Higgs boson 17

3. Aharonov-Bohm Effect and Hosotani Mechanism 23

3.1 Aharonov-Bohm effect in quantum mechanics 23
3.2 Gauge theory on $M^n \times S^1$ 25
3.3 Aharonov-Bohm phases in non-Abelian gauge theory . . . 27
3.4 Hosotani mechanism I – gauge symmetry breaking 28

4. Quantum Effects and Hosotani Mechanism 33

4.1 Effective potential in quantum mechanics 33
4.2 Effective potential in field theory 34
4.3 Effective potential for AB phases $V_{\rm eff}(\theta_H)$ 37

4.4 Hosotani mechanism II – dynamical gauge symmetry breaking 38
4.5 Masses of Higgs modes (A_y excitations) 42
4.6 Equivalence classes of boundary conditions and AB phases 44
4.7 Finiteness of $V_{\rm eff}(\theta_H)$ 45
4.8 QED and QCD at finite temperature 47
4.9 Gauge symmetry breaking on the lattice 49
4.9.1 Polyakov loop and distribution density of AB phases 50
4.9.2 Gauge-invariant gauge symmetry breaking 52

5. Gauge-Higgs Unification in Flat Space 57

5.1 $SU(5)$ grand unified theory on $M^4 \times S^1$ 57
5.2 Fermions in higher dimensions 59
5.3 Orbifold $M^4 \times (S^1/Z_2)$ 61
5.4 Chiral fermions 63
5.5 Gauge fields 64
5.6 Non-Abelian gauge fields and orbifold boundary conditions 65
5.7 AB phases and equivalence classes of boundary conditions 67
5.8 Model of universal extra dimensions (UED) 70
5.9 Kawamura model of $SU(5)$ grand unification 73
5.10 Electroweak unification – $SU(3)$ gauge-Higgs unification . 75
5.11 Twisted gauge in flat space 81

6. Randall-Sundrum Warped Space 85

6.1 Einstein equations 85
6.2 Randall-Sundrum solution 87
6.3 Scalar fields in the RS space 89
6.4 Gauge fields in the RS space 93
6.5 Fermion fields in curved spacetime 99
6.6 Fermion fields in the RS space 101

7. Gauge-Higgs Unification in the RS Space 109

7.1 Goal of gauge-Higgs unification 109
7.2 $SO(5) \times U(1)_X \times SU(3)_C$ GHU 111
7.3 Higgs boson and AB phase θ_H 115
7.4 Quarks 117
7.5 Leptons 120
7.6 Brane scalar and brane interactions 121

7.7 Dark fermions 124
7.8 Twisted gauge in the RS space 125
7.9 Mass spectrum of gauge fields 127
7.9.1 Gauge fields: A_μ and B_μ 127
7.9.2 Gauge fields: A_z and B_z 131
7.10 Mass spectrum of fermion fields 133
7.10.1 Quarks 133
7.10.2 Leptons 138
7.10.3 Dark fermions 141
7.11 Effective potential $V_{\rm eff}(\theta_H)$ and electroweak symmetry breaking 146

8. New Particles and Interactions 153

8.1 θ_H universality 153
8.2 Effective interactions of the Higgs field 157
8.3 Wave functions 158
8.3.1 Gauge fields 158
8.3.2 Quarks 164
8.3.3 Leptons 166
8.3.4 Dark fermions 168
8.3.5 Chiral structure in the fermion zero modes 170
8.4 Gauge couplings 173
8.4.1 Self-couplings of gauge fields 173
8.4.2 Gauge couplings of fermions 178
8.5 Higgs couplings 186
8.5.1 Gauge-Higgs couplings 186
8.5.2 Yukawa couplings 188

9. Predictions 191

9.1 The Fermi constant and W boson mass 191
9.2 Decay of the Higgs boson 196
9.3 Self-couplings of the Higgs boson 199
9.4 Production of W' and Z' bosons 199
9.5 Z' interference effect in e^-e^+ collisions 201
9.6 Coupling sum rules 207
9.6.1 Unitarity at the tree level 207
9.6.2 Oblique corrections at the 1-loop level 209
9.6.3 Gauge anomalies 213

9.7 Anomaly flow . . . 213
9.7.1 Anomaly coefficients . . . 216
9.7.2 Anomaly flow . . . 218
9.7.3 Anomaly formula . . . 219
9.8 Gauge-Higgs grand unification . . . 222
9.8.1 $SU(6)$ unification . . . 223
9.8.2 $SO(11)$ unification . . . 224
9.8.3 Fermions and a brane scalar . . . 225

Appendix A Natural Units, Dirac Matrices and $SO(N)$ 229

A.1 Metric, natural units, Dirac matrices . . . 229
A.2 $SO(N)$ and Clifford algebra . . . 230
A.3 $SU(N)$ . . . 236

Appendix B Effective Potential 237

B.1 Regularization . . . 237
B.2 Infinite sum in flat space . . . 239
B.3 Integral representation of $V_{\rm eff}(\theta_H)$ . . . 240
B.4 Finiteness of $V_{\rm eff}^{1\,\rm loop}(\theta_H)$ . . . 243

Appendix C Bessel Functions 245

C.1 Properties of Bessel functions . . . 245
C.2 Basis functions in the RS space . . . 248
C.3 Modified Bessel functions . . . 253

Bibliography 255

Index 261

About the Author 265

Chapter 1

Introduction

What are the smallest building blocks of matter – elementary particles? How do they interact with each other? What kinds of forces are there? Particle physics tries to answer to these questions. Now unification of forces is in sight.

1.1 Gauge theory and forces

There are at least four kinds of forces which elementary particles have. They are gravitational force (gravity), electromagnetic force, strong force, and weak force. Among them gravitational and electromagnetic forces are observed in daily life as forces among macroscopic objects. On the other hand strong and weak forces are not recognized at the macroscopic level, but they are important in controlling the world of elementary particles. Particle physics clarifies and explains all these forces at the most fundamental level.

Gravity is described by Newton's law of universal gravitation and by Einstein's theory of general relativity. Electromagnetic forces are described by Maxwell's theory, mediated by electric fields $\vec{E}(x)$ and magnetic fields $\vec{B}(x)$. Electric and magnetic fields combine to form electromagnetic fields, which evolve in time, following the Maxwell equations. Further, by expressing electromagnetic fields $\vec{E}, \vec{B}$ in terms of gauge potential A_μ as $E_j = F_{0j}, B_j = \frac{1}{2}\epsilon^{jk\ell}F_{k\ell}$ where $F_{\mu\nu} = \partial_\mu A_\nu - \partial_\nu A_\mu$, Maxwell's theory becomes a gauge theory. That is, the theory becomes invariant under a gauge transformation $A_\mu(x) \to A'_\mu(x) = A_\mu(x) + \partial_\mu \Lambda(x)$. By treating A_μ as dynamical variables, Maxwell's equations and Lorentz equation are derived from the principle of the least action. The gauge invariance guarantees the conservation of electric charges.

In quantum theory it becomes an absolute necessity to introduce gauge potentials A_μ. It is not possible to write equations of motion for matter in electromagnetic fields such as Schödinger equation and Dirac equation without A_μ. The electromagnetic force in quantum theory is formulated as gauge interactions. Let $\psi(x)$ be a matter field with an electric charge e. Under a gauge transformation the matter field is transformed as $\psi(x) \to \psi'(x) = \Omega(x)\psi(x)$ where $\Omega(x) = \exp\{i(e/\hbar c)\Lambda(x)\}$. $\Omega^\dagger\Omega = 1$ and $\Omega(x) \in U(1)$ in the terminology of group theory. We express that electromagnetic interactions are described by an $U(1)$ gauge theory.

Quantum electrodynamics (QED), which is an $U(1)$ gauge theory describing relativistic electrons and protons with electromagnetic interactions, has been confirmed experimentally to high accuracy. How about the strong and weak interactions? In the 1970's and 1980's it was established that the strong and weak interactions are described in terms of gauge theory as well. The strong force is described by an $SU(3)$ gauge theory. The weak force and electromagnetic force are unified and are described by an $SU(2)\times U(1)$ gauge theory. All these three forces are described as gauge theory.

1.2 Unification of forces

Formulation of gauge theory of strong and weak interactions was simultaneously carried out from the late 1960's to the early 1980's. It is based on non-Abelian gauge theory (Yang-Mills theory) which Yang and Mills proposed in 1954. [1] As explained in the previous section, the theory of QED is invariant under $U(1)$ gauge transformations of $\psi(x) \to \Omega(x)\psi(x)$ where $\Omega(x) \in U(1)$. Yang and Mills proposed a theory in which $\Omega(x) \in SU(2)$. At the time of the proposal the theory was not accepted as a physics theory describing nature. In the early 1970's the renormalizability of the Yang-Mills theory was proven. Since then the Yang-Mills theory has attracted more attention as a potential fundamental theory in particle physics.

Quite independently, it was established in the 1960's that hadrons such as protons, neutrons and mesons, which strongly interacts with each other, are made up of quarks. It turned out that quarks come in three colors, or as color triplets in $SU(3)$ whereas hadrons are all color singlets. In the late 1960's people started to suspect that strong interactions are gauge interactions coupling to color degrees of freedom. Furthermore it was shown in the 1970's that the Yang-Mills theory not only is renormalizable, but also has the property of asymptotic freedom with which the gauge coupling strength is getting weaker at short distances and at high energies. With these observations $SU(3)$ gauge theory (quantum chromodynamics, QCD) was formulated as a theory of strong interactions. The QCD property of asymptotic freedom was confirmed in electron-proton collision experiments at high energies. With further progress in experiments and advancement in lattice gauge theory there is no doubt that strong interactions are described by QCD.

One needs an additional device to formulate a theory of weak interactions. The reason why weak interactions are weak is that the mass of vector bosons mediating weak interactions is very large. The photon, which is the vector boson mediating electromagnetic interactions, is massless. On the other hand, the mass of the vector boson (W boson) mediating β-disintegration of nuclei and muons is about 90 times heavier than the mass of the proton. In gauge theory, however, gauge bosons (vector bosons) are massless.

A method to resolve this difficulty emerged from the study of strong interactions from the late 1950's to the early 1960's. Nambu proposed a scenario in which protons and neutrons are originally massless, and acquire finite masses as a consequence of strong interactions. Theory of massless fermions has chiral symmetry. Left-handed and right-handed fermions move as independent particles. As far as the chiral symmetry is preserved, the fermions remain massless. By analogy with the superconductivity Nambu foresaw that the chiral symmetry is spontaneously broken by strong interactions. As a consequence the masses of protons and neutrons are dynamically generated, and pions (π mesons) behave as nearly massless particles. It is known, at present, that 98% of the masses of protons and neutrons are explained by Nambu's mechanism. [2, 3]

The viewpoint that fundamental symmetry is spontaneously broken brought a revolution in particle physics. After Nambu and Jona-Lasinio, and later Goldstone, showed that symmetry is spontaneously broken in relativistic field theory, Higgs, Englert and Brown showed in gauge theory with scalar fields that gauge symmetry is spontaneously broken and gauge bosons become massive vector bosons. [4–6] This phenomenon is called the Higgs mechanism. It is applied to weak interactions.

There is one more step necessary to formulate a gauge theory of weak interactions. Instead of treating electromagnetic and weak interactions independently, one must unify these two kinds of the interactions in one single electroweak gauge theory. In the theory formulated by Glashow, Weinberg and Salam $SU(2)_L \times U(1)_Y$ gauge symmetry spontaneously breaks to $U(1)_{\rm EM}$ where $U(1)_{\rm EM}$ is the $U(1)$ gauge symmetry of the electromagnetic interactions whose gauge boson is the photon. In the $SU(2)_L \times U(1)_Y$ gauge theory there are four gauge bosons, three of which become the $W^\pm$ and Z bosons. [7–9]

The electromagnetic and weak forces are unified. The unification is achieved by starting from the $SU(2)_L \times U(1)_Y$ gauge theory. The $SU(2)_L \times U(1)_Y$ gauge symmetry spontaneously breaks down to the $U(1)_{\rm EM}$ gauge symmetry, thereby resulting in the observed electromagnetic and weak forces. In other words, the unification of forces and the spontaneous symmetry breaking are inseparably related to each other.

1.3 Mechanisms of spontaneous symmetry breaking

The model defined with the combination of the $SU(3)_C$ color gauge theory of strong interactions and the $SU(2)_L \times U(1)_Y$ electroweak unified gauge theory is called the standard model (SM). The SM has been confirmed by experiments and observation to high accuracy. The $SU(2)_L \times U(1)_Y$ gauge symmetry is spontaneously broken by the Higgs mechanism, with which the W and Z bosons acquire large masses and one neutral scalar, the Higgs boson, appears. The Higgs boson was discovered at the Large Hadron Collider (LHC) at CERN in Switzerland in 2012. [10, 11] The discovery of the Higgs boson confirms the scenario of the unification of forces.

However, the detail of the Higgs mechanism in the SM is yet to be scrutinized further. It is necessary to determine how the Higgs boson interact with itself and other particles more accurately. From a theoretical point of view, the part of the gauge interactions in the SM is beautiful as it is controlled by the gauge principle, but the part of the Higgs interactions is afflicted with arbitrariness and lacks a principle. It is far from the ultimate theory.

It is one of the most important subjects in contemporary physics to clarify the mechanism of spontaneous gauge symmetry breaking. Nambu and Jona-Lasinio found spontaneous symmetry breaking by fermion-pair condensation in analogy with the superconductivity. In the Higgs mechanism condensation of the Higgs scalar field replaces fermion-pair condensation.

There is an alternative mechanism of spontaneous gauge symmetry breaking by gauge fields in higher dimensions. It is supposed that there exists the fifth dimension in spacetime. The fifth-dimensional component of the gauge fields induces gauge symmetry breaking by an Aharonov-Bohm effect in the fifth dimension. [12] This mechanism is called the Hosotani mechanism, which is controlled by the gauge principle. By applying it to the electroweak unified theory, the Higgs boson becomes a part of the gauge fields. The scenario leads to gauge-Higgs unification, with which the gauge hierarchy problem in the SM is naturally resolved as well.

In this book we introduce the Hosotani mechanism as an alternative to the Higgs mechanism, and apply it to the electroweak interactions to formulate gauge-Higgs unification (GHU) models replacing the SM. The Hosotani mechanism is a mechanism of gauge symmetry breaking by gauge invariant quantities. GHU models formulated in the Randall-Sundrum warped space yield nearly the same phenomenology at low energies as the SM. It predicts new particles at high energies which can be confirmed in future collider experiments. The W boson mass is also predicted to be heavier than that in the SM.

Chapter 2

Higgs mechanism

The Higgs mechanism is a mechanism of spontaneous gauge symmetry breaking, and is applied to the standard model (SM). We overview the Higgs mechanism and the SM.

2.1 Spontaneous symmetry breaking

As a simple relativistic model of spontaneous symmetry breaking, let us consider a complex scalar field $\Phi(x)$. The action in the unit system of $\hbar = c = 1$, is given by

$$I_1 = \int d^4x \, \mathcal{L}_1 \; , \quad \mathcal{L}_1 = -\partial_\mu \Phi^\dagger \partial^\mu \Phi - V(\Phi) \; ,$$
$$V(\Phi) = m^2 \Phi^\dagger \Phi + \lambda (\Phi^\dagger \Phi)^2 \; , \quad m^2 < 0 \; , \quad \lambda > 0 \; . \tag{2.1}$$

The mass squared m^2 is taken to be negative. Here $\partial_\mu \Phi^\dagger \partial^\mu \Phi = \eta^{\mu\nu} \partial_\mu \Phi^\dagger \partial_\nu \Phi$ and $\eta_{\mu\nu} = \eta^{\mu\nu} = \text{diag}\,(-1,1,1,1)$. (See Appendix A.1.) With the conjugate momentum field $\Pi = \dot{\Phi} = \partial_t \Phi, \Pi^\dagger = \dot{\Phi}^\dagger$ the Hamiltonian is given by

$$H = \int d^3x \left\{ \Pi^\dagger \Pi + \nabla \Phi^\dagger \nabla \Phi + V(\Phi) \right\} . \tag{2.2}$$

As $m^2 < 0$, the potential becomes

$$V(\Phi) = \lambda \left\{ \left(\Phi^\dagger \Phi - \frac{v^2}{2} \right)^2 - \frac{v^4}{4} \right\} , \quad v = \sqrt{\frac{-m^2}{\lambda}} \; , \tag{2.3}$$

which takes the shape of the bottom of a wine bottle. The energy (H) is minimized when $\Phi(x)$ is constant and $|\Phi| = v/\sqrt{2}$.

The action and Hamiltonian are invariant under a phase transformation (a $U(1)$ transformation) of $\Phi(x)$.

$$\Phi(x) \to \Phi'(x) = e^{i\alpha} \, \Phi(x) \; ,$$
$$I_1[\Phi'] = I_1[\Phi] \; , \quad H[\Phi', \Pi'] = H[\Phi, \Pi] \; . \tag{2.4}$$

This fact is phrased as the Hamitonian has $U(1)$ symmetry. In physics the configuration of the state which is an eigenstate of the Hamitonian with the smallest eigenvalue (enegy) is called the ground state, or the vacuum in particle physics.

The vacuum is degenerate. Let us write $\Phi(x) = \rho(x)e^{i\theta(x)}$. Classically the vacuum corresponds to $\rho(x) = v$. However the value of $\theta(x) = \beta$ is undetermined. Under a $U(1)$ transformation of Eq. (2.4), $\beta \to \beta' = \beta + \alpha$. Pick one vacuum ($\beta = 0$) with $\Phi(x) = v$. Then the vacuum spontaneously breaks the $U(1)$ symmetry.

When the Hamiltonian is invariant under a transformation of symmetry $\mathcal{G}$ whereas the vacuum (ground state) is not invariant, we say that the symmetry $\mathcal{G}$ is spontaneously broken. This phenomenon is called **spontaneous symmetry breaking**.

Is any trace of the symmetry left once a vacuum is chosen? When a vacuum $|0\rangle$ is chosen such that $\langle\Phi\rangle \equiv \langle 0|\Phi(x)|0\rangle = v/\sqrt{2}$, we write

$$\Phi(x) = \frac{1}{\sqrt{2}}\{v + h(x)\}\, e^{i\varphi(x)/v} \ . \tag{2.5}$$

Inserting (2.5) into the action (2.1), one finds that

$$\begin{aligned} I_1 = \int d^4x \Big\{ &-\frac{1}{2}\partial_\mu h \partial^\mu h - \lambda v^2 h^2 - \frac{1}{2}\partial_\mu\varphi\partial^\mu\varphi + \frac{\lambda}{4}v^4 \\ &- \Big(\frac{h}{v} + \frac{h^2}{2v^2}\Big)\partial_\mu\varphi\partial^\mu\varphi - \lambda v h^3 - \frac{\lambda}{4}h^4 \Big\} \ . \end{aligned} \tag{2.6}$$

It is seen in the first line that the mass of the $h(x)$ field is $\sqrt{2\lambda v^2} = \sqrt{2}\,|m|$, whereas the $\varphi(x)$ field is massless. The second line represents cubic and quartic interaction terms among $h(x)$ and $\varphi(x)$.

The appearance of the massless field $\varphi(x)$ is a necessary consequence of spontaneous breaking of the $U(1)$ symmetry. As a result of the $U(1)$ symmetry in the theory, the phase of $\langle\Phi\rangle$ in the vacuum is not uniquely fixed. The $\varphi(x)$ mode represents a fluctuation of this phase.

In general, when global symmetry, symmetry which does not depend on spacetime coordinates x^μ, is spontaneously broken, there appears an associated massless field. In quantum theory it becomes a massless particle. This is called the **Nambu-Goldstone theorem**. The massless particle is called the **Nambu-Goldstone (NG) boson**. [3, 4]

2.2 Gauge invariance

What happens to the spontaneous symmetry breaking discussed above if the complex scalar field $\Phi(x)$ has an electric charge and interacts with electromagnetic fields? The electromagnetic interaction is a $U(1)$ gauge interaction described by a gauge potential $A_\mu(x)$. The action is given by

$$I_2 = \int d^4x \, \mathcal{L}_2 \, , \quad \mathcal{L}_2 = -(D_\mu \Phi)^\dagger D^\mu \Phi - V(\Phi) - \frac{1}{4} F_{\mu\nu} F^{\mu\nu} \, ,$$
$$D_\mu = \partial_\mu - ieA_\mu \, , \quad F_{\mu\nu} = \partial_\mu A_\nu - \partial_\nu A_\mu \, . \tag{2.7}$$

$F_{\mu\nu}$ is the field strength. Electric and magnetic fields are $E_k = F_{0k}$ and $B_k = \frac{1}{2}\epsilon_{k\ell m} F^{\ell m}$. D_μ is called a covariant derivative. The equations of motion are given by

$$D_\mu D^\mu \Phi - \frac{\delta V}{\delta \Phi^\dagger} = \left\{ D_\mu D^\mu - m^2 - 2\lambda \Phi^\dagger \Phi \right\} \Phi = 0 \, ,$$
$$\partial_\mu F^{\mu\nu} + ie\left\{ (D^\nu \Phi)^\dagger \Phi - \Phi^\dagger D^\nu \Phi \right\} = 0 \, . \tag{2.8}$$

The action I_2 is invariant under a gauge transformation

$$\Phi(x) \to \Phi'(x) = e^{i\Lambda(x)} \Phi(x) \, ,$$
$$A_\mu(x) \to A'_\mu(x) = A_\mu(x) + \frac{1}{e} \partial_\mu \Lambda(x) \, , \tag{2.9}$$

as $(D_\mu \Phi)' = e^{i\Lambda(x)} D_\mu \Phi$ and $F'_{\mu\nu} = F_{\mu\nu}$. $\Omega(x) = e^{i\Lambda(x)}$ satisfies $\Omega^\dagger = \Omega^{-1}$ so that $\Omega(x) \in U(1)$. The action I_2 is $U(1)$ gauge invariant.

With $V(\Phi)$ ($m^2 < 0$) in (2.1), the $U(1)$ symmetry is spontaneously broken. Choosing a vacuum with $\langle \Phi \rangle = v/\sqrt{2}$ and inserting (2.5) into $\mathcal{L}_2$, one finds that

$$\mathcal{L}_2 = -\frac{1}{2} \partial_\mu h \partial^\mu h - \lambda v^2 h^2 - \lambda v h^3 - \frac{\lambda}{4}(h^4 - v^4)$$
$$- \frac{1}{4} F_{\mu\nu} F^{\mu\nu} - \frac{1}{2} e^2 (v+h)^2 B_\mu B^\mu \, ,$$
$$B_\mu = A_\mu - \frac{1}{ev} \partial_\mu \varphi \, . \tag{2.10}$$

The $h(x)$ part in the first line of $\mathcal{L}_2$ is the same as that in Eq. (2.6), and therefore $h(x)$ describes a particle of a mass $\sqrt{2\lambda v^2}$. The situation for the Nambu-Goldstone boson field $\varphi(x)$ is slightly different, though. The $\varphi(x)$ field appears with A_μ solely in the combination of B_μ. What does it mean?

2.3 Higgs mechanism

Bilinear terms in the fluctuation modes φ and A_μ in $\mathcal{L}_2$ are

$$\mathcal{L}_3 = -\frac{1}{4}F_{\mu\nu}F^{\mu\nu} - \frac{1}{2}e^2v^2\Big(A_\mu - \frac{1}{ev}\partial_\mu\varphi\Big)\Big(A^\mu - \frac{1}{ev}\partial^\mu\varphi\Big) . \qquad (2.11)$$

Under a gauge transformation (2.9), $\varphi \to \varphi' = \varphi + v\Lambda$ so that the combination $B_\mu = A_\mu - (ev)^{-1}\partial_\mu\varphi$ is gauge-invariant, i.e. $\mathcal{L}_3$ is gauge-invariant.

(i) Unitary gauge

Theory is gauge-invariant. With a gauge transformation $\Lambda(x) = -\varphi(x)/v$, the fields become $\varphi' = 0$ and $A'_\mu = B_\mu$. In the new gauge it becomes equivalent to a theory of a vector field (called Procca field) described by

$$\mathcal{L}'_3 = -\frac{1}{4}F_{\mu\nu}F^{\mu\nu} - \frac{1}{2}m_V^2 A_\mu A^\mu \ , \quad m_V = ev \ . \qquad (2.12)$$

The Euler equation derived from (2.12) is

$$\partial_\mu F^{\mu\nu} - m_V^2 A^\nu = 0 \ . \qquad (2.13)$$

The equation can be obtained by inserting $\Phi = v/\sqrt{2}$ into (2.8) as well. By taking a divergence ∂_ν of (2.13), one obtains $\partial_\mu A^\mu = 0$ for $m_V \neq 0$. Hence Eq. (2.13) is reduced to $(\partial_\mu\partial^\mu - m_V^2)A^\nu = 0$, and the A_μ field describes a particle of spin 1 and mass m_V. A_μ has four components. With the condition $\partial_\mu A^\mu = 0$ the number of independent degrees of freedom is three. A_μ describes a massive spin-1 particle.

For $m^2 > 0$, A_μ is a massless gauge field with two degrees of freedom. Φ is a complex scalar field with two degrees of freedom. For $m^2 < 0$ one degree of freedom of the NG boson field $\varphi(x)$ is absorbed by A_μ, and A_μ becomes a massive vector field with three degrees of freedom. The total number of degrees of freedom remains unchanged.

When gauge symmetry is spontaneously broken, the corresponding gauge field absorbs the NG boson and becomes massive. This mechanism is called the **Higgs mechanism**. [5, 6]

(ii) General gauge

The Euler equations obtained by treating A_μ and φ as independent degrees of freedom in $\mathcal{L}_3$ are

$$\partial_\mu F^{\mu\nu} - m_V^2\Big(A^\nu - \frac{1}{m_V}\partial^\nu\varphi\Big) = 0 \ ,$$

$$\partial_\mu\Big(A^\mu - \frac{1}{m_V}\partial^\mu\varphi\Big) = 0 \ . \qquad (2.14)$$

The second equation results from the first equation as well. By writing $B_\mu = A_\mu - m_V^{-1}\partial_\mu \varphi$, the equation becomes the same as the equation in the unitary gauge, Eq. (2.13), as $F_{\mu\nu} = \partial_\mu B_\nu - \partial_\nu B_\mu$. $B_\mu(x)$ is a vector field of mass m_V with three degrees of freedom. $\varphi(x)$ corresponds to a gauge degree of freedom. Unless the gauge is fixed, $\varphi(x)$ is not determined by the equations of motion, remaining arbitrary.

2.4 Higgs mechanism in non-Abelian gauge theory

In Sections 2.2 and 2.3, $\Omega(x) = e^{i\Lambda(x)}$ in the gauge transformation (2.9) is an element of $U(1)$, and gauge symmetry associated with electromagnetic fields is symmetry of the Abelian group $U(1)$. **Non-Abelian gauge theory** (Yang-Mills theory) is a gauge theory in which gauge symmetry is promoted to a non-Abelian group. [1]

Let us take $SU(N)$ as a gauge group. An element of $SU(N)$ is a $N \times N$ matrix Ω satisfying $\Omega^\dagger = \Omega^{-1}$ and $\det\Omega = 1$. It can be expressed as

$$\Omega = \exp\left\{ i \sum_{a=1}^{N^2-1} \alpha_a T^a \right\} , \quad (T^a)^\dagger = T^a , \quad \mathrm{Tr}\, T^a = 0 , \tag{2.15}$$

where $\alpha_a = \alpha_a^*$ and $N \times N$ generators $\{T^a\}$ satisfy

$$[T^a, T^b] = i f^{abc} T^c , \quad \mathrm{Tr}\, T^a T^b = \frac{1}{2}\, \delta^{ab} . \tag{2.16}$$

f^{abc} is called the structure constant of the group. In the case of $SU(2)$, $T^a = \frac{1}{2}\tau^a$ and $f^{abc} = \epsilon^{abc}$.

$SU(N)$ gauge fields $A_\mu^a(x)$ $(a = 1, \cdots, N^2 - 1)$ are expressed in the matrix notation as

$$A_\mu(x) = \sum_{a=1}^{N^2-1} A_\mu^a(x) T^a , \quad A_\mu^\dagger = A_\mu , \quad \mathrm{Tr}\, A_\mu = 0 . \tag{2.17}$$

Let $\psi(x)$ be a field in the fundamental representation of $SU(N)$ (N components). Under an $SU(N)$ gauge transformation

$$\begin{aligned} &\psi(x) \to \psi'(x) = \Omega(x)\psi(x) , \\ &A_\mu(x) \to A'_\mu(x) = \Omega(x) A_\mu(x) \Omega(x)^{-1} + \frac{i}{g}\Omega(x)\partial_\mu \Omega(x)^{-1} , \\ &\Omega(x) = \exp\left\{ i \sum_{a=1}^{N^2-1} \alpha_a(x) T^a \right\} . \end{aligned} \tag{2.18}$$

In the case of $U(1)$, $\Omega = e^{i\Lambda}$ and the transformation reduces to (2.9). The inhomogeneous term $(i/g)\Omega\partial_\mu\Omega^{-1}$ in A'_μ above is determined by the requirement that the covariant derivative of ψ, $D_\mu\psi = (\partial_\mu - igA_\mu)\psi$, should transform as

$$D_\mu\psi \to (D_\mu\psi)' = (\partial_\mu - igA'_\mu)\psi' = \Omega\, D_\mu\psi \ . \tag{2.19}$$

g is the gauge coupling constant. The field strength $F_{\mu\nu}$ is found from $[D_\mu, D_\nu]\psi = -igF_{\mu\nu}\psi$ to be

$$\begin{aligned} F_{\mu\nu} &= \sum_{a=1}^{N^2-1} F^a_{\mu\nu}T^a = \partial_\mu A_\nu - \partial_\nu A_\mu - ig[A_\mu, A_\nu] \ , \\ F^a_{\mu\nu} &= \partial_\mu A^a_\nu - \partial_\nu A^a_\mu + gf^{abc}A^b_\mu A^c_\nu \ . \end{aligned} \tag{2.20}$$

Under a gauge transformation $F_{\mu\nu} \to F'_{\mu\nu} = \Omega F_{\mu\nu}\Omega^{-1}$. The action for the gauge fields is given by

$$\mathcal{L}_g = -\frac{1}{2}\,\mathrm{Tr}\, F_{\mu\nu}F^{\mu\nu} = \sum_{a=1}^{N^2-1} -\frac{1}{4}F^a_{\mu\nu}F^{a\mu\nu} \ . \tag{2.21}$$

The Higgs mechanism is extended to non-Abelian gauge theory. Take $SU(2)$ as an example. We consider an $SU(2)$ doublet scalar field Φ. The Lagrangian density takes the same form as in (2.7);

$$\mathcal{L}_4 = -(D_\mu\Phi)^\dagger D^\mu\Phi - m^2\Phi^\dagger\Phi - \lambda(\Phi^\dagger\Phi)^2 - \frac{1}{2}\,\mathrm{Tr}\, F_{\mu\nu}F^{\mu\nu} \ . \tag{2.22}$$

The action is invariant under $SU(2)$ gauge transformations. For $m^2 < 0$, the potential is minimized at $\Phi^\dagger\Phi = v^2/2$ as in (2.3). Choosing a vacuum with $\langle\Phi\rangle^t = (0, v/\sqrt{2})$, we write

$$\Phi(x) = \frac{1}{\sqrt{2}}\begin{pmatrix} \varphi_2(x) + i\varphi_1(x) \\ v + h(x) - i\varphi_3(x) \end{pmatrix} \tag{2.23}$$

and insert it into (2.22) to obtain

$$\begin{aligned} \mathcal{L}_4 &= \frac{\lambda}{4}v^4 - \frac{1}{2}\partial_\mu h\partial^\mu h - \lambda v^2h^2 - \frac{1}{4}\sum_{a=1}^{3} F^a_{\mu\nu}F^{a\mu\nu} \\ &\quad - \frac{g^2v^2}{8}\sum_{a=1}^{3}\Big(A^a_\mu - \frac{2}{gv}\partial_\mu\varphi_a\Big)\Big(A^{a\mu} - \frac{2}{gv}\partial^\mu\varphi_a\Big) + \cdots \end{aligned} \tag{2.24}$$

where the last part $\cdots$ represents terms cubic or quartic with respect to A_μ, h and φ_a. $h(x)$ describes a field of mass $\sqrt{2\lambda v^2}$ as in (2.10). A^a_μ and φ_a are appearing in the same form as in $\mathcal{L}_3$ in (2.11). All three components

$(a = 1, 2, 3)$ of gauge fields acquire a mass of $\frac{1}{2}gv$. By the nonvanishing expectation value of the doublet scalar field $\langle \Phi \rangle \neq 0$ the $SU(2)$ gauge symmetry is spontaneously broken. There is no massless gauge field left over. Three real components of Φ, φ_a $(a = 1, 2, 3)$, are absorbed by three components of gauge fields, A_μ^a $(a = 1, 2, 3)$.

As a next example, we consider a model in which $SU(2)$ gauge symmetry is broken by an $SU(2)$ triplet real scalar field $\vec{\phi}(x) = (\phi_1, \phi_2, \phi_3)^t$. Generators of $SU(2)$ in 3×3 matrix representation are $(T^a_{\rm vec})_{bc} = -i\epsilon^{abc}$. The covariant derivative of $\vec{\phi}$ is given by $D_\mu \vec{\phi} = (\partial_\mu - ig \sum_{a=1}^3 A_\mu^a T^a_{\rm vec})\vec{\phi}$. Under the $SU(2)$ gauge transformation specified in Eq. (2.18), $\vec{\phi} \to \vec{\phi}' = \Omega_{\rm vec}\phi$ and $D_\mu \vec{\phi} \to (D_\mu \vec{\phi})' = \Omega_{\rm vec} D_\mu \vec{\phi}$ where $\Omega_{\rm vec} = \exp\{i \sum \alpha_a(x) T^a_{\rm vec}\}$. Alternatively one may write $\hat{\phi} \equiv \sum_{a=1}^3 \phi_a T^a = \frac{1}{2}\vec{\phi}\vec{\tau}$. Then the covariant derivative is expressed as $D_\mu \hat{\phi} = \partial_\mu \hat{\phi} - ig[A_\mu, \hat{\phi}]$. Under a gauge transformation

$$\hat{\phi} \to \hat{\phi}' = \Omega\, \hat{\phi}\, \Omega^{-1} \ , \quad D_\mu \hat{\phi} \to (D_\mu \hat{\phi})' = \Omega\, D_\mu \hat{\phi}\, \Omega^{-1} \ . \tag{2.25}$$

We take a Lagrangian density given by

$$\mathcal{L}_5 = -\mathrm{Tr}\, D_\mu \hat{\phi}\, D^\mu \hat{\phi} - m^2 \mathrm{Tr}\, \hat{\phi}^2 - \lambda (\mathrm{Tr}\, \hat{\phi}^2)^2 - \frac{1}{2} \mathrm{Tr}\, F_{\mu\nu} F^{\mu\nu} \ . \tag{2.26}$$

$\mathcal{L}_5$ is gauge invariant. For $m^2 < 0$ the potential is minimized at $\mathrm{Tr}\, \hat{\phi}^2 = \frac{1}{2}v^2$, or $\vec{\phi}^{\,2} = v^2$ $(v^2 = -m^2/\lambda)$. With $\langle \vec{\phi} \rangle^t = (0, 0, v)$, we write as

$$\vec{\phi}\,(x) = \begin{pmatrix} -\varphi_2(x) \\ \varphi_1(x) \\ v + h(x) \end{pmatrix} \tag{2.27}$$

and insert (2.27) into $\mathcal{L}_5$ to obtain

$$\begin{aligned} \mathcal{L}_5 = \frac{\lambda}{4} v^4 - \frac{1}{2} \partial_\mu h \partial^\mu h - \lambda v^2 h^2 - \frac{1}{4} \sum_{a=1}^3 F^a_{\mu\nu} F^{a\mu\nu} \\ - \frac{g^2 v^2}{2} \sum_{a=1}^2 \Big(A_\mu^a - \frac{1}{gv} \partial_\mu \varphi_a \Big) \Big(A^{a\mu} - \frac{1}{gv} \partial^\mu \varphi_a \Big) + \cdots \ . \end{aligned} \tag{2.28}$$

As before $h(x)$ describes a particle of mass $\sqrt{2\lambda v^2}$. Two components of gauge fields, A_μ^1 and A_μ^2, become massive vector fields of mass gv by absorbing φ_1 and φ_2. The component A_μ^3 remains as a massless gauge field. We say that $SU(2)$ gauge symmetry is spontaneously broken to $U(1)$ gauge symmetry by the expectation value of the triplet scalar field $\langle \vec{\phi} \rangle \neq 0$.

In the absence of gauge interactions ($g = 0$) $SU(2)$ symmetry is spontaneously broken to $U(1)$ symmetry, thereby $\varphi_1(x)$ and $\varphi_2(x)$ being the Nambu-Goldstone bosons.

It is observed that how the gauge symmetry is broken in non-Abelian gauge theory depends on the representation of a scalar field responsible for symmetry breaking. In this section we have examined the cases in which $SU(2)$ gauge symmetry is spontaneously broken by a doublet scalar field Φ and by a triplet scalar field $\vec{\phi}$. The vacuum expectation value of each field is

$$\text{(a)} \;\; \langle\Phi(x)\rangle = \frac{1}{\sqrt{2}}\begin{pmatrix}0\\ v\end{pmatrix}, \quad \text{(b)} \;\; \langle\vec{\phi}\,(x)\rangle = \begin{pmatrix}0\\ 0\\ v\end{pmatrix}. \tag{2.29}$$

Under a gauge transformation (2.18), $\langle\Phi'\rangle = \Omega(x)\langle\Phi\rangle \neq \langle\Phi\rangle$ so that the vacuum expectation value is not $SU(2)$ invariant. On the other hand, under a $U(1)$ gauge transformation $\Omega_{\text{vec}}^{U(1)}(x) = \exp\left\{i\alpha_3(x)T_{\text{vec}}^3\right\}$, $\langle\vec{\phi}'\rangle = \Omega_{\text{vec}}^{U(1)}(x)\langle\vec{\phi}\rangle = \langle\vec{\phi}\rangle$ so that the vacuum expectation value of $\vec{\phi}$ remains invariant. The gauge field A_μ^3 corresponds to this $U(1)$ gauge invariance.

Suppose that theory has symmetry of a continuous group $\mathcal{G}$, and that a scalar field Φ develops a non-vanishing expectation value $\langle\Phi\rangle \neq 0$, spontaneously breaking the symmetry to a subgroup $\mathcal{H} \subset \mathcal{G}$. Let $n_\mathcal{G}$ be the dimension of $\mathcal{G}$ (the number of independent generators), and $n_\mathcal{H}$ be the dimension of $\mathcal{H}$, respectively. Then $n_\mathcal{G} - n_\mathcal{H}$ independent components of Φ appear massless Nambu-Goldstone bosons (**Nambu-Goldstone theorem**).

Further, when theory has gauge symmetry of a continuous group $\mathcal{G}$, $n_\mathcal{G} - n_\mathcal{H}$ components of gauge fields A_μ^a absorb the Nambu-Goldstone bosons to become massive vector fields. Gauge symmetry $\mathcal{G}$ spontaneously breaks to gauge symmetry $\mathcal{H}$. $n_\mathcal{H}$ components of gauge fields corresponding to $\mathcal{H}$ remain massless. This is the general **Higgs mechanism**.

Spontaneous symmetry breaking can take place even without scalar fields. In the context of particle physics Nambu, in analogy with the BCS theory of superconductivity, showed in 1960 that chiral symmetry is spontaneously broken by fermion-pair condensation and massless bosons appear. It is now confirmed that most (98%) of the masses of protons and neutrons are generated by Nambu's mechanism. Nambu's theory is highly sophisticated in comparison to scalar field theory discussed here. Nambu's theory laid the ground for unified theory of forces in the coming years.

2.5 Standard model

There are four kinds of forces which elementary particles have. They are gravitational force, electromagnetic force, strong force and weak force. The gravitational force (gravity) results as a consequence of distortion of space-time in general theory of relativity by Einstein. The electromagnetic, strong and weak forces are all gauge interactions. The standard model (SM) is formulated as $SU(3)_C \times SU(2)_L \times U(1)_Y$ gauge theory. $SU(3)_C$ gauge theory describes strong interactions (color interactions) and $SU(2)_L \times U(1)_Y$ gauge theory describes electromagnetic and weak interactions in a unified manner. [7–9]

Let G_μ^a $(a = 1 \sim 8)$, A_μ^b $(b = 1 \sim 3)$ and B_μ be gauge fields of $SU(3)_C$, $SU(2)_L$ and $U(1)_Y$, respectively. The Lagrangian density of the gauge fields is, with gauge coupling constants g_s, g, g', given by

$$\begin{aligned}
\mathcal{L}^{\rm gauge} &= -\frac{1}{2}\mathrm{Tr}\, G_{\mu\nu}G^{\mu\nu} - \frac{1}{2}\mathrm{Tr}\, F_{\mu\nu}F^{\mu\nu} - \frac{1}{4}B_{\mu\nu}B^{\mu\nu} ~, \\
G_{\mu\nu} &= \partial_\mu G_\nu - \partial_\nu G_\mu - ig_s[G_\mu, G_\nu] ~, \quad G_\mu = \sum_{a=1}^{8} G_\mu^a T_{SU(3)}^a ~, \\
F_{\mu\nu} &= \partial_\mu A_\nu - \partial_\nu A_\mu - ig[A_\mu, A_\nu] ~, \quad A_\mu = \sum_{a=1}^{3} A_\mu^a T_{SU(2)}^a ~, \\
B_{\mu\nu} &= \partial_\mu B_\nu - \partial_\nu B_\mu ~.
\end{aligned} \tag{2.30}$$

Quarks and leptons come in three generations. The first generation is (u, d, ν_e, e). The second and third generations are (c, s, ν_μ, μ) and (t, b, ν_τ, τ). Quarks are $SU(3)_C$ color triplets and couple to gluons G_μ^a, whereas leptons are color singlets and do not couple to gluons. As for the electroweak $SU(2)_L \times U(1)_Y$ interactions, the left-handed and right-handed components of quarks and leptons interact differently. For a Dirac field ψ, $\psi_R = \frac{1}{2}(1 + \gamma^5)\psi$ is right-handed and $\psi_L = \frac{1}{2}(1 - \gamma^5)\psi$ is left-handed. The left-handed components

$$\begin{aligned}
q_{1L} &= \begin{pmatrix} u_L \\ d_L \end{pmatrix}, \quad q_{2L} = \begin{pmatrix} c_L \\ s_L \end{pmatrix}, \quad q_{3L} = \begin{pmatrix} t_L \\ b_L \end{pmatrix}, \\
\ell_{1L} &= \begin{pmatrix} \nu_{eL} \\ e_L \end{pmatrix}, \quad \ell_{2L} = \begin{pmatrix} \nu_{\mu L} \\ \mu_L \end{pmatrix}, \quad \ell_{3L} = \begin{pmatrix} \nu_{\tau L} \\ \tau_L \end{pmatrix}
\end{aligned} \tag{2.31}$$

Table 2.1 Charges of quarks and leptons.

	T_L^3	Y	$Q_{\rm EM}$		T_L^3	Y	$Q_{\rm EM}$
u_{jL}	$\frac{1}{2}$	$\frac{1}{6}$	$\frac{2}{3}$	u_{jR}	0	$\frac{2}{3}$	$\frac{2}{3}$
d_{jL}	$-\frac{1}{2}$	$\frac{1}{6}$	$-\frac{1}{3}$	d_{jR}	0	$-\frac{1}{3}$	$-\frac{1}{3}$
ν_{jL}	$\frac{1}{2}$	$-\frac{1}{2}$	0	ν_{jR}	0	0	0
e_{jL}	$-\frac{1}{2}$	$-\frac{1}{2}$	-1	e_{jR}	0	-1	-1

are $SU(2)_L$ doublets, whereas the right-handed components

$$\begin{aligned} &u_{1R} = u_R \ , \ u_{2R} = c_R \ , \ u_{3R} = t_R \ , \\ &d_{1R} = d_R \ , \ d_{2R} = s_R \ , \ d_{3R} = b_R \ , \\ &\nu_{1R} = \nu_{eR} \ , \ \nu_{2R} = \nu_{\mu R} \ , \ \nu_{3R} = \nu_{\tau R} \ , \\ &e_{1R} = e_R \ , \ e_{2R} = \mu_R \ , \ e_{3R} = \tau_R \end{aligned} \tag{2.32}$$

are $SU(2)_L$ singlets. Let T_L^a ($a = 1, 2, 3$) be the generators of $SU(2)_L$ group. Then $T_L^3 = \frac{1}{2}$ for u_L, ν_{eL} and $T_L^3 = -\frac{1}{2}$ for d_L, e_L. All right-handed components have $T_L^3 = 0$.

The generator of $U(1)_Y$ is called weak hypercharge, and is related to the electric charge $eQ_{\rm EM}$ by

$$Q_{\rm EM} = T_L^3 + Y \ . \tag{2.33}$$

The values of $(T_L^3, Y, Q_{\rm EM})$ of quarks and leptons are summarized in Table 2.1.

Gauge interactions of quarks and leptons are given by

$$\begin{aligned} \mathcal{L}^{\rm fermion} = \sum_{j=1}^{3} \Big\{ &\bar{q}_{jL}\gamma^\mu D_\mu q_{jL} + \bar{u}_{jR}\gamma^\mu D_\mu u_{jR} + \bar{d}_{jR}\gamma^\mu D_\mu d_{jR} \\ &+ \bar{\ell}_{jL}\gamma^\mu D_\mu \ell_{jL} + \bar{\nu}_{jR}\gamma^\mu \partial_\mu \nu_{jR} + \bar{e}_{jR}\gamma^\mu D_\mu e_{jR} \Big\} , \end{aligned} \tag{2.34}$$

where covariant derivatives are

$$\begin{aligned} &D_\mu q_{jL} = \big(\partial_\mu - ig_s G_\mu - igA_\mu - iYg'B_\mu\big)\, q_{jL} \ , \\ &D_\mu \begin{pmatrix} u_{jR} \\ d_{jR} \end{pmatrix} = \big(\partial_\mu - ig_s G_\mu - iYg'B_\mu\big) \begin{pmatrix} u_{jR} \\ d_{jR} \end{pmatrix} , \\ &D_\mu \ell_{jL} = \big(\partial_\mu - igA_\mu - iYg'B_\mu\big)\, \ell_{jL} \ , \\ &D_\mu e_{jR} = \big(\partial_\mu - iYg'B_\mu\big)\, e_{jR} \ . \end{aligned} \tag{2.35}$$

Y is the weak hypercharge of each particle.

In the SM, in addition to gauge fields, quarks and leptons, we introduce the Higgs scalar field. The Higgs field Φ is an $SU(2)_L$ doublet and $SU(3)_C$ singlet. Its $U(1)_Y$ hypercharge is $Y = \frac{1}{2}$. The Lagrangian density is given by

$$\begin{aligned}
\mathcal{L}^{\text{Higgs}} &= -(D_\mu \Phi)^\dagger D^\mu \Phi - m^2 \Phi^\dagger \Phi - \lambda(\Phi^\dagger \Phi)^2 \ , \\
D^\mu \Phi &= (\partial_\mu - igA_\mu - iYg' B_\mu)\Phi \ , \\
m^2 &< 0 \ , \quad \lambda > 0 \ .
\end{aligned} \tag{2.36}$$

There are additional $SU(3)_C \times SU(2)_L \times U(1)_Y$ invariant interactions connecting the Higgs field Φ with quarks and leptons. They are Yukawa interactions given by

$$\begin{aligned}
\mathcal{L}^{\text{Yukawa}} = -\Big\{ & y^u_{jk} \bar{u}_{jR} \tilde{\Phi}^\dagger q_{kL} + y^d_{jk} \bar{d}_{jR} \Phi^\dagger q_{kL} \\
& + y^\nu_{jk} \bar{\nu}_{jR} \tilde{\Phi}^\dagger \ell_{kL} + y^e_{jk} \bar{e}_{jR} \Phi^\dagger \ell_{kL} + \text{h.c.} \Big\} \ .
\end{aligned} \tag{2.37}$$

Here $\tilde{\Phi} \equiv i\tau_2 \Phi^*$ transforms as an $SU(2)_L$ doublet.

The Lagrangian density of the SM is given by

$$\mathcal{L}^{\text{SM}} = \mathcal{L}^{\text{gauge}} + \mathcal{L}^{\text{fermion}} + \mathcal{L}^{\text{Higgs}} + \mathcal{L}^{\text{Yukawa}} \ , \tag{2.38}$$

which is also called the Glashow-Salam-Weinberg model. In the early stage of the model construction it was thought that neutrinos are massless and only left-handed neutrinos exist, and therefore the $y^\nu_{jk} \bar{\nu}_{jR} \tilde{\Phi}^\dagger \ell_{kL}$ term in $\mathcal{L}^{\text{Yukawa}}$ is absent. Now neutrino oscillation has been observed and the existence of right-handed neutrino has been established. As seen in Table 2.1, ν_{jR} is completely neutral in any of $SU(3)_C$, $SU(2)_L$ and $U(1)_Y$ gauge interactions. Therefore there can be Majorana mass terms consisting of only ν_{jR}.

2.6 *W* boson, *Z* boson and Higgs boson

As $m^2 < 0$ in the Higgs potential (2.36), $\Phi(x)$ develops a non-vanishing vacuum expectation value:

$$\langle \Phi(x) \rangle = \frac{1}{\sqrt{2}} \begin{pmatrix} 0 \\ v \end{pmatrix} , \quad v = \sqrt{\frac{-m^2}{\lambda}} \ . \tag{2.39}$$

$SU(2)_L \times U(1)_Y$ gauge symmetry is spontaneously broken to $U(1)_{\text{EM}}$ gauge symmetry, and W and Z bosons become massive.

Inserting the expression (2.23) into $\mathcal{L}^{\rm gauge} + \mathcal{L}^{\rm Higgs}$ and retaining bilinear terms in the $SU(2)_L \times U(1)_Y$ gauge bosons and the Higgs boson, one finds that

$$
\begin{aligned}
&\mathcal{L}^{\rm gauge} + \mathcal{L}^{\rm Higgs} \\
&= -\frac{1}{2}\partial_\mu h \partial^\mu h - \lambda v^2 h^2 - \frac{1}{4}\sum_{a=1}^{3} F^a_{\mu\nu} F^{a\mu\nu} - \frac{1}{4} B_{\mu\nu} B^{\mu\nu} \\
&\quad - \frac{g^2 v^2}{8} \sum_{a=1}^{2} \Big(A^a_\mu - \frac{2}{gv}\partial_\mu \varphi_a \Big)\Big(A^{a\mu} - \frac{2}{gv}\partial^\mu \varphi_a \Big) \\
&\quad - \frac{v^2}{8}\Big(g A^3_\mu - g' B_\mu - \frac{2}{v}\partial_\mu \varphi_3 \Big)\Big(g A^{3\mu} - g' B^\mu - \frac{2}{v}\partial^\mu \varphi_3 \Big) + \cdots . \qquad (2.40)
\end{aligned}
$$

The part of A^1_μ and A^2_μ is the same as in Eq. (2.24) so that A^1_μ and A^2_μ become vector fields with a mass $\frac{1}{2}gv$. For A^3_μ and B_μ, only the combination $gA^{3\mu} - g'B^\mu$ acquires a mass. Considering the normalization of the kinetic terms, we define $W^\pm_\mu$, Z_μ and A^γ_μ by

$$
\begin{aligned}
W^\pm_\mu &= \frac{1}{\sqrt{2}}(A^1_\mu \mp i A^2_\mu) , \\
Z_\mu &= \cos\theta_W A^3_\mu - \sin\theta_W B_\mu , \\
A^\gamma_\mu &= \sin\theta_W A^3_\mu + \cos\theta_W B_\mu , \\
\cos\theta_W &= \frac{g}{\sqrt{g^2 + g'^2}} , \quad \sin\theta_W = \frac{g'}{\sqrt{g^2 + g'^2}} . \qquad (2.41)
\end{aligned}
$$

θ_W is called the Weinberg angle, or the **weak mixing angle**. The bilinear part of (2.40) becomes

$$
\begin{aligned}
&\mathcal{L}^{\rm gauge} + \mathcal{L}^{\rm Higgs} \\
&= -\frac{1}{2}\partial_\mu h \partial^\mu h - \frac{1}{2} m_H^2 h^2 \\
&\quad - \frac{1}{2} W^+_{\mu\nu} W^{-\mu\nu} - m_W^2 \Big(W^+_\mu - \frac{\partial_\mu \varphi_+}{m_W} \Big)\Big(W^{-\mu} - \frac{\partial^\mu \varphi_-}{m_W} \Big) \\
&\quad - \frac{1}{4} Z_{\mu\nu} Z^{\mu\nu} - m_Z^2 \Big(Z_\mu - \frac{\partial_\mu \varphi_3}{m_Z} \Big)\Big(Z^\mu - \frac{\partial^\mu \varphi_3}{m_Z} \Big) \\
&\quad - \frac{1}{4} F^\gamma_{\mu\nu} F^{\gamma\mu\nu} . \qquad (2.42)
\end{aligned}
$$

Here $W^{\pm}_{\mu\nu} = \partial_\mu W^{\pm}_\nu - \partial_\nu W^{\pm}_\mu$, $\varphi_\pm = (\varphi_1 \mp i\varphi_2)/\sqrt{2}$, $Z_{\mu\nu} = \partial_\mu Z_\nu - \partial_\nu Z_\mu$, $F^\gamma_{\mu\nu} = \partial_\mu A^\gamma_\nu - \partial_\nu A^\gamma_\mu$, and the masses are given by

$$m_H = \sqrt{2\lambda v^2} \ , \quad m_W = \frac{gv}{2} \ , \quad m_Z = \frac{m_W}{\cos\theta_W} \ . \tag{2.43}$$

h, $W^\pm_\mu$, Z_μ and A^γ_μ are the Higgs boson, $W^\pm$ bosons, Z boson and photon (γ), respectively. The W boson and Z boson acquire finite masses by the Higgs mechanism. The photon, on the other hand, remains massless. It is due to the fact that the vacuum expectation value (2.39) is invariant under a gauge transformation

$$\Phi(x) \to \Phi'(x) = \Omega^{\rm EM}(x)\Phi(x) \ ,$$

$$\Omega^{\rm EM}(x) = \exp\Big\{ i\alpha_{\rm EM}(x) Q_{\rm EM} \Big\} \ . \tag{2.44}$$

It is said that the $SU(2)_L \times U(1)_Y$ symmetry is spontaneously broken to the $U(1)_{\rm EM}$ symmetry.

W and Z interact with quarks and leptons through the covariant derivative terms. For $SU(2)_L$ doublet fields $D_\mu = \partial_\mu - i(gA_\mu + Yg'B_\mu)$ and

$$\begin{aligned} gA_\mu + Yg'B_\mu &= \frac{g}{\sqrt{2}} \begin{pmatrix} & W^+_\mu \\ W^-_\mu & \end{pmatrix} \\ &\quad + \frac{g}{\cos\theta_W}(T^3_L - \sin^2\theta_W Q_{\rm EM}) Z_\mu + eQ_{\rm EM} A^\gamma_\mu \ , \\ e &= g\sin\theta_W \ , \end{aligned} \tag{2.45}$$

and for $SU(2)_L$ singlet fields $T^3_L = 0$, $Q_{\rm EM} = Y$ and

$$Yg'B_\mu = -\frac{g}{\cos\theta_W}\sin^2\theta_W Q_{\rm EM} Z_\mu + eQ_{\rm EM}A^\gamma_\mu \ . \tag{2.46}$$

Inserting these expressions into $\mathcal{L}^{\rm fermion}$, one finds

$$\begin{aligned} \mathcal{L}^{\rm fermion}_{\rm EW} &= eA^\gamma_\mu J^\mu_{\rm EM} + \frac{g}{\sqrt{2}}(W^+_\mu J^\mu_W + W^-_\mu J^{\mu\dagger}_W) + \frac{g}{\cos\theta_W} Z_\mu J^\mu_Z \ , \\ J^\mu_{\rm EM} &= \sum_{j=1}^{3} \Big\{ \tfrac{2}{3}\bar{u}_j\gamma^\mu u_j - \tfrac{1}{3}\bar{d}_j\gamma^\mu d_j - \bar{e}_j\gamma^\mu e_j \Big\}, \\ J^\mu_W &= \sum_{j=1}^{3} \Big\{ \bar{u}_{jL}\gamma^\mu d_{jL} + \bar{\nu}_{jL}\gamma^\mu e_{jL} \Big\}, \\ J^\mu_Z &= \sum_{j=1}^{3} \Big\{ \tfrac{1}{2}\bar{u}_{jL}\gamma^\mu u_{jL} - \tfrac{1}{2}\bar{d}_{jL}\gamma^\mu d_{jL} \\ &\quad + \tfrac{1}{2}\bar{\nu}_{jL}\gamma^\mu \nu_{jL} - \tfrac{1}{2}\bar{e}_{jL}\gamma^\mu e_{jL} \Big\} - \sin^2\theta_W J^\mu_{\rm EM} \ . \end{aligned} \tag{2.47}$$

From the cubic and quartic terms in $\mathcal{L}^{\text{gauge}}$,

$$ig\text{Tr}\,(\partial_\mu A_\nu - \partial_\nu A_\mu)[A^\mu, A^\nu] + \frac{g^2}{2}\text{Tr}\,[A_\mu, A_\nu][A^\mu, A^\nu]\ , \tag{2.48}$$

self-interaction terms among gauge fields are found to be

$$\begin{aligned}\mathcal{L}_{\text{EW}}^{\text{gauge}} &= -ie\,K[W^\dagger, W, A^\gamma] - \frac{1}{2}e^2 M[A^\gamma, W^\dagger, A^\gamma, W] \\ &\quad - \frac{1}{2}eg\cos\theta_W\Big\{M[A^\gamma, W^\dagger, Z, W] + M[Z, W^\dagger, A^\gamma, W]\Big\} \\ &\quad - ig\cos\theta_W\,K[W^\dagger, W, Z] - \frac{1}{4}g^2 M[W^\dagger, W, W^\dagger, W] \\ &\quad - \frac{1}{2}g^2\cos^2\theta_W M[Z, W^\dagger, Z, W]\ ,\end{aligned} \tag{2.49}$$

where

$$\begin{aligned}K[A,B,C] &= A^{\mu\nu}B_\mu C_\nu + B^{\mu\nu}C_\mu A_\nu + C^{\mu\nu}A_\mu B_\nu\ , \\ &\qquad (A_{\mu\nu} = \partial_\mu A_\nu - \partial_\nu A_\mu)\ , \\ M[A,B,C,D] &= (A_\mu B_\nu - A_\nu B_\mu)(C^\mu D^\nu - C^\nu D^\mu)\ .\end{aligned} \tag{2.50}$$

There is the Yukawa interaction $\mathcal{L}^{\text{Yukawa}}$ in (2.37). With $\langle\Phi\rangle \neq 0$ mass terms of quarks and leptons are generated.

$$\begin{aligned}\mathcal{L}^{\text{Yukawa}} &= -\sum_{j,k=1}^{3}\big\{\bar{u}_{jR}M_{jk}^u u_{kL} + \bar{d}_{jR}M_{jk}^d d_{kL} \\ &\qquad + \bar{\nu}_{jR}M_{jk}^\nu \nu_{kL} + \bar{e}_{jR}M_{jk}^e e_{kL} + h.c.\big\} + \cdots\ , \\ (M_{jk}^u, M_{jk}^d, M_{jk}^\nu, M_{jk}^e) &= \frac{v}{\sqrt{2}}(y_{jk}^u, y_{jk}^d, y_{jk}^\nu, y_{jk}^e)\end{aligned} \tag{2.51}$$

In general the Yukawa couplings y_{jk}^u etc. form 3-by-3 matrices with off-diagonal elements in the generation space so that the mass matrices M_{jk}^u etc. also have off-diagonal elements. u_j, d_j etc. written above are diagonal in gauge interactions, and are called gauge eigenstates. On the other hand, eigenstates of the Hamiltonian are states with diagonal mass terms. The eigenvalues of these mass eigenstates are the observed masses of quarks and leptons. These two sets of the eigenstates are related by unitary transformations. When the Lagrangian is expressed in terms of the mass eigenstates, the kinetic terms remain invariant, but the charged (W) current in the electroweak interaction (2.47) becomes

$$J_W^\mu = \sum_{j,k=1}^{3} \left\{ \bar{u}_{jL} \gamma^\mu K_{jk}^{\text{CKM}} d_{kL} + \bar{\nu}_{jL} \gamma^\mu K_{jk}^{\text{PMNS}} e_{kL} \right\} . \tag{2.52}$$

K^{CKM} is called the Cabbibo-Kobayashi-Maskawa matrix, and K^{PMNS} is called the Pontecorvo-Maki-Nakagawa-Sakata matrix. The neutral (Z) current J_Z^μ remains invariant, i.e. J_Z^μ remains diagonal at the tree level. This property is expressed as flavor changing neutral currents (FCNC) are naturally suppressed.

The gauge interaction part of the SM has been confirmed to high accuracy by numerous experiments and observation at low energies. However, there remains arbitrariness in the Higgs interaction part of the SM. There is no guiding principle for determining the Higgs potential and Yukawa interactions. Even if the parameter λ in the potential (2.36) is adjusted such that the observed Higgs boson mass m_H is reproduced, there arise large quantum corrections so that fine-tuning of the parameters is necessary to cancel it. This is called the gauge hierarchy problem.

Chapter 3

Aharonov-Bohm effect and Hosotani mechanism

In quantum theory new degrees of freedom emerge which cannot be expressed solely by field strengths of electromagnetic fields if space is not simply-connected. This is called the Aharonov-Bohm effect. In non-Abelian gauge theory gauge symmetry can be partially broken by the Aharonov-Bohm effect.

3.1 Aharonov-Bohm effect in quantum mechanics

Is the gauge potential $A^\mu(x) = (A^0, \vec{A})$ necessary to describe electromagnetic fields? Isn't it sufficient to use only electric and magnetic field $F_{\mu\nu} = (\vec{E}, \vec{B})$ to describe electromagnetic interactions? At the classical level only $\vec{E}$ and $\vec{B}$ appear in the Maxwell equations and Lorentz equations describing the motion of charged particles. Although the introduction of A^μ is necessary in order to derive these equations from the action principle, it merely is subsidiary.

The situation is completely different in quantum theory. The gauge potential $A^\mu(x)$ is absolutely necessary to write down the fundamental equation of motion for the electron in electromagnetic fields, namely the Schrödinger equation:

$$i\hbar \frac{\partial \psi(\vec{x}, t)}{\partial t} = \bigg\{ -\frac{\hbar^2}{2m} \Big(\nabla + i\frac{e}{\hbar c}\vec{A} \Big)^2 + eA_0 \bigg\} \psi(\vec{x}, t) \ . \tag{3.1}$$

The equation is invariant under a gauge transformation

$$\begin{aligned} A_\mu(x) &\to A'_\mu(x) = A_\mu(x) + \frac{1}{e}\, \partial_\mu \Lambda(x) \ , \\ \psi(x) &\to \psi'(x) = e^{-i\Lambda(x)/\hbar c}\, \psi(x) \ . \end{aligned} \tag{3.2}$$

The legitimate question is whether or not $A_\mu(x)$ is uniquely determined by $F_{\mu\nu}(x)$ up to the arbitrariness of (3.2). Aharonov and Bohm showed that depending on the topology of the space where the electron moves, there is a physical degree of freedom of electromagnetic fields which cannot be expressed in terms of the field strength $F_{\mu\nu}(x)$. This phenomenon occurs only in quantum theory, and is called the **Aharonov-Bohm effect.** [13]

Aharonov and Bohm considered the motion of an electron around a magnetic coil as depicted in Fig. 3.1. The electron cannot move into the coil. Inside the coil there is a magnetic field $\vec{B}$ in the upward direction in Fig. 3.1. Outside the coil there is no electromagnetic field, $\vec{B} = \vec{E} = 0$. The electron moves only in a region where $\vec{B} = \vec{E} = 0$. Does the magnetic field inside the coil affect the motion of the electron?

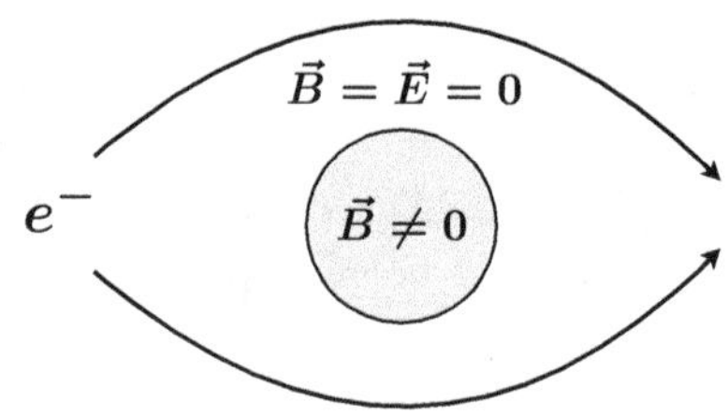

Fig. 3.1 Aharonov-Bohm effect. Electrons move around a magnetic coil.

In classical theory the electron moves by following the Lorentz equation so that its motion is not affected by the magnetic field inside the coil. In quantum theory the motion of the electron is determined by the Schödinger equation (3.1). Let us take the cylindrical coordinates (r, ϕ, z) with the center of the coil as the origin $(r = 0)$. Outside the coil $(r > R_{\rm coil})$

$$A_\phi = \frac{\Phi}{2\pi r} , \quad A_r = A_z = A_0 = 0 . \tag{3.3}$$

As $\Phi = \int_0^{2\pi} r d\phi \, A_\phi = \int d\vec{S} \cdot (\nabla \times \vec{A}) = \int d\vec{S} \cdot \vec{B}$, Φ is the total magnetic flux penetrating the coil from the bottom to the top.

Let us place a slit on the left of the coil and inject an electron beam from the left to the right. The Schrödinger waves of the electron propagate above and below the coil, and superpose on the right-side of the coil with interference. The magnitude of the interference is determined by the difference of the phases of the electron waves. It follows from Eqs. (3.1) and (3.3) that the phase difference is given by

$$\frac{e}{\hbar c} \int_{-\pi}^{0} r d\phi \, A_\phi - \frac{e}{\hbar c} \int_{\pi}^{0} r d\phi \, A_\phi = \frac{e\Phi}{\hbar c} . \tag{3.4}$$

In general the phase difference can be expressed as

$$\theta_{AB} = \frac{e}{\hbar c} \oint d\vec{x}\, \vec{A} \; . \tag{3.5}$$

Here the integration path in $\oint$ is any loop surrounding the coil. θ_{AB} is called the **Aharonov-Bohm phase (AB phase)**.

Is the AB phase θ_{AB} gauge invariant? Under the transformation (3.2)

$$\theta_{AB} \to \theta'_{AB} = \theta_{AB} + \frac{1}{\hbar c} \oint d\vec{x}\, \nabla \Lambda \tag{3.6}$$

so that θ_{AB} is invariant provided $\Lambda(x)$ is single-valued. The Schrödinger wave function $\psi(x)$ must be single-valued so that $\psi'(x)$ after the gauge transformation must be single-valued. However, this does not necessarily implies that $\Lambda(x)$ is single-valued. Indeed, when one applies a gauge transformation given in the cylindrical coordinates by

$$\Lambda(x) = 2\pi n\, \hbar c\, \phi \;, \quad (n : \text{an integer}) \;, \tag{3.7}$$

$\psi'(x)$ remains single-valued, but the AB phase changes as

$$\theta_{AB} \to \theta'_{AB} = \theta_{AB} + 2\pi n \; . \tag{3.8}$$

We stress that $e^{i\theta_{AB}}$ is gauge invariant.

The magnetic flux Φ penetrating the coil changes as

$$\Phi \to \Phi' = \Phi + n\, \Phi_0 \;, \quad \Phi_0 = \frac{2\pi \hbar c}{e} \; . \tag{3.9}$$

In other words, when one changes the magnetic flux by a multiple of Φ_0, the new electron wave function is related to the original one by a gauge transformation and physical results such as the interference pattern remain the same as the original ones. This has been confirmed experimentally.

3.2 Gauge theory on $M^n \times S^1$

In Fig. 3.1 the electron moves outside the coil $r > R_{\rm coil}$. What would happens if the electron can move only on the ring, namely on $r = R > R_{\rm coil}$? In this case also the motion of the electron is affected by the magnetic flux penetrating the ring through A_ϕ. The situation is the same as in the case of the electron moving in the spacetime $M^n \times S^1$ where M^n is the n-dimensional Minkowski spacetime and S^1 is a circle.

Consider the case $n = 1$, namely two-dimensional spacetime $(x^0, x^1) = (t, x)$ with the space S^1 of a radius R. The Schrödinger equation is

$$\Big(i\hbar \frac{\partial}{\partial t} - eA_0 \Big) \psi(t,x) = -\frac{\hbar^2}{2m} \Big(\frac{\partial}{\partial x} - i\frac{e}{\hbar c} A_x \Big)^2 \psi(t,x) \; . \tag{3.10}$$

The correspondence to (ϕ, A_ϕ) in the previous section is $(x = R\phi, A_x = -A_\phi)$. For the vanishing electric field $F_{01} = 0$,

$$A_0 = 0 \ , \quad A_x = \frac{\hbar c}{2\pi eR}\,\theta_H \ , \quad \theta_H = \frac{e}{\hbar c}\int_0^{2\pi R} dx\, A_x \ . \tag{3.11}$$

The wave function is single-valued; $\psi(t, x + 2\pi R) = \psi(t, x)$. By expanding it in a Fourier series $\psi(t,x) = \sum_{n=-\infty}^{\infty} \psi_n(t)\, e^{inx/R}$, one finds that

$$\begin{aligned} i\hbar \frac{\partial}{\partial t}\, \psi_n(t) &= \varepsilon_n\, \psi_n(t) \ , \\ \varepsilon_n &= \frac{\hbar^2}{2mR^2}\Big(n - \frac{\theta_H}{2\pi}\Big)^2 \ . \end{aligned} \tag{3.12}$$

Under a gauge transformation $\psi \to \psi' = \Omega(x)\psi$, $\Omega(x) = e^{ix/R}$, which retains the single-value property of $\psi(x)$,

$$\theta_H \to \theta'_H = \theta_H - 2\pi \ , \quad \varepsilon_n \to \varepsilon_{n+1} \tag{3.13}$$

so that the energy spectrum $\{\varepsilon_n\}$ remains invariant as a whole. Since the physical quantity $\{\varepsilon_n\}$ depends on θ_H, θ_H is a physical degree of freedom.

The extension to a relativistic theory on $M^4 \times S^1$ is straightforward. Let us denote the coordinates of M^4 and S^1 as $x^M = (x^\mu, y)$. The Lagrangian density for a complex massless scalar field Φ is given, in the natural units $\hbar = c = 1$, by

$$\mathcal{L} = -(D_M\Phi)^\dagger D^M \Phi \ , \quad D_M = \partial_M - ieA_M \ . \tag{3.14}$$

The Lagrangian density $\mathcal{L}$ must be single-valued on S^1. With the boundary conditions

$$\begin{aligned} A_M(x, y + 2\pi R) &= A_M(x,y) \ , \\ \Phi(x, y + 2\pi R) &= e^{i\beta}\Phi(x,y) \ , \end{aligned} \tag{3.15}$$

one expands Φ as

$$\Phi(x,y) = \sum_{n=-\infty}^{\infty} \Phi_n(x) \exp\Big\{\frac{iy}{R}\Big(n + \frac{\beta}{2\pi}\Big)\Big\} \ . \tag{3.16}$$

For $A_\mu = 0$ and $A_y = \theta_H/2\pi eR$, the field strengths vanish: $F_{MN} = 0$. Inserting (3.16) into the Euler equation $D_M D^M \Phi = 0$, one obtains

$$\begin{aligned} &(\partial_\mu \partial^\mu - m_n^2)\, \Phi_n(x) = 0 \ , \\ &m_n^2 = m_n(\theta_H; \beta)^2 = \frac{1}{R^2}\Big(n - \frac{\theta_H - \beta}{2\pi}\Big)^2 \ . \end{aligned} \tag{3.17}$$

$\Phi_n(x)$ represents a 4D scalar field with a mass m_n. $\{m_n\}$ depends on the Aharonov-Bohm phase θ_H.

3.3 Aharonov-Bohm phases in non-Abelian gauge theory

The arguments in Sections 3.1 and 3.2 are generalized to non-Abelian gauge theory. As in the case of electromagnetic fields ($U(1)$ gauge theory), physical degrees of freedom of A_μ in the absence of field strengths $F_{\mu\nu} = \partial_\mu A_\nu - \partial_\nu A_\mu - ig[A_\mu, A_\nu]$ are called Aharonov-Bohm (AB) phases in non-Abelian gauge theory. To make the definition precise we first define a **Wilson line integral** along a path C connecting two points x_i^μ and x_f^μ.

$$\begin{aligned} W(x_i, x_f; C) &= P \exp\Big\{ -ig \int_C dx^\mu A_\mu(x) \Big\} \\ &\equiv \lim_{N\to\infty} W(x_0, x_1) W(x_1, x_2) \cdots W(x_{N-1}, x_N) , \\ W(x_{j-1}, x_j) &= 1 - ig\Delta x_{j-1}^\mu A_\mu(x_j) . \end{aligned} \tag{3.18}$$

Here we have divided the path C into N segments $x_0 = x_i \to x_1 \to x_2 \to \cdots \to x_N = x_f$ with $\Delta x_j = x_j - x_{j-1}$. In non-Abelian gauge theory A_μ is a matrix, and the order of the product must be treated carefully. In the limit $N \to \infty$, $\Delta x_j \to 0$. In an $SU(N)$ gauge theory, under a gauge transformation (2.18)

$$W(x_{j-1}, x_j) \to \Omega(x_{j-1}) W(x_{j-1}, x_j) \Omega(x_j)^{-1}$$

to $O(\Delta x)$ so that the Wilson line integral (3.18) transforms as

$$W(x_i, x_f; C) \to \Omega(x_i) W(x_i, x_f; C) \Omega(x_f)^{-1} . \tag{3.19}$$

Let us consider an $SU(N)$ gauge theory on $M^4 \times S^1$ with a Dirac field $\psi(x, y)$ in the fundamental representation of $SU(N)$. The Lagrangian density is

$$\mathcal{L} = -\frac{1}{2} \mathrm{Tr}\, F_{MN} F^{MN} + \overline{\psi} \Gamma^M D_M \psi \tag{3.20}$$

where $x^M = (x^\mu, y)$ and $D_M = \partial_M - igA_M$. ψ transforms as a spinor under Lorentz transformations and has four components. Γ^M is, in terms of γ^μ in four dimensions, given by $(\Gamma^\mu, \Gamma^5) = (\gamma^\mu, \gamma^5)$. (See Appendix A.)

$\mathcal{L}$ must be single-valued on S^1. We take

$$\begin{aligned} A_M(x, y + 2\pi R) &= U A_M(x, y) U^{-1} , \\ \psi(x, y + 2\pi R) &= e^{i\beta} U \psi(x, y) , \quad U \in SU(N) . \end{aligned} \tag{3.21}$$

A_M becomes periodic on S^1 for $U = 1$, but in general can be twisted with an $SU(N)$ matrix U. The theory is specified with the Lagrangian density (3.20) and boundary conditions (3.21).

Under a gauge transformation (2.18), A'_M and ψ' after the transformation satisfy the boundary conditions given by

$$\begin{aligned} &A'_M(x,y+2\pi R) = U' A'_M(x,y) U'^{-1} + \frac{i}{g} U' \partial_M U'^{-1} , \\ &\psi'(x,y+2\pi R) = e^{i\beta} U' \psi'(x,y) , \\ &U' = \Omega(x,y+2\pi R) U \Omega(x,y)^{-1} . \end{aligned} \tag{3.22}$$

The residual gauge invariance is given by $U' = U$, namely by

$$\mathcal{G}_{\rm res} = \{ \Omega(x,y) ; \ \Omega(x,y+2\pi R) U \Omega(x,y)^{-1} = U \} . \tag{3.23}$$

Note that $\Omega(x,y+2\pi R) = U\Omega(x,y)U^{-1}$ so that $\Omega(x,y)$ is not periodic on S^1 in general.

Consider the Wilson line integral going around S^1 and multiplied by U:

$$\begin{aligned} &\hat{W}(x) = W(x)\, U , \\ &W(x) = P\exp\bigg\{ -ig \int_0^{2\pi R} dy A_y(x,y) \bigg\} . \end{aligned} \tag{3.24}$$

Under a gauge transformation $\Omega(x,y) \in \mathcal{G}_{\rm res}$ in (3.23),

$$\begin{aligned} \hat{W}(x) &\to \Omega(x,0) W(x) \Omega(x,2\pi R)^{-1} U \\ &= \Omega(x,0)\, \hat{W}(x)\, \Omega(x,0)^{-1} \end{aligned} \tag{3.25}$$

so that the eigenvalues of $\hat{W}(x)$ are gauge invariant. In $SU(N)$ gauge theory there is one relation among N eigenvalues $\{e^{i\theta_j(x)}, \ j = 1 \sim N\}$ given by $\sum \theta_j(x) = 0 \pmod{2\pi}$. In particular, configurations with constant θ_j have vanishing field strengths $F_{MN} = 0$. These $\{\theta_j, \ j = 1 \sim N, \ \sum \theta_j = 0 \pmod{2\pi}\}$ are called **Aharonov-Bohm phases in $SU(N)$ gauge theory**.

3.4 Hosotani mechanism I – gauge symmetry breaking

In the next chapter we will find that the values of the AB phases $\{\theta_j\}$ are dynamically determined by quantum effects. Here we examine the meaning of $\theta_j \neq 0$. When $U = 1$ in (3.21), the boundary condition is $SU(N)$ symmetric. $\{\theta_j\}$ are constant modes of A_y. As a classical background field we take

$$A^{\rm c}_\mu = 0, \ A^{\rm c}_y = \frac{1}{2\pi g R} \begin{pmatrix} \theta_1 & & \\ & \ddots & \\ & & \theta_N \end{pmatrix} , \ \sum_{j=1}^N \theta_j = 0 \pmod{2\pi} , \tag{3.26}$$

and examine how the fluctuation $A^{\rm q}_M$ around it ($A_M = A^{\rm c}_M + A^{\rm q}_M$) behaves.

In general, the procedure of gauge fixing becomes necessary to quantize gauge fields. It is due to the fact that gauge degrees of freedom are not dynamical variables. In the Fadeev-Popov quantization one takes, for gauge fields in (2.21),

$$\mathcal{L}^{\rm gauge} = -\frac{1}{2} \,{\rm Tr}\, F_{MN}F^{MN} - \frac{1}{\xi}\,{\rm Tr}\,[f_{\rm gf}(A)]^2 + g{\rm Tr}\,\bar{c}\,\frac{\delta f_{\rm gf}(A^\alpha)}{\delta\alpha}\,c\,. \tag{3.27}$$

$f_{\rm gf}(A)$ is a gauge-fixing function. The third term is the associated ghost term, where A^α represents the gauge configuration obtained by an infinitesimal gauge transformation $\Omega = e^{i\alpha}$ on A_M. c and $\bar{c}$ are unphysical ghost and anti-ghost fields which are scalar fields obeying the Fermi statistics. $c = \sum_a c^a T^a$ and $\bar{c}$ are in the matrix form. ξ is an arbitrary constant.

In the case under consideration it is convenient to take

$$\begin{aligned} &f_{\rm gf}(A) = \eta^{\mu\nu}\mathcal{D}^{\rm c}_\mu A^{\rm q}_\nu + \xi\mathcal{D}^{\rm c}_y A^{\rm q}_y\,, \\ &g\frac{\delta f_{\rm gf}(A^\alpha)}{\delta\alpha} = \eta^{\mu\nu}\mathcal{D}^{\rm c}_\mu \mathcal{D}^{\rm c+q}_\nu + \xi\mathcal{D}^{\rm c}_y \mathcal{D}^{\rm c+q}_y\,. \end{aligned} \tag{3.28}$$

Here $\mathcal{D}^{\rm c}_M B = \partial_M B - ig[A^{\rm c}_M, B]$ and $\mathcal{D}^{\rm c+q}_M B = \partial_M B - ig[A_M, B]$, and we have made use of $(A^{\rm q}_M)^\alpha = A^{\rm q}_M + (1/g)\mathcal{D}^{\rm c+q}_M\alpha$. Inserting (3.26) and (3.28) into (3.27) and integrating in part, one obtains, up to terms quadratic in $A^{\rm q}_M$, c and $\bar{c}$,

$$\begin{aligned} \mathcal{L}^{\rm gauge}_2 = {\rm Tr}\Bigg[&A^{\rm q}_\mu\Big\{\eta^{\mu\nu}\Big(\Box_4 + \mathcal{D}^{\rm c}_y\mathcal{D}^{\rm c}_y\Big) - \Big(1 - \frac{1}{\xi}\Big)\partial^\mu\partial^\nu\Big\}A^{\rm q}_\nu \\ &+ A^{\rm q}_y\Big(\Box_4 + \xi\mathcal{D}^{\rm c}_y\mathcal{D}^{\rm c}_y\Big)A^{\rm q}_y + \bar{c}\Big(\Box_4 + \xi\mathcal{D}^{\rm c}_y\mathcal{D}^{\rm c}_y\Big)c\Bigg]\,, \end{aligned} \tag{3.29}$$

where $\Box_4 = \eta^{\mu\nu}\partial_\mu\partial_\nu$. The gauge-fixing function $f_{\rm gf}(A)$ has been introduced such that $A^{\rm q}_\mu$ does not intertwine with $A^{\rm q}_y$. It is the extension of the R_ξ gauge in the SM introduced by 't Hooft.

With $U = 1$ in the boundary condition (3.21) A_M, c and $\bar{c}$ are periodic on S^1, and therefore can be expanded as

$$\begin{pmatrix} A_M(x,y) \\ c(x,y) \\ \bar{c}(x,y) \end{pmatrix} = \sum_{n=-\infty}^{\infty} \begin{pmatrix} A^{(n)}_M(x) \\ c^{(n)}(x) \\ \bar{c}^{(n)}(x) \end{pmatrix} \frac{e^{iny/R}}{\sqrt{2\pi R}}\,. \tag{3.30}$$

Expansion of fields in five-dimensional spacetime such as $A_M(x,y)$ in a series of four-dimensional fields ($A^{(n)}_M(x)$ etc.) is called the **Kaluza-Klein (KK) expansion.** $\{A^{(n)}_M(x)\}$ are KK modes of $A_M(x,y)$.

For the matrix element B_{jk} of B, $(\mathcal{D}^c_y B)_{jk} = [\partial_y - (i/2\pi R)(\theta_j - \theta_k)]B_{jk}$ so that each term in $\mathcal{L}_2^{\rm gauge}$ takes the diagonal form of $B^\dagger_{jk} \cdots B_{jk}$. The action of the $A_\mu^{(n)}$ mode is, in the momentum representation, given by $A_\mu^{(n)} K^{\mu\nu} A_\nu^{(n)}$ where $K^{\mu\nu} = \eta^{\mu\nu}(p^2 + M_n^2) - (1-\xi^{-1})p^\mu p^\nu$. The two-point function (propagator) $D_{\mu\nu}$ in four dimensions is given by $K^{\mu\nu}D_{\nu\rho} = \delta^\mu{}_\rho$. For each element one finds

$$
\begin{aligned}
&\langle A^{(n)}_{\mu,jk} A^{(n)}_{\nu,kj} \rangle : \frac{\eta_{\mu\nu}}{p^2 + M^2_{jk,n}} + \frac{p_\mu p_\nu}{M^2_{jk,n}} \left(\frac{1}{p^2 + M^2_{jk,n}} - \frac{1}{p^2 + \xi M^2_{jk,n}} \right), \\
&\langle A^{(n)}_{y,jk} A^{(n)}_{y,kj} \rangle : \frac{1}{p^2 + \xi M^2_{jk,n}} , \\
&\langle \, c^{(n)}_{jk} \, \bar{c}^{(n)}_{kj} \, \rangle \ : \frac{1}{p^2 + \xi M^2_{jk,n}} , \quad M_{jk,n} = \frac{1}{R}\Big(n - \frac{\theta_j - \theta_k}{2\pi} \Big) . \qquad (3.31)
\end{aligned}
$$

$\xi = 0, 1$ and ∞ correspond to the Landau, Feynman and unitary gauges, respectively. For $\xi = 1$ the two-point function of $A^{(n)}_{\mu,jk}$ becomes diagonal. The parts depending on the parameter ξ correspond to unphysical modes. Observable, physical quantities do not depend on ξ. For the $A^{(n)}_{\mu,jk}$ element

$$
\det D_\mu{}^\nu = \xi (p^2 + M^2_{jk,n})^{-3} (p^2 + \xi M^2_{jk,n})^{-1} , \qquad (3.32)
$$

and $c, \bar{c}$ are complex fields with two negative-norm unphysical degrees of freedom. Hence the number of degrees of freedom with ξ-dependent mass is $2 - 2 = 0$, and is cancelled. We have a spin 1 vector particle with the mass $M_{jk,n}$ carrying three physical degrees of freedom.

The Lagrangian density for a fermion field ψ in the fundamental representation of $SU(N)$ is

$$
\mathcal{L}^{\rm fermion} = \overline{\psi}(\Gamma^M D_M - m_f)\psi \ , \ \ D_M = \partial_M - igA_M \qquad (3.33)
$$

and with the boundary condition (3.21) ψ is expanded as

$$
\psi(x,y) = \sum_{n=-\infty}^{\infty} \psi^{(n)}(x) e^{i(n+\beta/2\pi)(y/R)} .
$$

By inserting $D^c_y = \partial_y - igA^c_y$ into $\mathcal{L}^{\rm fermion}$, the mass of $\psi_j^{(n)}(x)$ is found to be

$$
\psi_j^{(n)} \ : \ M^2_{j,n} = m_f^2 + \frac{1}{R^2}\Big(n + \frac{\beta - \theta_j}{2\pi} \Big)^2 . \qquad (3.34)
$$

The mass spectra $M_{jk,n}$ and $M_{j,n}$ depend on the AB phases $\{\theta_j\}$. In particular, if there is a pair (j,k) with $\theta_j \neq \theta_k$ (mod 2π), then the mass spectrum is not $SU(N)$ symmetric, that is, the $SU(N)$ gauge symmetry breaks down. If $\theta_j = \theta_k$, $A^{(0)}_{\mu,jk}(x)$ becomes the massless gauge field in four dimensions.

To what extent the gauge symmetry is broken by the AB phases can be deduced from $\hat{W}$ in (3.24) with $A^{\rm c}_y$ inserted as well:

$$\hat{W}_{\rm sym} = P\exp\left\{ - ig \int_0^{2\pi R} dy A^{\rm c}_y \right\} U \ . \tag{3.35}$$

Among the generators $\mathbf{g} = \{T^a\}$ of the gauge group $\mathcal{G}$, the set of the generators which commute with $\hat{W}_{\rm sym}$, $\mathbf{h} = \{T^a \,;\, [T^a, \hat{W}_{\rm sym}] = 0\}$, forms a subgroup $\mathcal{H}_{\rm sym}$. The original gauge symmetry $\mathcal{G}$ is broken to $\mathcal{H}_{\rm sym}$ by the AB phases. The proof for the general case $U \neq 1$ is given below after quantum effects are taken into account. The mechanism in which gauge symmetry is spontaneously broken by Aharonov-Bohm phases is called the **Hosotani mechanism.** [12, 14]

As an example, consider $SU(3)$ gauge theory. Among three AB phases there is a relation $\sum_{j=1}^3 \theta_j = 0 \pmod{2\pi}$. $\mathcal{H}_{\rm sym}$ is given by

$$\begin{array}{cl}
\mathcal{H}_{\rm sym} & (\theta_1, \theta_2, \theta_3) \\[6pt]
SU(3) & (0,0,0),\ (\pm\frac{2}{3}\pi, \pm\frac{2}{3}\pi, \pm\frac{2}{3}\pi) \\
SU(2)\times U(1) & (\alpha, \alpha, -2\alpha)\ + \text{permutations} \\
U(1)\times U(1) & (\beta_1, \beta_2, -\beta_1 - \beta_2)\ + \text{permutations.}
\end{array} \tag{3.36}$$

Here $\alpha \neq 0, \pm\frac{2}{3}\pi$ and $\beta_1 \neq \beta_2, -\frac{1}{2}\beta_2, -2\beta_2$.

In the next chapter we are going to see that in the Hosotani mechanism the values of the Aharonov-Bohm phases are determined by quantum effects once matter fields in the models are specified.

Chapter 4

Quantum effects and Hosotani mechanism

The values of Aharonov-Bohm phases are determined by the location of the global minimum of the effective potential which takes quantum effects into account. It is shown that in $SU(3)$ gauge theory on $M^4 \times S^1$ the gauge symmetry can be spontaneously broken by matter fermions.

4.1 Effective potential in quantum mechanics

Let H and $|\psi\rangle$ be the Hamiltonian and state vector respectively. The Schrödinger equation for stationary states can be obtained from the variational problem, namely the problem of finding $|\psi\rangle$ which minimizes $\langle\psi|H|\psi\rangle$ under the condition $\langle\psi|\psi\rangle = 1$. By using the method of a Lagrangian multiplier, one has

$$\delta\big\{\langle\psi|H|\psi\rangle - E(\langle\psi|\psi\rangle - 1)\big\} = \delta\langle\psi|H - E|\psi\rangle = 0\ . \tag{4.1}$$

For the ground state, $(H - E_0)|\psi_0\rangle = 0$ is derived. The energy eigenvalue E appears as a Lagrangian multiplier.

Let us impose an additional condition, $\langle\psi|Q|\psi\rangle = Q_c$ ($Q^\dagger = Q$). The variational problem then becomes

$$\delta\langle\psi|H - E - \lambda Q|\psi\rangle = 0 \quad \Rightarrow \quad (H - E_0 - \lambda Q)|\psi_0\rangle = 0\ . \tag{4.2}$$

At this stage $E_0 = E_0(\lambda)$ and $|\psi_0\rangle = |\psi_0(\lambda)\rangle$. The value of λ ($\equiv J$) is determined by $\langle\psi_0(J)|Q|\psi_0(J)\rangle = Q_c$.

$$\big\{H - JQ - E_0(J)\big\}|\psi_0(J)\rangle = 0\ . \tag{4.3}$$

Differentiating the whole of Eq. (4.3) with respect to J and taking an inner product with $\langle\psi_0(J)|$, one obtains

$$\frac{dE_0(J)}{dJ} = -Q_c\ . \tag{4.4}$$

The relation $Q_c = \langle\psi_0(J)|Q|\psi_0(J)\rangle$ can be written as $Q_c = Q_c(J)$ and $J = J(Q_c)$. The **effective potential** for Q_c, $V_{\rm eff}(Q_c)$, is defined as

$$V_{\rm eff}(Q_c) = \langle\psi_0(J)|H|\psi_0(J)\rangle = E_0(J) + JQ_c \ . \tag{4.5}$$

Differentiating the both sides of Eq. (4.5) with respect to Q_c and making use of the relation (4.4), we find

$$\frac{dV_{\rm eff}(Q_c)}{dQ_c} = J \ . \tag{4.6}$$

$E_0(J)$ is an eigenvalue of $H - JQ$, the smallest energy in the presence of an external source J for Q. $V_{\rm eff}(Q_c)$ is the smallest energy among the states with $\langle\psi|Q|\psi\rangle = Q_c$. These two are related to each other by (4.5). We are interested in the ground state of the Hamiltonian H, for which $J = 0$ and therefore the relation (4.6) implies that **the ground state corresponds to the global minimum of the effective potential**, $V_{\rm eff}(Q_c)$. [15]

$V_{\rm eff}(Q_c)$ is defined with all quantum corrections taken into account. In terms of $V_{\rm eff}(Q_c)$, a criterion for being the true ground state is given. When $V_{\rm eff}(Q_c)$ has a global minimum at $Q_c = Q_c^{\rm min}$, then in the ground state $|\psi_G\rangle$, $\langle\psi_G|Q|\psi_G\rangle = Q_c^{\rm min}$.

In the case of the Hamiltonian $H = p^2/2m + V(x)$, $V_{\rm eff}(x_c)$ for $x_c = \langle\psi|x|\psi\rangle$ differs from the original potential; $V_{\rm eff}(x_c) \neq V(x_c)$. The location of the minimum of $V_{\rm eff}(x_c)$ differs from that of $V(x)$ in general.

4.2 Effective potential in field theory

The above argument can be extended to field theory. Let us consider a scalar field $\phi(x)$ in four dimensions. Let $\mathcal{L}(\phi)$ and $\mathcal{H}(\phi)$ be the Lagrangian and Hamiltonian densities, respectively. We introduce an external source $J(x)$ coupling to $\phi(x)$ in the form $\mathcal{L}_J = \mathcal{L} + \phi\, J$. The corresponding Hamiltonian is $H_J = H - \int d^3x\, \phi\, J$. In the path-integral representation we define $W[J]$ by

$$\begin{aligned} e^{iW[J]} &= \lim_{T\to\infty} \langle\Psi_0|e^{-iTH_J}|\Psi_0\rangle/\langle\Psi_0|e^{-iTH}|\Psi_0\rangle \\ &= N\int \mathcal{D}\phi\, e^{i\int d^4x\, \{\mathcal{L}(\phi)+\phi J\}} \ . \end{aligned} \tag{4.7}$$

Here $|\Psi_0\rangle$ is the ground state (the vacuum) of H. We assume that $J(x) \to 0$ in the limit $t \to \pm\infty$, and $J(x)$ is turned on and off adiabatically. The normalization factor N is determined such that $W[0] = 0$. Rigorously speaking, it has been assumed in the second equality in (4.7) that the analytic continuation from real t to imaginary time $\tau = it$ is possible. Let $\mathcal{E}(J)$ be the energy density of the ground state of H_J defined adiabatically. Then

$$W[J] = \int d^4x \{ -\mathcal{E}(J) + \mathcal{E}(0) \} \tag{4.8}$$

and

$$\begin{aligned}\frac{\delta W[J]}{\delta J(x)} &= \langle \phi(x) \rangle = \langle \Psi_0(J_t) | \phi(t, \vec{x}) | \Psi_0(J_t) \rangle \\ &\equiv \phi_c(x) \ ,\end{aligned} \tag{4.9}$$

where $|\Psi_0(J_t)\rangle$ is the ground state of H_J at time t. In the adiabatic process of switching J on and off, the ground state continuously changes from $|\Psi_0\rangle$ to $|\Psi_0(J_t)\rangle$, and to $|\Psi_0\rangle$. Equation (4.9) takes the same form as Eq. (4.4) in quantum mechanics. Following the definition (4.5), we define the **effective action** $\Gamma[\phi_c]$ by

$$\Gamma[\phi_c] = W[J] - \int d^4x \, J \, \phi_c \tag{4.10}$$

to find

$$\frac{\delta \Gamma[\phi_c]}{\delta \phi_c(x)} = -J(x) \ . \tag{4.11}$$

When the x-dependence of $\phi_c(x)$ is small, one can expand $\Gamma[\phi_c]$ as

$$\Gamma[\phi_c] = \int d^4x \Big\{ -V_{\rm eff}(\phi_c) - \frac{1}{2} Z(\phi_c) \partial_\mu \phi_c \partial^\mu \phi_c + \cdots \Big\} \ . \tag{4.12}$$

The first term $V_{\rm eff}(\phi_c)$ is called the **effective potential.**

In particular, for constant ϕ_c, $V_{\rm eff}(\phi_c)$ is the smallest energy density $\langle \Psi | \mathcal{H}_J | \Psi \rangle$ among the states $|\Psi\rangle$ with the constraint $\langle \Psi | \phi(x) | \Psi \rangle = \phi_c$ and satisfies

$$\frac{dV_{\rm eff}(\phi_c)}{d\phi_c} = J \ . \tag{4.13}$$

$J = 0$ in the ground state (the vacuum) $|\Psi_0\rangle$ of the Hamiltonian H. Let $V_{\rm eff}(\phi_c)$ be minimized at $\phi_c = \phi_c^{\rm min}$. Then $\langle \Psi_0 | \phi(x) | \Psi_0 \rangle = \phi_c^{\rm min}$. In field theory as well **the ground state (the vacuum) of the Hamiltonian corresponds to the global minimum of the effective potential.**

How can $V_{\rm eff}(\phi_c)$ be evaluated? We describe here a perturbative method called the loop expansion ($\hbar$ expansion). [16,17] Consider a real scalar field $\hat{\phi}(x)$ with Lagrangian density $\mathcal{L}(\hat{\phi}) = -\frac{1}{2}\partial_\mu \hat{\phi} \, \partial^\mu \hat{\phi} - V_0(\hat{\phi})$ where $V_0 =$

$\frac{1}{2}m^2\hat{\phi}^2+(\lambda/4!)\,\hat{\phi}^4$. For $\langle\hat{\phi}\rangle = \langle\Psi|\hat{\phi}|\Psi\rangle = \phi_c = \text{constant}$, we write $\hat{\phi} = \phi_c+\phi$. Expanding $\mathcal{L}$ in ϕ, we obtain

$$\begin{aligned}
\mathcal{L} &= -V_0(\phi_c) - V_0'(\phi_c)\phi + \mathcal{L}_2 + \mathcal{L}_{int} \ , \\
\mathcal{L}_2 &= -\frac{1}{2}\partial_\mu\phi\partial^\mu\phi - \frac{1}{2}\Big(m^2 + \frac{\lambda}{2}\,\phi_c^2\Big)\phi^2 \ , \\
\mathcal{L}_{int} &= -\frac{\lambda}{6}\phi_c\phi^3 - \frac{\lambda}{4!}\,\phi^4 \ .
\end{aligned} \tag{4.14}$$

From $\mathcal{L}_2$ the propagator $\mathcal{D}$ is given by

$$\mathcal{D}(p;\phi_c) = \frac{-i}{p^2 + m^2 + \frac{1}{2}\lambda\phi_c^2 - i\epsilon} \ , \tag{4.15}$$

and the effective potential becomes

$$\begin{aligned}
V_{\rm eff}(\phi_c) =& V_0(\phi_c) - \frac{i}{2}\int\frac{d^4p}{(2\pi)^4}\,\ln\, i\mathcal{D}^{-1}(p;\phi_c) \\
&+ i\Big\langle \exp\Big\{ i\int d^4x\,\mathcal{L}_{int}(\phi;\phi_c)\Big\}\Big\rangle_{\rm 1PI} \ .
\end{aligned} \tag{4.16}$$

The last term represents the sum of 1PI (1-Particle Irreducible) bubble diagrams with two or more loops, namely vacuum diagrams which cannot be separated by cutting one internal line, generated with $\mathcal{D}$ as the propagator and $\mathcal{L}_{int}$ as interaction vertices. The second term is called the 1-loop effective potential $V_{\rm eff}^{\rm 1\,loop}$.

The 1-loop effective potential is interpreted as the sum of zero-point energies. Indeed

$$\begin{aligned}
V_{\rm eff}(\phi_c)^{\rm 1\,loop} &= -\frac{i}{2}\int\frac{d^4p}{(2\pi)^4}\,\ln\,\Big(p^2 + m^2 + \frac{\lambda}{2}\phi_c^2 - i\epsilon\Big) \\
&= -\frac{i}{2}\int\frac{d^3p}{(2\pi)^3}\int_{-\infty}^{\infty}\frac{dp_0}{2\pi}\,\ln\big\{ -p_0^2 + E(\vec{p},\phi_c)^2 - i\epsilon\big\} \ , \\
E(\vec{p},\phi_c;a) &= \sqrt{\vec{p}^{\,2} + a\big(m^2 + \tfrac{1}{2}\lambda\phi_c^2\big)} \ , \quad E(\vec{p},\phi_c) \equiv E(\vec{p},\phi_c;1) \ ,
\end{aligned} \tag{4.17}$$

and by making use of

$$\frac{d}{da}\int_{-\infty}^{\infty}\frac{dp_0}{2\pi}\,\ln\big\{ -p_0^2 + E(\vec{p},\phi_c;a)^2 - i\epsilon\big\} = i\frac{d}{da}E(\vec{p},\phi_c;a) \ , \tag{4.18}$$

we find that up to a ϕ_c-independent constant term

$$(vol)V_{\rm eff}(\phi_c)^{\rm 1\,loop} = \int\frac{d^3x d^3p}{(2\pi\hbar)^3}\,\frac{1}{2}E(\vec{p},\phi_c) \tag{4.19}$$

where $\hbar$ has been restored. In other words, $(vol)V_{\rm eff}(\phi_c)^{1\,\rm loop}$ represents the variation in the zero-point energies when $\langle\hat{\phi}\rangle = \phi_c$.

The formula (4.19) is for a boson field. In the case of a fermion field a zero-point energy is given by $-\frac{1}{2}E(\vec{p})$. In general

$$V_{\rm eff}(\phi_c)^{1\,\rm loop} = \sum \mp\frac{i}{2}\int\frac{d^4p}{(2\pi)^4}\ln\left\{p^2+m(\phi_c)^2-i\epsilon\right\}. \tag{4.20}$$

The sign is $-$ $(+)$ for bosons (fermions). $m(\phi_c)$ is the effective mass when $\langle\hat{\phi}\rangle = \phi_c$. The sum extends over all degrees of freedom. For a Dirac fermion, for instance, a factor 4 should be multiplied as it has four degrees of freedom.

4.3 Effective potential for AB phases $V_{\rm eff}(\theta_H)$

In the previous chapter we have seen that gauge symmetry in non-Abelian gauge theory on $M^4\times S^1$ is spontaneously broken by Aharonov-Bohm (AB) phases. The pattern of symmetry breaking depends on the values of the AB phases, θ_H, which are determined to minimize the effective potential $V_{\rm eff}(\theta_H)$. At the classical level $V_{\rm eff}(\theta_H)$ vanishes, as the field strengths F_{MN} vanish for the AB phases. One sees that $V_{\rm eff}(\theta_H) = V_{\rm eff}^{1\,\rm loop} + V_{\rm eff}^{2\,\rm loop} + \cdots$.

In $SU(N)$ gauge theory with $U=1$ in the boundary conditions (3.21) the AB phases θ_j ($j = 1 \sim N$) is related to $\langle A_y\rangle = A_y^c$ by Eq. (3.26). The effective masses of gauge fields, ghost fields and fermion fields are given by (3.31), (3.32) and (3.34), respectively. From the formula (4.20) the 1-loop potential $V_{\rm eff}^{1\,\rm loop}$ in the presence of $N_{\rm f}$ fermion fields in the fundamental representation is given by

$$\begin{aligned} V_{\rm eff}^{1\,\rm loop}(\theta_j) = &-3\sum_{j,k=1}^{N}\sum_{n=-\infty}^{\infty}\frac{i}{2}\int\frac{d^4p}{(2\pi)^4}\ln\left\{p^2+M_{jk,n}^2-i\epsilon\right\} \\ &+4N_{\rm f}\sum_{j=1}^{N}\sum_{n=-\infty}^{\infty}\frac{i}{2}\int\frac{d^4p}{(2\pi)^4}\ln\left\{p^2+M_{j,n}^2-i\epsilon\right\}. \end{aligned}\tag{4.21}$$

Here we have taken into account that two degrees of freedom of the gauge fields with the effective mass $\xi M_{jk,n}$ are cancelled by two degrees of freedom of the ghost fields so that three degrees of freedom of the gauge fields with the effective mass $M_{jk,n}$ contribute to $V_{\rm eff}^{1\,\rm loop}$ and that each fermion field has four degrees of freedom. We also have utilized the fact that in the sum $\sum_{j,k}$ in the $SU(N)$ space the diagonal components $j=k$ are independent of θ_j.

The contribution to $V_{\rm eff}^{\rm 1\, loop}$ from one degree of freedom in gauge theory on $M^d \times S^1$ is expressed in terms of

$$I(\theta, m; d) = -\frac{i}{2} \int \frac{d^d p}{(2\pi)^d} \sum_{n=-\infty}^{\infty} \ln \left\{ p^2 + m^2 + \frac{1}{R^2}\Big(n - \frac{\theta}{2\pi}\Big)^2 - i\epsilon \right\}. \tag{4.22}$$

Although the integration and infinite sum in this expression diverges, the θ-dependent part of $I(\theta, m; d)$ is finite. The proof is given in Appendix B. The result is

$$\begin{aligned} & I(\theta, m; d) - I(0, m; d) \\ & \quad = \frac{\Gamma\big[\frac{d+1}{2}\big]}{\pi^{(d+1)/2}(2\pi R)^d} \sum_{n=1}^{\infty} \frac{1-\cos n\theta}{n^{d+1}} B_{(d+1)/2}(2\pi R\, nm) \ , \\ & B_\nu(z) = \frac{z^\nu K_\nu(z)}{2^{\nu-1}\Gamma(\nu)} \ , \quad B_\nu(0) = 1 \ . \end{aligned} \tag{4.23}$$

Here $K_\nu(z)$ is the modified Bessel function. The convergence in the infinite sum is very fast, and the expression can be well approximated by the $n=1$ term. $V_{\rm eff}^{\rm 1\, loop}$ in $SU(N)$ gauge theory is

$$\begin{aligned} V_{\rm eff}^{\rm 1\, loop}(\theta_j) = \frac{3}{64\pi^6 R^4} \bigg\{ & -3 \sum_{j,k=1}^{N} \sum_{n=1}^{\infty} \frac{\cos n(\theta_j - \theta_k)}{n^5} \\ & + 4N_{\rm f} \sum_{j=1}^{N} \sum_{n=1}^{\infty} \frac{\cos n(\theta_j - \beta)}{n^5} B_{5/2}(2\pi R\, nm) \bigg\} + ({\rm const}) \end{aligned} \tag{4.24}$$

where $\sum_{j=1}^{N} \theta_j = 0 \pmod{2\pi}$. The fermion contribution can be neglected for $mR \gg 1$. β is the phase in the boundary condition $\psi(x, y + 2\pi R) = e^{i\beta}\psi(x,y)$.

4.4 Hosotani mechanism II – dynamical gauge symmetry breaking

In pure $SU(N)$ gauge theory without fermion fields the effective potential in (4.24) takes the minimum value when $\theta_j = \theta_k$ and $\theta_j = 2\pi\ell/N$ ($\ell = 0, 1, \cdots, N-1$). $SU(N)$ gauge symmetry is preserved. In pure $SU(N)$ gauge theory there are N degenerate vacua, reflecting the Z_N symmetry.

The symmetry structure becomes rich in the presence of fermions. Let us consider the case in which there are $N_{\rm f}$ massless fermions in the fundamental representation ($\psi_{\rm f}(x, y + 2\pi R) = e^{i\beta_{\rm f}}\psi_{\rm f}(x,y)$) and $N_{\rm ad}$ massless fermions in the adjoint representation ($\psi_{\rm ad}(x, y + 2\pi R) = e^{i\beta_{\rm ad}}\psi_{\rm ad}(x, y)$). $V_{\rm eff}^{\rm 1\,loop}(\theta_j)$ is given by

$$
\begin{aligned}
V_{\rm eff}^{\rm 1\,loop}(\theta_j) &\equiv \frac{3}{64\pi^6 R^4}\hat{V}_{\rm eff}^{\rm 1\,loop}(\theta_j) \ , \\
\hat{V}_{\rm eff}^{\rm 1\,loop}(\theta_j) &= -3\sum_{j,k=1}^{N} h(\theta_j - \theta_k) + 4N_{\rm f}\sum_{j=1}^{N} h(\theta_j - \beta_{\rm f}) \\
&\qquad + 4N_{\rm ad}\sum_{j,k=1}^{N} h(\theta_j - \theta_k - \beta_{\rm ad}) \ , \\
h(\theta) &= h_5(\theta) \ , \quad h_d(\theta) = \sum_{n=1}^{\infty}\frac{\cos n\theta}{n^d} \ .
\end{aligned}
\tag{4.25}
$$

At the one loop level the pattern of symmetry breaking is determined by the value of $(N_{\rm f}, \beta_{\rm f}, N_{\rm ad}, \beta_{\rm ad})$.

Let us take $SU(3)$ gauge theory as an example. $\theta_1 + \theta_2 + \theta_3 = 0 \pmod{2\pi}$ so that there are two independent AB phases. The behavior of $\hat{V}_{\rm eff}^{\rm 1\,loop}$ is depicted in Fig. 4.1.

(i) $N_{\rm f} = N_{\rm ad} = 0$

The degenerate global minima are located at

$$
(\theta_1, \theta_2, \theta_3) = (0,0,0), (\pm\tfrac{2}{3}\pi, \pm\tfrac{2}{3}\pi, \pm\tfrac{2}{3}\pi) \quad : SU(3) \ . \tag{4.26}
$$

$SU(3)$ gauge symmetry is preserved.

(ii) $N_{\rm f} \neq 0,\ N_{\rm ad} = 0,\ \beta_{\rm f} = 0,\ m_{\rm f} = 0$

We take θ_1, θ_2 as independent variables. $\hat{V}_{\rm eff}^{\rm 1\,loop}$ is expressed as

$$
\begin{aligned}
\hat{V}_{\rm eff}^{\rm 1\,loop} &= -6\big\{h(\theta_1 - \theta_2) + h(\theta_1 + 2\theta_2) + h(2\theta_1 + \theta_2)\big\} \\
&\quad + 4N_{\rm f}\big\{h(\theta_1 - \beta_{\rm f}) + h(\theta_2 - \beta_{\rm f}) + h(\theta_1 + \theta_2 + \beta_{\rm f})\big\}.
\end{aligned}
\tag{4.27}
$$

The location of the global minimum of $\hat{V}_{\rm eff}^{\rm 1\,loop}$ depends on the parameter $\beta_{\rm f}$:

$$
(\theta_1, \theta_2, \theta_3) = \begin{cases}
(-\frac{2}{3}\pi, -\frac{2}{3}\pi, -\frac{2}{3}\pi) & \text{for } 0 < \beta_{\rm f} < \frac{2}{3}\pi, \\
(0,0,0) & \text{for } \frac{2}{3}\pi < \beta_{\rm f} < \frac{4}{3}\pi, \\
(\frac{2}{3}\pi, \frac{2}{3}\pi, \frac{2}{3}\pi) & \text{for } \frac{4}{3}\pi < \beta_{\rm f} < 2\pi.
\end{cases}
\tag{4.28}
$$

For $\beta_{\rm f} = 0, \pm\frac{2}{3}\pi$ two minima become degenerate. In all cases $SU(3)$ gauge symmetry is preserved.

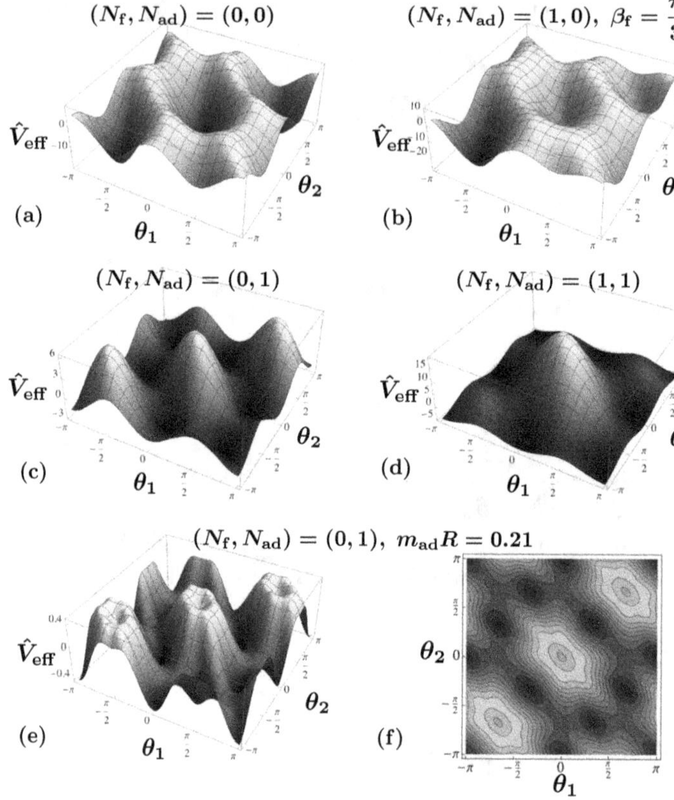

Fig. 4.1 Effective potential $\hat{V}_{\text{eff}}^{\text{1 loop}}(\theta_1, \theta_2)$ in $SU(3)$ gauge theory on $M^4 \times S^1$ is plotted with various fermion content. (a) $(N_{\text{f}}, N_{\text{ad}}) = (0, 0)$. (b) $(N_{\text{f}}, N_{\text{ad}}) = (1, 0)$, $m_{\text{f}} = 0$ and $\beta_{\text{f}} = \frac{1}{3}\pi$. (c) $(N_{\text{f}}, N_{\text{ad}}) = (0, 1)$, $m_{\text{ad}} = 0$ and $\beta_{\text{ad}} = 0$. (d) $(N_{\text{f}}, N_{\text{ad}}) = (1, 1)$, $m_{\text{f}} = m_{\text{ad}} = 0$ and $\beta_{\text{f}} = \beta_{\text{ad}} = 0$. (e) $(N_{\text{f}}, N_{\text{ad}}) = (0, 1)$, $m_{\text{ad}}R = 0.21$ and $\beta_{\text{ad}} = 0$. (f) Contour plot of $\hat{V}_{\text{eff}}^{\text{1 loop}}$ in the case (e). Gauge symmetry at the global minima is $SU(3)$ for (a), (b), $U(1) \times U(1)$ for (c), and $SU(2) \times U(1)$ for (d), (e).

(iii) $N_{\text{f}} = 0,\ N_{\text{ad}} \neq 0, \beta_{\text{ad}} = 0, m_{\text{ad}} = 0$

In this case

$$\hat{V}_{\text{eff}}^{\text{1 loop}} = -2(3 - 4N_{\text{ad}})\big\{h(\theta_1 - \theta_2) + h(\theta_1 + 2\theta_2) + h(2\theta_1 + \theta_2)\big\}\ . \quad (4.29)$$

The sign of $\hat{V}_{\text{eff}}^{\text{1 loop}}$ is flipped compared to that in the case $N_{\text{f}} = N_{\text{ad}} = 0$. The global minimum is located at

$$(\theta_1, \theta_2, \theta_3) = (0, \tfrac{2}{3}\pi, -\tfrac{2}{3}\pi) \text{ and permutations } : \ U(1) \times U(1)\ . \quad (4.30)$$

$SU(3)$ gauge symmetry breaks down to $U(1) \times U(1)$ gauge symmetry.

(iv) $N_{\rm f} = 1,\ N_{\rm ad} = 1, \beta_{\rm f} = \beta_{\rm ad} = 0, m_{\rm f} = m_{\rm ad} = 0$

In this case $\hat{V}_{\rm eff}^{\rm 1\,loop}$ is minimized at

$$(\theta_1, \theta_2, \theta_3) = (\pi, \pi, 0), \text{ and permutations } : SU(2) \times U(1) . \tag{4.31}$$

$SU(3)$ gauge symmetry breaks down to $SU(2) \times U(1)$ gauge symmetry.

(v) $N_{\rm f} = 0,\ N_{\rm ad} \neq 0, \beta_{\rm ad} = 0, m_{\rm ad} \neq 0$

When the adjoint fermion has a mass $m_{\rm ad}$, there arises a situation intermediate between the cases (i) and (iii) above. If $m_{\rm ad} R$ is sufficiently large, it becomes the same as in the case (i) and $SU(3)$ gauge symmetry is preserved. If $m_{\rm ad} R$ is sufficiently small, it becomes the same as in the case (iii) and the symmetry spontaneously breaks down to $U(1) \times U(1)$ gauge symmetry. For $N_{\rm ad} = 1$ and $0.188 < m_{\rm ad} R < 0.223$, the global minima of $\hat{V}_{\rm eff}^{\rm 1\,loop}$ are located at

$$(\theta_1, \theta_2, \theta_3) = (\pi, \pi, 0),\ (\pm\tfrac{1}{3}\pi, \pm\tfrac{1}{3}\pi, \mp\tfrac{2}{3}\pi),$$
$$\text{and permutations } : SU(2) \times U(1) . \tag{4.32}$$

$SU(3)$ gauge symmetry breaks down to $SU(2) \times U(1)$ gauge symmetry. We add that in $SU(3)$ gauge theory on $M^3 \times S^1$ with $N_{\rm ad} = 1$, the global minima of $\hat{V}_{\rm eff}^{\rm 1\,loop}$ are located at $SU(2) \times U(1)$ symmetric points for $0.145 < m_{\rm ad} R < 0.187$.

It is seen that gauge symmetry can be spontaneously broken at the quantum level, depending on the content of matter fields (fermion fields in the present case). In the Hosotani mechanism the values of the AB phases θ_j are determined at the quantum level. The AB phases are gauge-invariant quantities. **The Hosotani mechanism is a mechanism for dynamical gauge symmetry breaking by gauge invariant AB phases.** [12, 14, 18–25]

One comment on $U(1)$ gauge theory on $M^4 \times S^1$ is appropriate. In $U(1)$ gauge theory the AB phase θ_H is related to the constant $A_y = A_y^c$ by $\theta_H = 2\pi g R A_y^c$. In $U(1)$ theory the gauge symmetry is not broken, but $V_{\rm eff}(\theta_H)$ has nontrivial θ_H dependence. When there are $N_{\rm f}$ fermions with charge g and mass m, and they satisfy boundary condition $\psi(x, y + 2\pi R) = e^{i\beta}\psi(x, y)$, $V_{\rm eff}^{\rm 1\,loop}(\theta_H)$ becomes

$$V_{\rm eff}^{\rm 1\,loop}(\theta_H) = \frac{3}{64\pi^6 R^4} \cdot 4N_{\rm f} \sum_{n=1}^{\infty} \frac{\cos n(\theta_H - \beta)}{n^5} B_{5/2}(2\pi R n m) . \tag{4.33}$$

When fermions are periodic ($\beta = 0$), $V_{\rm eff}^{\rm 1\,loop}$ has the minimum at $\theta_H = \pi$. On the other hand, when fermions are anti-periodic ($\beta = \pi$), the minimum of $V_{\rm eff}^{\rm 1\,loop}$ is located at $\theta_H = 0$.

4.5 Masses of Higgs modes (A_y excitations)

In the examples discussed above the AB phases θ_j correspond to the diagonal components of the background field A_y^c of A_y. As seen in Eq. (3.31) fluctuation modes around A_y^c have masses $\sqrt{\xi}M_{jk,n}$. In particular, the $n = 0$ and $j = k$ modes are massless. It is due to the fact that the energy at the classical level does not depend on the values of the AB phases. At the quantum level these zero modes acquire finite masses.

Let us denote the zero mode part of $A_y(x, y)$ as $A_y^{(0)}(x) = A_y^c + H(x)$. The AB phase is then given by $\theta_H = \sqrt{2\pi R}\, agA_y^c$, where the coefficient a depends on gauge group and gauge components. In $U(1)$ gauge theory $a = 1$. The fluctuation mode $H(x)$ is a field with the normalization $\int_0^{2\pi R} dy\, \mathcal{L} = -\frac{1}{2}\partial_\mu H \partial^\mu H + \cdots$. This mode becomes the 4D Higgs field when the scenario is applied to electroweak interactions below. We shall call these $H(x)$ modes **Higgs modes**. We define $\hat{\theta}_H(x)$ by

$$\hat{\theta}_H(x) = \theta_H + \frac{H(x)}{f_H}\,, \quad f_H = \frac{1}{\sqrt{2\pi R}\, ag} = \frac{1}{2\pi aRg_4}\,. \tag{4.34}$$

Here $g_4 = g\sqrt{2\pi R}$ is the 4D gauge coupling constant. The effective potential for the Higgs mode $H(x)$ is $V_{\rm eff}(\theta_H)|_{\theta_H \to \hat{\theta}_H}$. The mass m_H of the Higgs mode is given by

$$m_H^2 = \frac{1}{f_H^2} \frac{d^2 V_{\rm eff}}{d\theta_H^2}\bigg|_{\rm min}\,. \tag{4.35}$$

With the use of (4.25) and (4.34), one finds that $m_H = O(g_4)$ at the one loop level.

In $SU(3)$ gauge theory on $M^4 \times S^1$ there are two independent AB phases, and therefore there appear two independent Higgs modes. The diagonal components of A_y^c are decomposed as

$$\begin{aligned}
&\frac{1}{\sqrt{2\pi R}} \left(A_y^{c(3)} T^3 + A_y^{c(8)} T^8 \right) \\
&= \frac{1}{2\sqrt{2\pi R}} \begin{pmatrix} A_y^{c(3)} + \frac{1}{\sqrt{3}} A_y^{c(8)} & & \\ & -A_y^{c(3)} + \frac{1}{\sqrt{3}} A_y^{c(8)} & \\ & & -\frac{2}{\sqrt{3}} A_y^{c(8)} \end{pmatrix} \\
&= \frac{1}{2\pi Rg} \begin{pmatrix} \theta_1 & & \\ & \theta_2 & \\ & & \theta_3 \end{pmatrix},
\end{aligned} \tag{4.36}$$

so that

$$(\theta_1, \theta_2) = g\sqrt{\frac{\pi R}{2}}\Big(A_y^{c(3)} + \frac{1}{\sqrt{3}}A_y^{c(8)}, -A_y^{c(3)} + \frac{1}{\sqrt{3}}A_y^{c(8)}\Big). \tag{4.37}$$

Denoting the fluctuation modes of $A_y^{(3)}$ and $A_y^{(8)}$ by $H_3(x)$ and $H_8(x)$, one finds that

$$\begin{aligned} &\hat{\theta}_1(x) = \theta_1 + \frac{H_3(x)}{f_{H_3}} + \frac{H_8(x)}{f_{H_8}} ~, \quad \hat{\theta}_2(x) = \theta_2 - \frac{H_3(x)}{f_{H_3}} + \frac{H_8(x)}{f_{H_8}} ~, \\ &f_{H_3} = \frac{1}{\pi R g_4} ~, \quad f_{H_8} = \frac{\sqrt{3}}{\pi R g_4} ~. \end{aligned} \tag{4.38}$$

The effective potential for $H_3(x), H_8(x)$ is given by $V_{\rm eff}(\theta_1,\theta_2)|_{\theta_j\to\hat{\theta}_j}$. It follows that the masses m_3, m_8 of $H_3(x), H_8(x)$ are given by

$$\begin{aligned} m_3^2 &= \frac{\partial^2}{\partial H_3^2} V_{\rm eff}\Big|_{\rm min} = \frac{1}{f_{H_3}^2}\Big(\frac{\partial}{\partial\theta_1} - \frac{\partial}{\partial\theta_2}\Big)^2 V_{\rm eff}\Big|_{\rm min} ~, \\ m_8^2 &= \frac{\partial^2}{\partial H_8^2} V_{\rm eff}\Big|_{\rm min} = \frac{1}{f_{H_8}^2}\Big(\frac{\partial}{\partial\theta_1} + \frac{\partial}{\partial\theta_2}\Big)^2 V_{\rm eff}\Big|_{\rm min} ~. \end{aligned} \tag{4.39}$$

Note that $\partial^2 V_{\rm eff}/\partial H_3 \partial H_8|_{\rm min} = 0$ as $V_{\rm eff}(\theta_1,\theta_2) = V_{\rm eff}(\theta_2,\theta_1)$.

In pure gauge theory without fermions ($N_{\rm f} = N_{\rm ad} = 0$) $SU(3)$ gauge symmetry is preserved and it follows from (4.27) and (4.39) that

$$m_3^2 = m_8^2 = \frac{27\zeta_R(3) g_4^2}{16\pi^4 R^2} ~. \tag{4.40}$$

Here $\zeta_R(z) = \sum_{n=1}^\infty 1/n^z$ is Riemann's zeta function.

In the case $N_{\rm f} = 0$, $N_{\rm ad} \neq 0$, $m_{\rm ad} = 0$ and $\beta_{\rm ad} = 0$, the gauge symmetry is broken to $U(1) \times U(1)$. With (4.29) and (4.30) one finds that

$$\begin{aligned} &m_3^2 = m_8^2 = \frac{9(3 - 4N_{\rm ad}) g_4^2}{16\pi^4 R^2}\, h^{(2)}(\tfrac{2}{3}\pi) ~, \\ &h^{(2)}(\tfrac{2}{3}\pi) = \sum_{n=1}^\infty \frac{\cos\frac{2}{3}\pi n}{n^3} \sim -0.5343 ~. \end{aligned} \tag{4.41}$$

In the case $N_{\rm f} = 0$, $N_{\rm ad} = 1$, $0.188 < m_{\rm ad}R < 0.223$, ($N_{\rm ad} = 2$, $0.406 < m_{\rm ad}R < 0.456$) and $\beta_{\rm ad} = 0$, the gauge symmetry is broken to $SU(2)\times U(1)$. In this case $m_3^2 \neq m_8^2$.

4.6 Equivalence classes of boundary conditions and AB phases

In Sections 4.3–4.5 we analyzed $SU(N)$ gauge theory with gauge fields obeying the periodic boundary condition $U = 1$ in (3.21). Does $U \neq 1$ imply gauge symmetry breaking? In this respect dynamics of AB phases and its effective potential $V_{\rm eff}(\theta_H)$ play an important role. **Physical symmetry is determined by the combination of boundary conditions and dynamics of AB phases.**

An unitary matrix $U \in SU(N)$ can be diagonalized in the form $U = V^{-1}\Lambda V$ $[\Lambda_{jk} = e^{i\lambda_j}\delta_{jk}, V \in SU(N)]$. $A'_M = VA_MV^{-1}$ and $\psi' = V\psi$ satisfy the boundary condition with $U' = \Lambda$ in (3.22). That is to say, without loss of generality one may assume $U_{jk} = e^{i\lambda_j}\delta_{jk}$, $\det U = 1$.

Next we make a gauge transformation

$$\Omega(y) = \begin{pmatrix} e^{i\alpha_1 y/R} & & & \\ & e^{i\alpha_2 y/R} & & \\ & & \ddots & \\ & & & e^{i\alpha_N y/R} \end{pmatrix}, \quad \sum_{j=1}^{N} \alpha_j = 0 \ . \tag{4.42}$$

Then $U' = \Omega(y + 2\pi R)U\Omega(y)^{-1}$ in (3.22) becomes

$$U' = \begin{pmatrix} e^{i(\lambda_1+2\pi\alpha_1)} & & & \\ & e^{i(\lambda_2+2\pi\alpha_2)} & & \\ & & \ddots & \\ & & & e^{i(\lambda_N+2\pi\alpha_N)} \end{pmatrix}. \tag{4.43}$$

In other words the boundary condition changes unless all α_j's are integers. Ω in (4.42) does not belong to $\mathcal{G}_{\rm res}$ in (3.23). Such Ω is called a large gauge transformation. AB phases $\{\theta_j\}$ are related to the background field A^c_y by (3.26). After the gauge transformation $A'^c_y = \Omega A^c_y \Omega^{-1} + (i/g)\Omega\partial_y\Omega^{-1}$ so that the AB phases change to

$$\theta'_j = \theta_j + 2\pi\alpha_j \ . \tag{4.44}$$

When two sets of boundary conditions $BC = \{U, \beta\}$ and $BC' = \{U', \beta\}$ are connected by a large gauge transformation, we write $BC \sim BC'$. The relation $\sim$ is an equivalence relation, which defines **equivalence classes of boundary conditions.** [14]

It is seen in Sections 4.1–4.4 that the values of the AB phases are determined by the location of the global minimum of the effective potential

$V_{\rm eff}(\theta_j; U, \beta)$ once the boundary conditions are specified. When $\{U, \beta\}$ and $\{U', \beta\}$ belong to the same equivalence class, it is confirmed that

$$V_{\rm eff}(\theta_j; U, \beta) = V_{\rm eff}(\theta'_j; U', \beta) \tag{4.45}$$

where $\{\theta'_j\}$ is related to $\{\theta_j\}$ by the gauge transformation in (4.44). To see it, first note that when A^c_y satisfies the equations of motion, as does the corresponding, A'^c_y. In the path-integral quantization of gauge fields, the terms which are not manifestly gauge-invariant are gauge-fixing terms and ghost terms. In the current case the gauge-fixing function $f_{\rm gf}(A; A^c) = \partial^\mu A^q_\mu + \xi(\partial_y A^q_y - ig[A^c_y, A^q_y])$ satisfies $f_{\rm gf}(A'; A'^c) = \Omega f_{\rm gf}(A; A^c)\Omega^{-1}$, and $\mathcal{L}^{\rm gauge}$ in (3.27) is invariant under the transformation. Consequently, the relation (4.45) is derived.

The relation (4.45) implies that the **physics depends solely on the equivalence class of boundary conditions**, and that even for two different sets of boundary conditions, the physics remains the same provided the two sets belong to the same equivalence class. When $V_{\rm eff}(\theta_j; U, \beta)$ is minimized at $\{\theta_j^{\rm min}\}$, $V_{\rm eff}(\theta'_j; U', \beta)$ is minimized at $\{\theta'^{\rm min}_j\}$. With the boundary condition $\{U, \beta\}$ one formulates a theory around $\{\theta_j^{\rm min}\}$. With the boundary condition $\{U', \beta\}$ one formulates a theory around $\{\theta'^{\rm min}_j\}$. These two procedures yield the same physics. The dynamics of the AB phases guarantees this equivalence.

In $SU(N)$ gauge theory on $M^n \times S^1$, the choice of $\alpha_j = -\lambda_j/2\pi$ in (4.42) leads to $U' = 1$. Apart from the twist phases $\{\beta\}$ for matter fields, there is only one equivalence class of boundary conditions. The situation becomes richer in $SU(N)$ gauge theory on $M^n \times T^2$. There are boundary conditions in two extra dimensions (T^2) and there appear AB phases along the two extra dimensions. The boundary conditions are classified into a finite number of equivalence classes. In the later chapters we need to consider gauge theory with the orbifold structure in the fifth dimension. We shall see that there appear a finite number of equivalence classes of boundary conditions there, too.

4.7 Finiteness of $V_{\rm eff}(\theta_H)$

It is seen in Sections 4.3–4.5 that the effective potential for the AB phase, $V_{\rm eff}(\theta_H)$, is finite at the one loop level, and the mass of the Higgs mode corresponding to quantum fluctuations of the AB phase also becomes finite. Gauge theory in five or more dimensions is not renormalizable, and radiative corrections at higher orders are in general divergent and cannot be controlled in an orderly manner. Nevertheless $V_{\rm eff}(\theta_H)$ is finite. There must be a reason for that.

We are going to show that the finiteness of $V_{\rm eff}(\theta_H)$ at the one loop level is guaranteed by gauge invariance. As seen in Section 4.3, $V_{\rm eff}^{\rm 1\,loop}(\theta_H)$ is summarized in the form of Eq. (4.21). In gauge theory on $M^4 \times S^1$, after making the Wick rotation in the p_0 integration ($p_0 \to ip_4$), we have the expression

$$V_{\rm eff}^{\rm 1\,loop}(\theta_H) = \sum \pm \frac{1}{2} \int \frac{d^4p_E}{(2\pi)^4} \sum_{n=-\infty}^{\infty} \ln\left\{p_E^2 + m_n(\theta_H)^2\right\} ,$$

$$m_n(\theta_H) = \frac{1}{R}\left(n + a + \ell\,\frac{\theta_H}{2\pi}\right) . \tag{4.46}$$

Here the sign is $+$ ($-$) for bosons (fermions). a is a constant and $\ell (= 0, 1, 2, \cdots)$ is an integer. The expression of $V_{\rm eff}^{\rm 1\,loop}(\theta_H)$ in (4.46) is apparently divergent due to the $\int d^4p_E$ integration and infinite sum $\sum_n$. We regularize it by first making the integration dimension to d. Then we perform the integration and sum, and take the limit $d \to 4$ at the end.

When $\{\alpha_j\}$ in the gauge transformation (4.42) are integers, the boundary condition remains invariant ($U' = U$) and each AB phase θ_j shifts by a multiple of 2π as in Eq. (4.44). The mass spectrum of each KK tower remains invariant as a whole. Hence the gauge invariance of the theory implies that

$$V_{\rm eff}(\theta_H + 2\pi) = V_{\rm eff}(\theta_H) , \tag{4.47}$$

which is valid to all orders in perturbation theory. We expand $V_{\rm eff}(\theta_H)$ in a Fourier series:

$$V_{\rm eff}(\theta_H) = \sum_{n=-\infty}^{\infty} v_n\, e^{in\theta_H} . \tag{4.48}$$

By differentiating it we have

$$\frac{d^q}{d\theta_H^q} V_{\rm eff}(\theta_H) = \sum_{n\neq 0} v_n^{(q)}\, e^{in\theta_H} ~,~~ v_n^{(q)} = (in)^q\, v_n . \tag{4.49}$$

We insert the 1-loop expression (4.46) into the left-hand side of (4.49). Except for special values of θ_H, the ultraviolet divergence degree of the expression becomes smaller as the order of the differentiation is increased, and becomes finite for $q \geq 6$. Since $v_n = v_n^{(q)}/(in)^q$ ($n \neq 0$), $V_{\rm eff}(\theta_H) - v_0$ is finite. In general v_0 diverges. For θ_H yielding $m_n(\theta_H) = 0$ there arises infrared divergence coming from the pole at $p_E^2 = 0$ in the p_E integration for $d^q V_{\rm eff}/d\theta_H^q$. For instance, divergence appears at $\theta_H = 0$ (π) for $a = 0$ $(\frac{1}{2})$. $V_{\rm eff}(\theta_H)$ has a cusp singularity there.

It is seen that the gauge invariance guarantees the finiteness of the θ_H-dependent part of $V_{\rm eff}(\theta_H)$ at least at the one loop level. It is expected that some of the physical quantities such as the mass of the Higgs mode remain finite even when higher order corrections are taken into account. [26, 27] This property is made use of for resolving the gauge hierarchy problem later.

4.8 QED and QCD at finite temperature

In evaluating $V_{\rm eff}(\theta_H)$ in gauge theory on $M^n \times S^1$, we have Wick-rotated the p_0 integration path to convert the momentum integration into that in the Euclidean space. Mathematically it is equivalent to evaluating the effective potential in field theory defined on $R^n \times S^1$. Field theory at finite temperature ($T > 0$) in four-dimensional spacetime is defined by replacing the time axis by a periodic imaginary time axis with a period $1/k_B T$ and by imposing periodic (anti-periodic) boundary conditions on boson (fermion) fields. Mathematically it is equivalent to field theory defined on $R^3 \times S^1$.

There is a close connection between the two. In gauge theory on $M^3 \times S^1$, S^1 is a space dimension so that fermion fields need not be anti-periodic on S^1 and may have a $e^{i\beta}$ factor in the boundary condition (3.21). The radius of S^1 and the temperature have the correspondence $2\pi R \leftrightarrow 1/k_B T$. Gauge theory at finite temperature corresponds to gauge theory on $M^3 \times S^1$ with the boundary condition $\{U = 1, e^{i\beta} = -1)\}$. The effective potential $V_{\rm eff}$ is nothing but the free energy F.

Take QED (Quantum Electrodynamics), namely $U(1)$ gauge theory, as an example. As gauge theory on $M^3 \times S^1$, the effective potential in four dimensions is given, instead of (4.33) evaluated on $M^4 \times S^1$, by

$$V_{\rm eff}^{\rm 1\,loop}(\theta_H) = \frac{1}{16\pi^6 R^4} \cdot 4N_{\rm f} \sum_{n=1}^{\infty} \frac{\cos n(\theta_H - \beta)}{n^4} B_2(2\pi R n m) \,. \qquad (4.50)$$

For $\beta = \pi$, which corresponds to QED at finite temperature, $V_{\rm eff}^{\rm 1\,loop}(\theta_H)$ is minimized at $\theta_H = 0$. The mass of the Higgs mode obtained by expanding around $\theta_H = 0$ corresponds to the plasmon frequency.

What would happen if $\beta = 0$? In this case $V_{\rm eff}^{\rm 1\,loop}(\theta_H)$ is minimized at $\theta_H = \pi$. One expands the theory around $\theta_H = \pi$ and evaluate physical quantities. Then it will be found that the mass of the Higgs mode and all other physical quantities become the same as those obtained in the case

$\beta = \pi$. By applying the argument in Section 4.6 one sees that the value of $e^{i\beta}$ in $U(1)$ gauge theory on $M^n \times S^1$ can be changed by large gauge transformations, and that there is only one equivalence class of boundary conditions. Physics does not depend on the value of $e^{i\beta}$ as a result of the dynamics of the AB phase θ_H.

As another example let us consider $SU(N_c)$ gauge theory on $M^3 \times S^1$. In QCD (Quantum Chromodynamics) $N_c = 3$. Let light quarks are massless and ignore contributions from heavy quarks. Then the effective potential is given, in place of (4.25), by

$$
\begin{aligned}
V_{\rm eff}^{\rm 1\,loop}(\theta_j) &\equiv \frac{1}{16\pi^6 R^4} \hat{V}_{\rm eff}^{\rm 1\,loop}(\theta_j) ~, \\
\hat{V}_{\rm eff}^{\rm 1\,loop}(\theta_j) &= -2 \sum_{j,k=1}^{N_c} h_4(\theta_j - \theta_k) + 4N_{\rm f}^{\rm light} \sum_{j=1}^{N_c} h_4(\theta_j - \beta_{\rm f}) ~, \\
h_4(\theta) &= \sum_{n=1}^{\infty} \frac{\cos n\theta}{n^4} \\
&= \frac{\pi^4}{90} - \frac{1}{48}\theta^2(\theta - 2\pi)^2 \quad \text{for } 0 \le \theta \le 2\pi ~,
\end{aligned}
\tag{4.51}
$$

where $\theta_{N_c} = -(\theta_1 + \cdots + \theta_{N_c - 1})$. In QCD at finite temperature $\beta_{\rm f} = \pi$ and $V_{\rm eff}^{\rm 1\,loop}$ takes the minimum at $\theta_j = 0$. The configuration $\{\theta_j = 0\}$ is $SU(N_c)$ symmetric. The mass of the Higgs mode is given by the one corresponding to m_3 in Eq. (4.39). In $SU(N_c)$ gauge theory we take $\theta_1, \theta_2, \cdots, \theta_{N_c - 1}$ as independent variables, and find

$$
\begin{aligned}
m_3^2 &= (\pi R g)^2 \Big(\frac{\partial}{\partial\theta_1} - \frac{\partial}{\partial\theta_2} \Big)^2 V_{\rm eff} \bigg|_{\theta_j = 0, \beta_{\rm f} = \pi} \\
&= \frac{g^2}{3(2\pi R)^2} \Big(N_c + \frac{1}{2} N_{\rm f}^{\rm light} \Big) ~.
\end{aligned}
\tag{4.52}
$$

The result (4.52) implies that in QCD at finite temperature, the longitudinal components of gluons acquire (mass)^2 of $\frac{1}{3} g^2 (k_B T)^2 (N_c + \frac{1}{2} N_{\rm f}^{\rm light})$, which leads to the Debye screening.

What would happen if quarks are periodic on S^1? Along the same line of the manipulation one finds

$$
\begin{aligned}
m_3^2 &= (\pi R g)^2 \Big(\frac{\partial}{\partial\theta_1} - \frac{\partial}{\partial\theta_2} \Big)^2 V_{\rm eff} \bigg|_{\theta_j = 0, \beta_{\rm f} = 0} \\
&= \frac{g^2}{3(2\pi R)^2} \big(N_c - N_{\rm f}^{\rm light} \big) ~.
\end{aligned}
\tag{4.53}
$$

If $N_f^{\rm light} > N_c$, the mass squared, m_3^2, becomes negative, and therefore the configuration $\{\theta_j = 0\}$ turns out unstable. Where is the true vacuum (ground state) then? This question led to the discovery of the Hosotani mechanism.

The true vacuum is the configuration which minimizes $V_{\rm eff}$. The analysis in Section 4.4 applies as it is. For $N_c = 3$, $N_{\rm f} \geq 1$ and $\beta_{\rm f} = 0$, $V_{\rm eff}$ is minimized at $(\theta_1, \theta_2, \theta_3) = (\pm\frac{2}{3}\pi, \pm\frac{2}{3}\pi, \pm\frac{2}{3}\pi)$. The energy density and the mass of the Higgs modes are

$$\begin{aligned} V_{\rm eff}\Big|_{\theta_j=\frac{2}{3}\pi} - V_{\rm eff}\Big|_{\theta_j=0} &= -\frac{N_{\rm f}^{\rm light}}{81\pi^2 R^4} < 0 , \\ m_3^2\Big|_{\theta_j=\frac{2}{3}\pi} &= \frac{g^2}{3(2\pi R)^2}\left(3 + \frac{1}{3}N_{\rm f}^{\rm light}\right) . \end{aligned} \tag{4.54}$$

4.9 Gauge symmetry breaking on the lattice

The Hosotani mechanism is a mechanism for dynamical gauge symmetry breaking by gauge invariant AB phases. It is contrary to the Higgs mechanism in which gauge symmetry is broken by a non-vanishing expectation value of a gauge non-invariant quantity such as the Higgs field. So far we have analyzed the Hosotani mechanism in perturbation theory.

It is desirable to investigate and confirm the Hosotani mechanism beyond the perturbation theory. **Lattice gauge theory** is non-perturbative formulation of gauge theory. Spacetime is discretized, and is represented by lattice sites. Gauge fields are placed on each link connecting neighbouring lattice sites, whereas matter fields such as quarks are defined on lattice sites. Theory is invariant under local gauge transformations defined on each lattice site. Applying this formulation to QCD of strong interactions, we have $SU(3)$ lattice gauge theory. With a lattice spacing a, we prepare N sites in each direction and impose periodic boundary conditions to define 4D gauge theory with a volume $(Na)^4$. The total number of degrees of freedom is finite so that one can explore various physical quantities nonperturbatively by numerical simulations such as the Monte Carlo method. In particular, in four dimensions the gauge coupling constant is dimensionless and $SU(3)$ gauge theory is asymptotically free so that the continuum limit $a \to 0$ can be safely taken and we can explore physics in the continuum QCD. The $SU(3)$ lattice gauge theory has been quite successful.

The Hosotani mechanism can be explored in lattice gauge theory. In Sections 4.3–4.5 we have seen that gauge symmetry can be dynamically broken by evaluating the effective potential $V_{\rm eff}(\theta_j)$ at the one loop level in $SU(3)$ gauge theory on $M^4 \times S^1$. The argument there can be extended to gauge theory on $M^\ell \times S^1$. One need to replace $h(\theta) = h_5(\theta)$ in $\hat{V}_{\rm eff}^{\rm 1\,loop}(\theta_j)$ in Eq. (4.25) by $h_{\ell+1}(\theta)$, and adjust the numbers of degrees of freedom of gauge and fermion fields in the $\ell+1$ dimensions. The Hosotani mechanism operates irrespective of the spacetime dimensions.

Let us consider $SU(3)$ gauge theory on $M^3 \times S^1$. By Wick-rotating the time axis it becomes $SU(3)$ gauge theory on $R^3 \times S^1$. The corresponding lattice gauge theory is defined on the $N_x^3 \times N_y$ lattice. N_y corresponds to S^1, and it is assumed that $N_x \gg N_y$. Concrete numerical simulations have been done, for instance, on $16^3 \times 4$ lattice. Let us label lattice sites as $n = (n_1, n_2, n_3, n_4)$. An unit vector in the μ direction is denoted as e_μ. For instance $e_1 = (1,0,0,0)$. A link variable $U_{n,\mu}$, corresponding to a gauge field in the continuum theory, is defined on a link connecting lattice sites n and $n+e_\mu$. $U_{n,\mu} \in SU(3)$ and it corresponds to $P \exp\left\{ig \int_{na}^{(n+e_\mu)a} dx^\mu A_\mu\right\} \simeq \exp\left\{igaA_\mu(x)\right\}$ in the continuum theory.

Under a gauge transformation Ω_n at each site,

$$U_{n,\mu} \to U'_{n,\mu} = \Omega_n U_{n,\mu} \Omega^\dagger_{n+e_\mu} \,. \tag{4.55}$$

The plaquette $P_{n,\mu\nu}$ is defined by

$$P_{n,\mu\nu} = {\rm Tr}\left(U_{n,\mu} U_{n+e_\mu,\nu} U^\dagger_{n+e_\nu,\mu} U^\dagger_{n,\nu}\right) , \tag{4.56}$$

and is gauge invariant. The periodic boundary condition is imposed on the lattice, that is, we identify $n + N_x e_j \sim n$ $(j = 1,2,3)$ and $n + N_y e_4 \sim n$. The action for gauge fields is given by

$$I_W = \frac{1}{g^2} \sum_n \sum_{\mu,\nu} \left\{3 - {\rm Re}\,(P_{n,\mu\nu})\right\} . \tag{4.57}$$

In the continuum limit $(a \to 0)$ $3 - {\rm Re}(P_{n,\mu\nu})$ becomes $\frac{1}{2} g^2 a^4 {\rm Tr}\, F_{\mu\nu}^2$, and the action of Yang-Mills fields is reproduced.

4.9.1 *Polyakov loop and distribution density of AB phases*

In the Hosotani mechanism the Aharonov-Bohm phase along S^1 plays an important role. The relevant quantity in lattice gauge theory is the **Polyakov loop** defined, with $n = (\vec{n}, n_4)$, by

$$W(\vec{n}) = \prod_{n_4=1}^{N_y} U_{(\vec{n},n_4),4} \,. \tag{4.58}$$

It corresponds to $P\exp\left\{ig\int_0^{N_y a} dy\, A_y\right\}$ in the continuum theory. Under a gauge transformation (4.55), $W(\vec{n}) \to \Omega_{(\vec{n},1)} W(\vec{n}) \Omega^\dagger_{(\vec{n},1)}$ so that the eigenvalues of $W(\vec{n})$ are gauge invariant. In $SU(3)$ theory there are three eigenvalues $\{e^{i\theta_1}, e^{i\theta_2}, e^{i\theta_3}\}$ $(e^{i(\theta_1+\theta_2+\theta_3)} = 1)$. Here, $\{\theta_1, \theta_2, \theta_3\}$ are nothing but the Aharonov-Bohm (AB) phases along S^1.

With matter fields (fermions) specified the model is fixed. By introducing only quark fields in the fundamental representation of $SU(3)$, one has lattice QCD theory. The lattice $N_y < N_x$ corresponds to QCD at finite temperature. By evaluating the effective potential $V_{\rm eff}(\theta_j)$ in perturbation theory in Section 4.4, we have seen that spontaneous gauge symmetry breaking is induced in the presence of fermion fields in the adjoint representation. A similar thing is expected to happen in lattice gauge theory.

In lattice gauge theory, typical configurations of gauge fields (link variables) are generated by numerical simulations in the Monte Carlo method, from which various physical quantities are evaluated. Given one typical configuration one evaluates $W(\vec{n})$ in (4.58), and determines its eigenvalues to find $\{\theta_j(\vec{n})\}$ at each space site $\vec{n}$. As the lattice gauge theory is a theory of a finite number of physical degrees of freedom, $\{\theta_j(\vec{n})\}$ necessarily takes various values in a distributed form. One can explore gauge symmetry breaking from how the values of $\{\theta_j(\vec{n})\}$ are distributed.

Let us first check what kind of behavior is expected from the result in the continuum theory. The effective potential $V_{\rm eff}(\theta_j)$ is the energy density of the state when an external source for the AB phase is introduced such that the expectation value of the AB phase becomes the given θ_j in the entire region of the space. The ground state, for which there is no external source, corresponds to the minimum of $V_{\rm eff}(\theta_j)$. The configuration of $\{\theta_j\}$ corresponding to the minimum of $V_{\rm eff}(\theta_j)$ is most favored to be realized even locally.

At the quantum level there are always local fluctuations in the values of $\{\theta_j\}$. The fluctuations become large for systems of a finite number of degrees of freedom. The probability of locally taking a configuration $\{\theta_j\}$ is larger for a smaller value of $V_{\rm eff}(\theta_j)$, and smaller for a larger value of $V_{\rm eff}(\theta_j)$.

The relation between the probability and $V_{\rm eff}(\theta_j)$ is inferred from the path-integral formula of physical quantities after the Wick rotation. For instance, the expectation value of gauge-invariant $P_3 = {\rm Tr}\, W_3$ $(W_3 =$

$P\exp\{ig\int_0^{N_y a} dy\, A_y\})$ is given in the continuum theory by

$$\langle P_3\rangle = \int [DA][D\bar{\psi}][D\psi]\; P_3\, e^{-S_E[A,\psi,\bar{\psi}]}$$
$$\sim \int [DW_3]\; P_3\, e^{-(\text{vol})V_{\text{eff}}(\theta_j)}\;. \tag{4.59}$$

Here $[DW_3]$ is the integration measure for $SU(3)$ matrices, namely the group-invariant Haar measure. There are two independent AB phases θ_1, θ_2 in the $SU(3)$ theory, and $[DW_3]$ can be written as

$$[DW_3] = d\theta_1 d\theta_2\, \rho_{\text{Haar}}(\theta_1,\theta_2)\;,$$
$$\rho_{\text{Haar}}(\theta_1,\theta_2) = \prod_{j<k} \sin^2 \tfrac{1}{2}(\theta_j - \theta_k)\Big|_{\theta_3=-\theta_1-\theta_2}\;. \tag{4.60}$$

In lattice gauge theory $\langle P_3\rangle$ is evaluated as

$$\langle P_3\rangle = \frac{1}{N_x^3}\sum_{n_1,n_2,n_3} \langle \text{Tr}\, W(\vec{n})\rangle \tag{4.61}$$

by using a typical gauge field configuration. At each site $\vec{n}$ one determines $\{\theta_j\}$ for this configuration, and therefore $\langle P_3\rangle$ is expressed in the form

$$\langle P_3\rangle = \int d\theta_1 d\theta_2\, P_3\, \rho(\theta_1,\theta_2)\;. \tag{4.62}$$

In other words, one can measure the distribution density $\rho(\theta_1,\theta_2)$ of the AB phases by numerical simulations. Comparing (4.59) and (4.62), we find that

$$\rho(\theta_1,\theta_2)^{\text{norm}} = \frac{\rho(\theta_1,\theta_2)}{\rho_{\text{Haar}}(\theta_1,\theta_2)} \Leftrightarrow e^{-(\text{vol})V_{\text{eff}}(\theta_1,\theta_2)}\;. \tag{4.63}$$

The **normalized distribution density** $\rho(\theta_1,\theta_2)^{\text{norm}}$ in lattice gauge theory carries the information of the effective potential $V_{\text{eff}}(\theta_1,\theta_2)$ in continuum theory.

4.9.2 *Gauge-invariant gauge symmetry breaking*

Simulations in lattice gauge theory are done by varying the values of the gauge coupling constant and fermion masses. One can learn how gauge symmetry is broken by seeing how the normalized distribution density $\rho(\theta_1,\theta_2)^{\text{norm}}$ changes with these parameters. [28, 29]

The result of simulations for the normalized distribution density $\rho(\theta_1,\theta_2)^{\text{norm}}$ in the presence of fermions in the adjoint representation is shown in Fig. 4.2. For $\beta = 6/g^2 = 5.30$, which corresponds to the strong gauge coupling, the distribution density is almost uniform (Fig. 4.2(a)). This is due to the fact that quantum fluctuations are large because of the strong gauge coupling and values of (θ_1, θ_2) not fixed at all. This corresponds to the confined phase (denoted as the X phase), and the $SU(3)$ symmetry is preserved. When the gauge coupling becomes a little bit smaller, namely for $\beta = 6/g^2 = 5.46$, the system moves in the deconfined phase (the A phase), and $\rho(\theta_1, \theta_2)^{\rm norm}$ becomes the largest at $(\theta_1, \theta_2, \theta_3) = (0, 0, 0)$ (Fig. 4.2(b)), which is one of the $SU(3)$ symmetric configurations in Eq. (3.36). The $SU(3)$ symmetry is unbroken in the deconfined phase.

When the gauge coupling is further reduced to $\beta = 6/g^2 = 5.95$, $\rho(\theta_1, \theta_2)^{\rm norm}$ becomes largest at $(\theta_1, \theta_2, \theta_3) = (\frac{1}{3}\pi, \frac{1}{3}\pi, -\frac{2}{3}\pi)$, $(\frac{1}{3}\pi, -\frac{2}{3}\pi, \frac{1}{3}\pi)$ and $(-\frac{2}{3}\pi, \frac{1}{3}\pi, \frac{1}{3}\pi)$ (Fig. 4.2(c)). This corresponds to the $SU(2) \times U(1)$ symmetric phase (the B phase) in Eq. (4.32). For $\beta = 6/g^2 = 6.50$, $\rho(\theta_1, \theta_2)^{\rm norm}$ becomes largest at $(\theta_1, \theta_2, \theta_3) = (0, \frac{2}{3}\pi, -\frac{2}{3}\pi)$ and its permutations (Fig. 4.2(d)), which corresponds to the $U(1) \times U(1)$ symmetric phase (the C phase) in Eq. (4.30). In Fig. 4.2(e) the Haar measure $\rho_{\rm Haar}(\theta_1, \theta_2)$ in (4.60) is plotted. In Figs. 4.2(c) and (d), there are black points in the vicinity of $(\theta_1, \theta_2) = (0, 0)$ and $(\pm\frac{2}{3}\pi, \pm\frac{2}{3}\pi)$. This is due to the fact that the Haar measure, which appears in the denominator of the definition of $\rho(\theta_1, \theta_2)^{\rm norm}$, vanishes at $\theta_1 = \theta_2$, $\theta_1 = \theta_3 = -\theta_1 - \theta_2$, and $\theta_2 = \theta_3 = -\theta_1 - \theta_2 \pmod{2\pi}$. Small number of samples emerging as statistical fluctuations there are enhanced to show up in $\rho(\theta_1, \theta_2)^{\rm norm}$.

From the non-perturbative analysis of the lattice gauge theory it is observed that the phase transition

$$\begin{aligned}
&\text{confined phase } [SU(3)] \text{ (X phase)} \\
&\to \text{ deconfined phase } [SU(3)] \text{ (A phase)} \\
&\to SU(2) \times U(1) \text{ (B phase)} \\
&\to U(1) \times U(1) \text{ (C phase)}
\end{aligned} \tag{4.64}$$

takes place as the gauge coupling g^2 becomes smaller (β becomes larger). Similar behavior is recognized from the perturbative analysis in the continuum. Of course the results in perturbation theory in the strong coupling region are not reliable. Nevertheless in the $SU(3)$ gauge theory on $R^3 \times S^1$ with fermions in the adjoint representation the transition A $\to$ B $\to$ C is seen at the one loop level as the fermion mass is decreased.

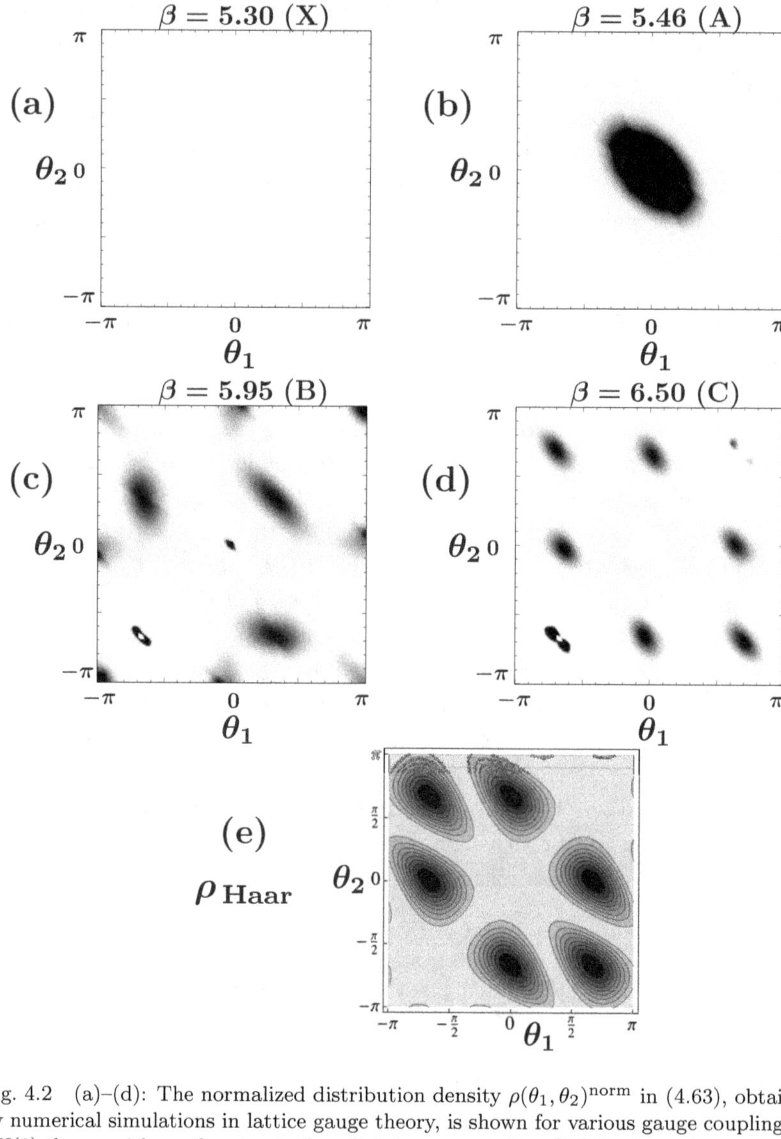

Fig. 4.2 (a)–(d): The normalized distribution density $\rho(\theta_1, \theta_2)^{\rm norm}$ in (4.63), obtained by numerical simulations in lattice gauge theory, is shown for various gauge couplings in $SU(3)$ theory with one fermion in the adjoint representation. [28] Darker points represent larger values of $\rho^{\rm norm}$. The lattice size is $16^3 \times 4$. The fermion mass m is taken to be $ma = 0.1$ where a is the lattice spacing. β is related to the gauge coupling constant by $\beta = 6/g^2$. (e): Contour plot of the Haar measure $\rho_{\rm Haar}(\theta_1, \theta_2)$ in (4.60). For strong coupling, namely for small β as in the case (a), $\rho(\theta_1, \theta_2)^{\rm norm}$ is almost constant and the distribution is uniform, implying $\rho(\theta_1, \theta_2) \propto \rho_{\rm Haar}(\theta_1, \theta_2)$.

In Fig. 4.3 contour plots of

$$\hat{\rho}(\theta_1, \theta_2) = e^{-\{\hat{V}_{\text{eff}}(\theta_1,\theta_2) - \hat{V}_{\text{eff}}|_{\text{min}}\}} \tag{4.65}$$

at the one loop level in $SU(3)$ gauge theory on $R^3 \times S^1$ with one fermion in the adjoint representation with $m_{\text{ad}}R = 0, 0.18$ and 0.21 are shown. $\hat{\rho}(\theta_1, \theta_2)$ corresponds to the normalized distribution density $\rho(\theta_1, \theta_2)^{\text{norm}}$ in lattice gauge theory as is seen in Eq. (4.63). For a sufficiently large fermion mass (the case of $m_{\text{ad}}R = 0.21$) the system approaches pure gauge theory without fermions, and $\hat{\rho}(\theta_1, \theta_2)$ becomes largest at the $SU(3)$ symmetric minimum of the effective potential V_{eff}, the system being in the A phase. As the fermion mass becomes smaller, say for $m_{\text{ad}}R = 0.18$, $\hat{\rho}$ becomes largest at the $SU(2) \times U(1)$ symmetric minimum of V_{eff}, the system being in the B phase. In the massless fermion limit ($m_{\text{ad}}R = 0$) $\hat{\rho}$ becomes largest at the $U(1)) \times U(1)$ symmetric minimum of V_{eff}, the system being in the C phase. We comment that similar behavior of the phase structure has been observed in the renormalization group analysis of five-dimensional gauge theory. [30]

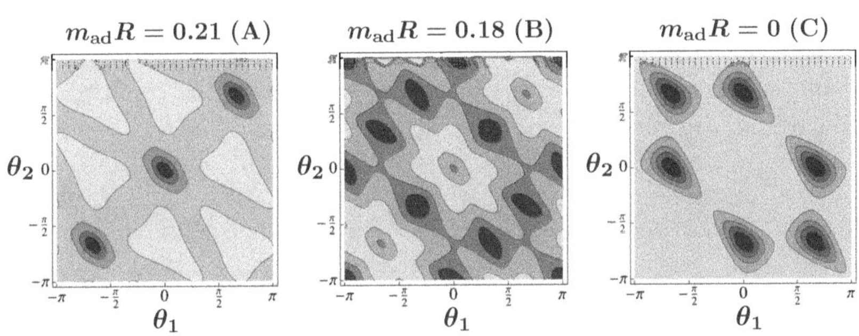

Fig. 4.3 Contour plots of the $\hat{\rho}(\theta_1, \theta_2)$ function in (4.65) at the one loop level in $SU(3)$ gauge theory on $R^3 \times S^1$ with one fermion in the adjoint representation are shown. The fermion mass is taken to be $m_{\text{ad}}R = 0.21, 0.18$ and 0 from the left to the right. Darker points represent larger values of $\hat{\rho}$.

Let us come back to the problem of how gauge symmetry is spontaneously broken in lattice gauge theory. There is Elitzur's theorem in lattice gauge theory. [31] It states that vacuum expectation values of gauge non-invariant quantities strictly vanish in lattice gauge theory. Consequently the expectation value of the Higgs field in the Higgs mechanism, for instance, vanishes, and spontaneous breaking of gauge symmetry by the Higgs mechanism does not take place.

The result of the simulations in lattice gauge theory described in Fig. 4.2 shows that the phase transition (4.64), namely spontaneous breaking of gauge symmetry takes place. This does not conflict with Elitzur's theorem. The important point is that the result in Fig. 4.2 represents the change in the distribution density of gauge invariant AB phases.

In the Hosotani mechanism, spontaneous gauge symmetry breaking is caused by the change in the distribution density of gauge invariant quantities (AB phases). In the Higgs mechanism and in the fermion pair condensation mechanism in the technicolor theory gauge symmetry is spontaneously broken by non-vanishing vacuum expectation values of gauge non-invariant quantities. In contrast, **the Hosotani mechanism is a mechanism for dynamical gauge symmetry breaking by gauge invariant quantities.**

As the lattice size gets bigger, the volume (vol) factor in (4.63) increases and the distribution density becomes sharper and the behavior of the phase transition should become more distinct. It is desirable to determine the masses of gauge bosons in the broken phase directly from simulations on the lattice. Although symmetry changes in the phase transition by the Hosotani mechanism, the structure of the phase transition seems different from that in the ordinary second-order phase transition. It may be possible to have a new way to characterize the transition from the confined phase to the deconfined phase by the Hosotani mechanism.

Chapter 5

Gauge-Higgs unification in flat space

We encounter the chiral-fermion problem in higher-dimensional theory. One of the promising scenarios for resolving this problem is gauge theory on orbifolds. We provide an overview of the universal extra-dimension model (UED) and the $SU(5)$ Kawamura model, and introduce $SU(3)$ gauge-Higgs unification models on $M^4 \times (S^1/Z_2)$. Aharonov-Bohm phases play an important role here.

5.1 $SU(5)$ grand unified theory on $M^4 \times S^1$

Gauge symmetry can be spontaneously broken by the Hosotani mechanism. How is this mechanism applied to electroweak unified theory or grand unification theory? As seen in Chapters 3 and 4, the extra-dimensional component of gauge fields A_M plays the role of the 4D Higgs field. Gauge fields belong to the adjoint representation of the gauge group $\mathcal{G}$ so that the Higgs field also belongs to the adjoint representation of $\mathcal{G}$. In 4D grand unified theory one introduces a Higgs field in the adjoint representation to reduce $\mathcal{G}$ to the SM gauge symmetry $\mathcal{G}_{\rm SM} = SU(3)_C \times SU(2)_L \times U(1)_Y$, and then introduces another Higgs field to reduce the $SU(2)_L \times U(1)_Y$ electroweak symmetry to $U(1)_{\rm EM}$. For instance, at least 2 kinds of Higgs fields are introduced in the $SU(5)$ grand unified theory, **24** in the adjoint representation, and **5** in the fundamental representation. With these circumstances taken into account, it may look reasonable to apply the Hosotani mechanism to grand unified theory. We first analyze $SU(5)$ gauge theory on $M^4 \times S^1$, to clarify a few problems to be resolved in constructing gauge-Higgs unification models.

In the 4D $SU(5)$ grand unified theory, fermions in each generation are introduced, in the form of left-handed components, in $\overline{\mathbf{5}}$ (fundamental representation) and in **10** (antisymmetric representation). [32] In the first generation one has

$$\overline{\mathbf{5}} : \psi_L = (d_1^c, d_2^c, d_3^c, e^-, -\nu_e)_L$$

$$\mathbf{10} : \chi_L = \frac{1}{\sqrt{2}} \begin{pmatrix} 0 & u_3^c & -u_2^c & u_1 & d_1 \\ -u_3^c & 0 & u_1^c & u_2 & d_2 \\ u_2^c & -u_1^c & 0 & u_3 & d_3 \\ -u_1 & -u_2 & -u_3 & 0 & e^+ \\ -d_1 & -d_2 & -d_3 & -e^+ & 0 \end{pmatrix}_L , \tag{5.1}$$

$(u_j^c)_L = (u_{jR})^c$ etc. Left-handed components of charge-conjugated fields $(u_j^c, d_j^c, e^+ = e^{-c})$ represent right-handed components of the original fields. The right-handed neutrino ν_{eR} is introduced in the singlet (**1**).

In Section 4.4 we have evaluated the 1-loop effective potential $V_{\rm eff}(\theta_j)$ in $SU(N)$ gauge theory on $M^4 \times S^1$. We consider $SU(5)$ gauge theory with $N_{\mathbf{5}}$ fermions $\Psi_{\mathbf{5}}$ in the **5** representation and $N_{\mathbf{10}}$ fermions $\Psi_{\mathbf{10}}$ in the **10** representation. As is seen in the next section, left-handed and right-handed fermions always appear in a pair in theory on $M^4 \times S^1$. The effective potential is given by

$$V_{\rm eff}^{\rm 1\,loop}(\theta_j) \equiv \frac{3}{64\pi^6 R^4} \hat{V}_{\rm eff}^{\rm 1\,loop}(\theta_j) ,$$

$$\hat{V}_{\rm eff}^{\rm 1\,loop}(\theta_j) = -3 \sum_{j,k=1}^{5} h(\theta_j - \theta_k) + 4N_{\mathbf{5}} \sum_{j=1}^{5} h(\theta_j - \beta_{\mathbf{5}}) + 4N_{\mathbf{10}} \sum_{1\le j<k\le 5} h(\theta_j + \theta_k - \beta_{\mathbf{10}}) , \tag{5.2}$$

where $\Psi_{\mathbf{5}/\mathbf{10}}(x, y + 2\pi R) = e^{i\beta_{\mathbf{5}/\mathbf{10}}} \Psi_{\mathbf{5}/\mathbf{10}}(x, y)$. There are four independent AB phases ($\sum_{j=1}^{5} \theta_j = 0 \bmod 2\pi$) in $SU(5)$ theory. If $V_{\rm eff}(\theta_j)$ is minimized at

$$e^{i\theta_1} = e^{i\theta_2} = e^{i\theta_3} \neq e^{i\theta_4} = e^{i\theta_5} , \tag{5.3}$$

then $SU(5)$ gauge symmetry is spontaneously broken to $SU(3) \times SU(2) \times U(1)$ gauge symmetry, namely the SM gauge symmetry $\mathcal{G}_{\rm SM}$.

Following the quark-lepton spectrum, we set $N_{\mathbf{5}} = N_{\mathbf{10}} = 3$. For $\beta_{\mathbf{5}} = \beta_{\mathbf{10}} = 0$, $V_{\rm eff}$ is minimized at $\theta_1 = \cdots = \theta_5 = \pm\frac{2}{5}\pi$. For $\beta_{\mathbf{5}} = \beta_{\mathbf{10}} = \pi$, $V_{\rm eff}$ is minimized at $\theta_1 = \cdots = \theta_5 = 0$. For $N_{\mathbf{5}} \neq 0, N_{\mathbf{10}} = 0, \beta_{\mathbf{5}} = 0$, $V_{\rm eff}$ is minimized at $\theta_1 = \cdots = \theta_5 = \pm\frac{4}{5}\pi$. In either case $SU(5)$ gauge symmetry remains unbroken. In general all 4D fermion fields in the Kaluza-Klein (KK) expansion have masses of $O(1/R)$ for $\theta_j \neq 0$ as in (3.34).

In gauge theory on $M^4 \times S^1$, once the matter content is specified, the effective potential $V_{\rm eff}(\theta_H)$ for the AB phases is determined, and it is also determined whether or not the gauge symmetry is broken. With the simple matter content discussed above, it is difficult to have realistic grand unification. By adopting different matter content and imposing a distinct β value on each field, it is not impossible to have gauge symmetry breaking to $\mathcal{G}_{\rm SM}$.

However, there remain many problems in this setup. The most severe one is that gauge interactions of 4D fermion fields obtained by the KK expansion of the 5D field necessarily become vector-like, that is, left-handed and right-handed fermions have the same gauge couplings. This is a generic problem in theories on $M^4 \times S^1$.

5.2 Fermions in higher dimensions

The free Dirac equation in five-dimensional $M^4 \times S^1$ is

$$\big(\Gamma^M \partial_M - m\big)\Psi(x,y) = 0 \ , \quad \{\Gamma^M, \Gamma^N\} = 2\eta^{MN} \ , \tag{5.4}$$

where $\eta^{MN} = \mathrm{diag}\,(-1,1,1,1,1)$. With the boundary condition on S^1

$$\Psi(x, y + 2\pi R) = e^{i\beta}\Psi(x,y) \tag{5.5}$$

the KK expansion is given by

$$\Psi(x,y) = \sum_{n=-\infty}^{\infty} \psi_n(x) \exp\left\{ i\Big(n + \frac{\beta}{2\pi}\Big)\frac{y}{R}\right\} . \tag{5.6}$$

The Dirac algebra $\{\Gamma^M, \Gamma^N\} = 2\eta^{MN}$ is satisfied by 4×4 matrices. Making use of Dirac matrices in four dimensions $\{\gamma^\mu, \gamma^\nu\} = 2\eta^{\mu\nu}$, one can take $\Gamma^\mu = \gamma^\mu$ ($\mu = 0 \sim 3$) and $\Gamma^5 = \gamma^5 = -i\gamma^0\gamma^1\gamma^2\gamma^3$. By adopting the spinor representation in four dimensions we have

$$\Gamma^\mu = \begin{pmatrix} 0 & \sigma^\mu \\ \bar\sigma^\mu & 0 \end{pmatrix} , \ \Gamma^5 = \begin{pmatrix} I_2 & 0 \\ 0 & -I_2 \end{pmatrix} , \ \begin{pmatrix} \sigma^\mu \\ \bar\sigma^\mu \end{pmatrix} = (\pm I_2, \vec\sigma\,) \ . \tag{5.7}$$

In this representation the upper two components of Ψ represent right-handed components Ψ_R ($\gamma^5 = +1$), whereas the lower two components represent left-handed components Ψ_L ($\gamma^5 = -1$). The Dirac equation becomes

$$\begin{pmatrix} \partial_y - m & \sigma^\mu \partial_\mu \\ \bar\sigma^\mu \partial_\mu & -\partial_y - m \end{pmatrix} \begin{pmatrix} \Psi_R \\ \Psi_L \end{pmatrix} = 0 \ . \tag{5.8}$$

Inserting the KK expansion (5.6) into (5.8), we have

$$
\begin{aligned}
\bar{\sigma}\partial\,\psi_{nR} &= \left\{ +\frac{i}{R}\Big(n+\frac{\beta}{2\pi}\Big) + m \right\} \psi_{nL} \ , \\
\sigma\partial\,\psi_{nL} &= \left\{ -\frac{i}{R}\Big(n+\frac{\beta}{2\pi}\Big) + m \right\} \psi_{nR} \ ,
\end{aligned}
\tag{5.9}
$$

so that the mass m_n of the $\psi_n(x)$ field becomes

$$
m_n = \left[\frac{1}{R^2}\Big(n+\frac{\beta}{2\pi}\Big)^2 + m^2 \right]^{1/2} . \tag{5.10}
$$

The KK mass scale $m_{\rm KK} = 1/R$ is assumed to be much larger than the masses of quarks and leptons. Quarks and leptons must correspond to the $n = 0$ modes with $\beta = 0$. Even if $m = 0$ and $m_0 = 0$, ψ_{0R} and ψ_{0R} appear in a pair. To have local-Lorentz invariant action and equations of motion in five dimensions, one needs both Ψ_R and Ψ_L as a set. Gauge interactions are introduced by replacing ∂_M by the covariant derivative $D_M = \partial_M - igA_M$. As a consequence massless fermion fields in four dimensions necessarily appear with both right-handed and left-handed components having the same gauge interactions. In weak interactions only left-handed quarks and leptons couple to the W boson. In a gauge theory on $M^4 \times S^1$ one cannot describe the observed weak interactions. This is called the **chiral fermion problem**.

The situation is not improved in gauge theory in higher dimensions. For instance, the Dirac equation and Dirac algebra in the six-dimensional spacetime $M^4 \times T^2$ take the same form as in Eq. (5.4). The Dirac algebra is satisfied with 8×8 matrices. In six dimensions Γ^7 satisfying $\{\Gamma^7, \Gamma^M\} = 0$ and $(\Gamma^7)^2 = 1$ can be defined, corresponding to γ^5 in four dimensions. For instance, $\Gamma^7 = \gamma^5\bar{\gamma}$ where $\bar{\gamma} = i\Gamma^5\Gamma^6$ satisfies the conditions. In the case $m = 0$, one can have a consistent theory by restricting $\Gamma^7\Psi_+ = +\Psi_+$ or $\Gamma^7\Psi_- = -\Psi_-$. $\Psi_\pm$ is called a 6D Weyl fermion, having four components. $\Psi_\pm$ is expanded in 4D KK modes. In these 4D KK modes, both right-handed and left-handed modes appear in a pair. For instance, in the KK expansion of Ψ_+, a pair of $(\gamma^5, \bar{\gamma}) = (+,+)$ and $(-,-)$ appears.

To have chiral gauge interactions in four dimensions, it is necessary that massless modes (zero modes) in four dimensions show up only in either Ψ_R or Ψ_L in the KK expansion of Ψ. There are a couple of methods of achieving it. Zero modes appear only in either right-handed or left-handed components if the extra-dimensional space is (i) two-dimensional S^2 with $U(1)$ magnetic fields of a magnetic monopole, (ii) two-dimensional T^2 with an uniform magnetic field, or (iii) four-dimensional S^4 with $SU(N)$ instanton fields. Or when the extra-dimensional space has special topology as in Calabi-Yau space, four dimensional fields can become chiral. An alternative possibility is the case where the extra-dimensional space is an orbifold, which turns out crucial to construct realistic gauge-Higgs unification models.

5.3 Orbifold $M^4 \times (S^1/Z_2)$

The simplest example of orbifolds is S^1/Z_2 which is obtained by identifying the upper half of a circle with its lower half. [33] Let y and R be the coordinate and radius of S^1, respectively. Then on S^1/Z_2

$$S^1/Z_2 : \; y \sim y + 2\pi R \; , \quad y \sim -y \; . \tag{5.11}$$

Two points $y = 0$ and $y = \pi R$ are the fixed points which remain invariant under the reflection $y \to -y$. We consider five-dimensional spacetime $M^4 \times (S^1/Z_2)$ with x^μ being the coordinates of M^4. There are two identification relations $(x^\mu, y) \sim (x^\mu, y + 2\pi R) \sim (x^\mu, -y)$. Three operations are defined by

$$\begin{aligned}
&\hat{P}_0 : (x^\mu, y) \to (x^\mu, -y) \; , \\
&\hat{P}_1 : (x^\mu, \pi R + y) \to (x^\mu, \pi R - y) \; , \\
&\hat{U} \; : (x^\mu, y) \to (x^\mu, y + 2\pi R) \; .
\end{aligned} \tag{5.12}$$

$\hat{P}_0$ and $\hat{P}_1$ are the reflections around $y = 0$ and $y = \pi R$ axes, and $\hat{U}$ is the translation going around the circle S^1. These three operations are not independent. There are relations among them:

$$(\hat{P}_0)^2 = (\hat{P}_1)^2 = 1 \; , \quad \hat{P}_1 \hat{P}_0 = \hat{U} \; . \tag{5.13}$$

The region $0 \le y \le \pi R$ is taken as the fundamental region of S^1/Z_2.

Let us consider a complex scalar field $\Phi(x, y)$ on $M^4 \times (S^1/Z_2)$. Physical quantities must be single-valued on $M^4 \times (S^1/Z_2)$. In particular, the Lagrangian density must satisfy $\mathcal{L}(x, y) = \mathcal{L}(x, y + 2\pi R) = \mathcal{L}(x, -y)$. With $\mathcal{L} = -\partial_M \Phi^\dagger \partial^M \Phi - m^2 \Phi^\dagger \Phi$, the field Φ needs to satisfy the boundary condition

$$\begin{aligned}
&\Phi(x, -y) = P_0 \Phi(x, y) \; , \\
&\Phi(x, \pi R - y) = P_1 \Phi(x, \pi R + y) \; .
\end{aligned} \tag{5.14}$$

It follows from (5.13) that $P_0^2 = P_1^2 = 1$ so that $P_0, P_1 = +1$ or -1. The field is characterized by the parity P_j, either even $+$ or odd $-$. There are four cases: $(P_0, P_1) = (+,+), (-,-), (+,-), (-,+)$. In the first two cases $U = P_1 P_0 = +1$ and Φ is periodic on S^1. In the last two cases $U = -1$ and Φ is anti-periodic. The KK expansion in each case is

(i) $(P_0, P_1) = (+,+)$

$$\Phi(x,y) = \frac{1}{\sqrt{\pi R}} \Phi_0(x) + \sqrt{\frac{2}{\pi R}} \sum_{n=1}^{\infty} \Phi_n(x) \cos \frac{ny}{R} ,$$

(ii) $(P_0, P_1) = (-,-)$

$$\Phi(x,y) = \sqrt{\frac{2}{\pi R}} \sum_{n=1}^{\infty} \Phi_n(x) \sin \frac{ny}{R} ,$$

(iii) $(P_0, P_1) = (+,-)$

$$\Phi(x,y) = \sqrt{\frac{2}{\pi R}} \sum_{n=0}^{\infty} \Phi_{n+\frac{1}{2}}(x) \cos \frac{(n+\frac{1}{2})y}{R} ,$$

(iv) $(P_0, P_1) = (-,+)$

$$\Phi(x,y) = \sqrt{\frac{2}{\pi R}} \sum_{n=0}^{\infty} \Phi_{n+\frac{1}{2}}(x) \sin \frac{(n+\frac{1}{2})y}{R} . \tag{5.15}$$

In either case the action is expressed as

$$I = \int d^4x \sum_{\ell} \Big\{ - \partial_\mu \Phi_\ell^\dagger \partial^\mu \Phi_\ell - m_\ell^2 \Phi_\ell^\dagger \Phi_\ell \Big\} ,$$

$$m_\ell^2 = m^2 + \frac{\ell^2}{R^2} , \tag{5.16}$$

where ℓ is an integer or half-odd integer appearing in the KK expansion (5.15). The zero mode ($\ell = 0$) exists only in the case $(P_0, P_1) = (+,+)$. The fundamental region of S^1/Z_2 is the interval $0 \le y \le \pi R$. Parity P_j at y_j $[(y_0, y_1) = (0, \pi R)]$ corresponds to the Neumann or Dirichlet condition at the edge point of the interval.

$$\begin{aligned} P_j = + : \ & \partial_y \Phi(x,y)|_{y=y_j} = 0 \quad \text{(Neumann)} \\ P_j = - : \ & \Phi(x,y)|_{y=y_j} = 0 \quad \text{(Dirichlet)} \end{aligned} \tag{5.17}$$

5.4 Chiral fermions

The action of a free massless fermion field $\Psi(x,y)$ on $M^4 \times (S^1/Z_2)$ is given by

$$I = \int d^4x dy \, \mathcal{L} \ , \quad \mathcal{L} = \overline{\Psi}\Gamma^M \partial_M \Psi = \overline{\Psi}(\Gamma^\mu \partial_\mu + \Gamma^5 \partial_y)\Psi \tag{5.18}$$

where $\overline{\Psi} = i\Psi^\dagger\Gamma^0$. Γ^M satisfies the Dirac algebra (5.4). It is seen that $(\overline{\Psi}_1\Psi_2)^\dagger = \overline{\Psi}_2\Psi_1$ and $(\overline{\Psi}_1\Gamma^M\Psi_2)^\dagger = -\overline{\Psi}_2\Gamma^M\Psi_1$. $\mathcal{L} - \mathcal{L}^\dagger = \partial_M(\overline{\Psi}\Gamma^M\Psi)$ so that the action is hermitian. To satisfy $\mathcal{L}(x, y_j - y) = \mathcal{L}(x, y_j + y)$ one needs

$$\Psi(x, y_j - y) = P_j \Gamma^5 \Psi(x, y_j + y) \ , \tag{5.19}$$

where $P_j = +1$ or -1. The factor Γ^5 is necessary as $\partial_y \to -\partial_y$ under the operation $\hat{P}_j$. Since $\overline{\Psi}(x, y_j - y) = -\overline{\Psi}(x, y_j + y)\Gamma^5 P_j$ and therefore $\overline{\Psi}\Psi(x, y_j - y) = -\overline{\Psi}\Psi(x, y_j + y)$, the ordinary mass term $-m\overline{\Psi}\Psi$ is forbidden on $M^4 \times (S^1/Z_2)$.

In the representation (5.7) of the Dirac matrices, Γ^5 is γ^5 in four dimensions so that the parity under $\hat{P}_j$ depends on whether the field component is right-handed or left-handed in four dimensions. For instance, in the case $(P_0, P_1) = (+1, +1)$ in (5.19)

$$\begin{aligned}
&\Psi_R(x, -y) = +\Psi_R(x, y) \ , \quad \Psi_R(x, \pi R - y) = +\Psi_R(x, \pi R + y) \ , \\
&\Psi_L(x, -y) = -\Psi_L(x, y) \ , \quad \Psi_L(x, \pi R - y) = -\Psi_L(x, \pi R + y) \ , \\
&\Psi_R(x, y) = \frac{1}{\sqrt{\pi R}} \psi_{0R}(x) + \sqrt{\frac{2}{\pi R}} \sum_{n=1}^{\infty} \psi_{nR}(x) \cos \frac{ny}{R} \ , \\
&\Psi_L(x, y) = \sqrt{\frac{2}{\pi R}} \sum_{n=1}^{\infty} \psi_{nL}(x) \sin \frac{ny}{R} \ .
\end{aligned} \tag{5.20}$$

In the case $(P_0, P_1) = (-1, -1)$, R and L are interchanged. The Dirac equations become

$$\begin{aligned}
&\bar{\sigma}\partial\psi_{0R} = 0 \ , \\
&\bar{\sigma}\partial\psi_{nR} = \frac{n}{R}\psi_{nL} \ , \quad \sigma\partial\psi_{nL} = \frac{n}{R}\psi_{nR} \quad (n \geq 1).
\end{aligned} \tag{5.21}$$

Depending on the boundary condition, a zero mode appears in either the right-handed or left-handed component. We thus obtain a **chiral fermion**. No zero mode appears in the case $(P_0, P_1) = (+1, -1)$ or $(-1, +1)$.

5.5 Gauge fields

Gauge fields on $M^4 \times (S^1/Z_2)$ are defined in a similar manner. For $U(1)$ gauge fields the Lagrangian density is $\mathcal{L} = -\frac{1}{4}F_{MN}F^{MN}$. The boundary conditions are

$$\begin{pmatrix} A_\mu \\ A_y \end{pmatrix} (x, y_j - y) = P_j \begin{pmatrix} A_\mu \\ -A_y \end{pmatrix} (x, y_j + y) \, . \tag{5.22}$$

A_y and A_μ have opposite parity in order for $F_{\mu y} = \partial_\mu A_y - \partial_y A_\mu$ to have a covariant relation $F_{\mu y}(x, y_j - y) = -P_j F_{\mu y}(x, y_j + y)$. The KK expansion is

$$\begin{aligned}
&\text{(i)} \quad (P_0, P_1) = (+,+) \\
&\qquad A_\mu(x,y) = \frac{1}{\sqrt{\pi R}}\, A_\mu^{(0)}(x) + \sqrt{\frac{2}{\pi R}} \sum_{n=1}^{\infty} A_\mu^{(n)}(x) \cos\frac{ny}{R} \, , \\
&\qquad A_y(x,y) = \sqrt{\frac{2}{\pi R}} \sum_{n=1}^{\infty} A_y^{(n)}(x) \sin\frac{ny}{R} \, , \\
&\text{(ii)} \quad (P_0, P_1) = (-,-) \\
&\qquad A_\mu(x,y) = \sqrt{\frac{2}{\pi R}} \sum_{n=1}^{\infty} A_\mu^{(n)}(x) \sin\frac{ny}{R} \, , \\
&\qquad A_y(x,y) = \frac{1}{\sqrt{\pi R}}\, A_y^{(0)}(x) + \sqrt{\frac{2}{\pi R}} \sum_{n=1}^{\infty} A_y^{(n)}(x) \cos\frac{ny}{R} \, , \\
&\text{(iii)} \quad (P_0, P_1) = (+,-) \\
&\qquad A_\mu(x,y) = \sqrt{\frac{2}{\pi R}} \sum_{n=0}^{\infty} A_\mu^{(n)}(x) \cos\frac{(n+\frac{1}{2})y}{R} \, , \\
&\qquad A_y(x,y) = \sqrt{\frac{2}{\pi R}} \sum_{n=0}^{\infty} A_y^{(n)}(x) \sin\frac{(n+\frac{1}{2})y}{R} \, , \\
&\text{(iv)} \quad (P_0, P_1) = (-,+) \\
&\qquad A_\mu(x,y) = \sqrt{\frac{2}{\pi R}} \sum_{n=0}^{\infty} A_\mu^{(n)}(x) \sin\frac{(n+\frac{1}{2})y}{R} \, , \\
&\qquad A_y(x,y) = \sqrt{\frac{2}{\pi R}} \sum_{n=0}^{\infty} A_y^{(n)}(x) \cos\frac{(n+\frac{1}{2})y}{R} \, .
\end{aligned} \tag{5.23}$$

Zero modes appear in A_μ in the case $(P_0, P_1) = (+,+)$, and in A_y in the case $(P_0, P_1) = (-,-)$.

Inserting the KK expansion into $\mathcal{L}^{4d} = \int_0^{\pi R} dy\, \mathcal{L}$, one finds in the case $(P_0, P_1) = (+,+)$, for instance, that

$$\mathcal{L}^{4d} = -\frac{1}{4} F_{\mu\nu}^{(0)} F^{(0)\mu\nu} + \sum_{n=1}^{\infty} \Big\{ -\frac{1}{4} F_{\mu\nu}^{(n)} F^{(n)\mu\nu} - \frac{1}{2} \Big(\frac{n}{R} A_\mu^{(n)} + \partial_\mu A_y^{(n)} \Big) \Big(\frac{n}{R} A^{(n)\mu} + \partial^\mu A_y^{(n)} \Big) \Big\} \tag{5.24}$$

where $F_{\mu\nu}^{(n)} = \partial_\mu A_\nu^{(n)} - \partial_\nu A_\mu^{(n)}$. $A_\mu^{(0)}(x)$ is the massless gauge field in four dimensions. There is $U(1)$ gauge invariance in four dimensions. On the other hand the Lagrangian density for the $n \geq 1$ mode is precisely the density $\mathcal{L}_3$ in Eq. (2.11) where the replacements $A_\mu \to A_\mu^{(n)}$, $\varphi \to -A_y^{(n)}$ and $ev \to n/R$ are made. In other words, for the KK excited modes ($n \geq 1$ modes) the Higgs mechanism in four dimensions takes place to make $A_\mu^{(n)}$ a vector field with mass n/R. In the case $(P_0, P_1) = (-,-)$ the zero mode $A_y^{(0)}(x)$ of A_y becomes a massless scalar field at the classical level.

5.6 Non-Abelian gauge fields and orbifold boundary conditions

Consider $SU(N)$ gauge theory on $M^4 \times (S^1/Z_2)$ with massless fermion fields in the fundamental representation. The Lagrangian density is given by $\mathcal{L} = \mathcal{L}^{\text{gauge}} + \mathcal{L}^{\text{fermion}}$ with $\mathcal{L}^{\text{gauge}}$ in (3.27) and $\mathcal{L}^{\text{fermion}}$ in (3.33) ($m_f = 0$).

The Lagrangian density on $M^4 \times (S^1/Z_2)$ must satisfy $\mathcal{L}(x,y) = \mathcal{L}(x, y + 2\pi R) = \mathcal{L}(x, -y)$. In non-Abelian gauge theory

$$\begin{aligned}
&\hat{P}_j A_M(x, y_j + y) = \begin{pmatrix} A_\mu \\ A_y \end{pmatrix} (x, y_j - y) = P_j \begin{pmatrix} A_\mu \\ -A_y \end{pmatrix} (x, y_j + y) P_j^{-1}, \\
&\hat{P}_j \Psi(x, y_j + y) = \Psi(x, y_j - y) = \eta_j P_j \Gamma^5 \Psi(x, y_j + y), \\
&\qquad P_j \in SU(N)\,, \quad P_j^2 = 1\,, \quad \eta_j = \pm 1\,.
\end{aligned} \tag{5.25}$$

Field strengths $F_{MN} = \partial_M A_N - \partial_N A_M - ig[A_M, A_N]$ satisfy $F_{\mu\nu}(x, y_j - y) = P_j F_{\mu\nu}(x, y_j + y) P_j^{-1}$ and $F_{\mu y}(x, y_j - y) = -P_j F_{\mu y}(x, y_j + y) P_j^{-1}$. P_j is a matrix, with non-vanishing off-diagonal elements in general. For instance, in $SU(2)$ gauge theory

$$P_j = \begin{pmatrix} \cos\alpha & e^{-i\beta} \sin\alpha \\ e^{i\beta} \sin\alpha & -\cos\alpha \end{pmatrix}$$

is possible. Recalling that $\hat{U} = \hat{P}_1\hat{P}_0$, we have from (5.25)

$$\begin{aligned}
&\hat{U} A_M(x,y) = A_M(x + 2\pi R) = U A_M(x,y) U^{-1} , \\
&\hat{U}\Psi(x,y) = \Psi(x + 2\pi R) = \eta_0\eta_1 U\Psi(x) , \\
&U = P_1 P_0 .
\end{aligned} \tag{5.26}$$

When either P_0 or P_1 is not proportional to the identity matrix, the gauge group $\mathcal{G}$ breaks down to a subgroup $\mathcal{H}^{\mathrm{BC}}$. Let $\hat{\mathcal{G}} = \{T^a\}$ be the generators of the group $\mathcal{G}$. Then the generators of $\mathcal{H}^{\mathrm{BC}}$ is given by $\hat{\mathcal{H}}^{\mathrm{BC}} = \{T^\alpha \in \hat{\mathcal{G}} \mid [T^\alpha, P_j] = 0\}$. The set $\{P_0, P_1, \eta_0, \eta_1\}$ is called **orbifold boundary condition** on $M^4 \times (S^1/Z_2)$.

When the group $\mathcal{G}$ breaks down to $\mathcal{H}^{\mathrm{BC}}$ by orbifold boundary condition, zero modes of A_μ appear in the $\mathcal{H}^{\mathrm{BC}}$ part. How about zero modes of A_y? It depends on P_0 and P_1. Take $\mathcal{G} = SU(3)$ as an example. We make use of the representation of the generators $\{T^a\}$ in Appendix A.3. In the case

$$\text{Case 1}: \; P_0 = P_1 = \begin{pmatrix} -1 & 0 & 0 \\ 0 & -1 & 0 \\ 0 & 0 & 1 \end{pmatrix} \tag{5.27}$$

$\mathcal{H}^{\mathrm{BC}} = SU(2) \times U(1)$, and the parity of A_μ, A_y is given by

$$A_\mu : \begin{pmatrix} (+,+) & (+,+) & (-,-) \\ (+,+) & (+,+) & (-,-) \\ (-,-) & (-,-) & (+,+) \end{pmatrix} , \; A_y : \begin{pmatrix} (-,-) & (-,-) & (+,+) \\ (-,-) & (-,-) & (+,+) \\ (+,+) & (+,+) & (-,-) \end{pmatrix} \tag{5.28}$$

so that zero modes of A_y appear in four components in $\mathcal{G}/\mathcal{H}^{\mathrm{BC}}$. Next in the case

$$\text{Case 2}: \; P_0 = \begin{pmatrix} -1 & 0 & 0 \\ 0 & -1 & 0 \\ 0 & 0 & 1 \end{pmatrix} , \; P_1 = \begin{pmatrix} -1 & 0 & 0 \\ 0 & 1 & 0 \\ 0 & 0 & -1 \end{pmatrix} \tag{5.29}$$

$\mathcal{H}^{\mathrm{BC}} = U(1) \times U(1)$, and the parity of A_μ, A_y is given by

$$A_\mu : \begin{pmatrix} (+,+) & (+,-) & (-,+) \\ (+,-) & (+,+) & (-,-) \\ (-,+) & (-,-) & (+,+) \end{pmatrix} , \; A_y : \begin{pmatrix} (-,-) & (-,+) & (+,-) \\ (-,+) & (-,-) & (+,+) \\ (+,-) & (+,+) & (-,-) \end{pmatrix} . \tag{5.30}$$

Zero modes of A_μ appear in the diagonal (T^3, T^8) components, and zero modes of A_y appear in the T^6, T^7 components. As another example, consider

$$\text{Case 3}: \; P_0 = \begin{pmatrix} -1 & 0 & 0 \\ 0 & -1 & 0 \\ 0 & 0 & 1 \end{pmatrix}, \quad P_1 = \begin{pmatrix} -1 & 0 & 0 \\ 0 & 0 & -i \\ 0 & i & 0 \end{pmatrix}. \tag{5.31}$$

In this case A_μ has a zero mode only in the $\frac{\sqrt{3}}{2}T^3 + \frac{1}{2}T^8$ component, and therefore $\mathcal{H}^{\mathrm{BC}} = U(1)$. A_y has a zero mode in the T^6 component. One sees that various patterns of symmetry breaking arise, depending on the choice of P_j.

5.7 AB phases and equivalence classes of boundary conditions

It is interesting that the symmetry $\mathcal{H}^{\mathrm{BC}}$ kept by the orbifold boundary conditions varies depending on the choice of boundary conditions. It also implies the arbitrariness in the choice of the boundary conditions. One needs to know the relationship between $\mathcal{H}^{\mathrm{BC}}$ and the physical symmetry of the system $\mathcal{H}^{\mathrm{phys}}$.

We have encountered this problem in gauge theory on $M^4 \times S^1$, and have discussed about it in Section 4.6. It is seen there that the Aharonov-Bohm (AB) phase along S^1 plays an important role. Its value is dynamically determined at the quantum level, and physical symmetry $\mathcal{H}^{\mathrm{phys}}$ is determined in the combination of the AB phase and boundary condition. Two apparently different boundary conditions may be connected by an AB phase, and may describe the same physics. Indeed, in $SU(N)$ gauge theory on $M^4 \times S^1$, the theory with boundary condition $\{U, \beta\}$ is equivalent to the theory with $\{U' = I, \beta\}$.

In gauge theory on the orbifold $M^4 \times (S^1/Z_2)$ the situation is more involved. Suppose that the orbifold boundary condition is given by (5.25). We perform a gauge transformation $A_M \to A'_M = \Omega A_M \Omega^{-1} - (i/g)\Omega \partial_M \Omega^{-1}$ and $\Psi \to \Psi' = \Omega\Psi$. The transformed fields, in general, do not satisfy the original boundary condition. It is easy to see that

$$\begin{pmatrix} A'_\mu \\ A'_y \end{pmatrix}(x, y_j - y) = P'_j \begin{pmatrix} A'_\mu \\ -A'_y \end{pmatrix}(x, y_j + y) P'^{-1}_j$$
$$- \frac{i}{g} P'_j \begin{pmatrix} \partial_\mu \\ -\partial_y \end{pmatrix} P'^{-1}_j ,$$
$$\Psi'(x, y_j - y) = \eta_j P'_j \Gamma^5 \Psi'(x, y_j + y) ,$$
$$P'_j = \Omega(x, y_j - y) P_j \Omega^{-1}(x, y_j + y) . \tag{5.32}$$

P'_j in general depends on (x, y). If $\partial_M P'_j = 0$ and P'_j are constant matrices, then the boundary condition changes from (P_0, P_1) to (P'_0, P'_1). In this case, by taking $y = -y_j$ in P'_j in (5.32), one has a relation $P'_j = \Omega(x, 2y_j) P_j \Omega^{-1}(x, 0)$. Therefore U in (5.26) becomes $U' = P'_1 P'_0 = \Omega(x, 2y_1) U \Omega^{-1}(x, 0)$, which takes the same form as the relation in (3.22).

The boundary condition remains invariant when $(P'_0, P'_1) = (P_0, P_1)$. The whole set of gauge transformations satisfying this condition is called the **residual gauge symmetry** surviving after the orbifold boundary condition is imposed.

$$\mathcal{G}_{\rm res} = \big\{ \Omega(x, y) \in G \,; \partial_M \big[\Omega(x, y_j - y) P_j \Omega^{-1}(x, y_j + y) \big] = 0,$$
$$P_j \Omega(x, y_j + y) = \Omega(x, y_j - y) P_j \,\big\} . \tag{5.33}$$

It is five-dimensional local gauge symmetry, and is larger than the four-dimensional gauge symmetry of $\mathcal{H}^{\rm BC}$. It will be seen below that many good properties of gauge-Higgs unification models follow from this five-dimensional local gauge symmetry.

When $(P'_0, P'_1) \neq (P_0, P_1)$, gauge transformations connecting these two sets of boundary conditions are called **large gauge transformations**. The theory with (P_0, P_1) is different from the theory with (P'_0, P'_1). But these two are connected by a large gauge transformation, and therefore physics must be the same in these two theories. The equivalence is guaranteed by the Aharonov-Bohm (AB) phase. Indeed, when $A_y(x, y)$ has a zero mode $A_y^{(0)}(x)$, eigenvalues of the line integral

$$\hat{W} = P \exp \bigg\{ -i \frac{g}{\sqrt{\pi R}} \int_0^{2y_1} dy \, A_y^{(0)} \bigg\} \cdot P_1 P_0 \tag{5.34}$$

are gauge invariant **AB phases** as in (3.24).

As an example we consider $SU(3)$ gauge theory with the boundary conditions (5.27) (Case 1) and (5.29) (Case 2). Zero modes of A_y appear in the four components in $SU(3)$ in Case 1 and two components in Case 2.

Inserting the common T^6 component

$$\frac{1}{\sqrt{\pi R}} A_y^{(0)} = -\frac{\theta_H}{g\pi R} T^6 \ , \quad T^6 = \frac{1}{2} \begin{pmatrix} 0 & 0 & 0 \\ 0 & 0 & 1 \\ 0 & 1 & 0 \end{pmatrix} \tag{5.35}$$

into (5.34), one finds that

$$\hat{W} = \begin{pmatrix} 1 & 0 & 0 \\ 0 & & \\ 0 & \multicolumn{2}{c}{\pm e^{i\theta_H \tau_1}} \end{pmatrix} . \tag{5.36}$$

Here + and − corresponds to Case 1 and Case 2 respectively. The phase θ_H of the eigenvalues is the AB phase. The value of θ_H is not arbitrary. It is determined by the condition that the effective potential $V_{\text{eff}}(\theta_H)$ is minimized at the quantum level as shown in Chapter 4.

We make a large gauge transformation

$$\Omega = \exp\left\{ i \frac{\beta y}{\pi R} T^6 \right\} . \tag{5.37}$$

Noting $\{T^6, P_j\} = 0$, one finds that

$$\frac{1}{\sqrt{\pi R}} A_y'^{(0)} = -\frac{\theta_H'}{g\pi R} T^6 \ , \quad \theta_H' = \theta_H - \beta \ ,$$

$$P_0' = P_0 \ , \quad P_1' = P_1 \begin{pmatrix} 1 & 0 & 0 \\ 0 & & \\ 0 & \multicolumn{2}{c}{e^{-i\beta\tau_1}} \end{pmatrix} . \tag{5.38}$$

Starting from (5.27) in Case 1 and making a large gauge transformation with $\beta = \pm\pi$, one finds that (P_0', P_1') becomes (P_0, P_1) in (5.29) in Case 2. That is, Case 1 and Case 2 become equivalent to each other at the quantum level. By making a large gauge transformation with $\beta = -\frac{1}{2}\pi$, (P_0', P_1') becomes (P_0, P_1) in (5.31) in Case 3. To summarize, the symmetry of boundary condition $\mathcal{H}^{\text{BC}}$ is different in Case 1, Case 2 and Case 3, but these three cases are connected by the AB phase so that physics in all cases is the same.

When two sets of boundary conditions are connected by a large gauge transformation, we write an equivalence relation

$$(P_0, P_1) \sim (P_0', P_1') \ . \tag{5.39}$$

The relation (5.39) defines **equivalence classes of boundary conditions.** In one equivalence class of boundary conditions, physics is the same. How can one characterize the physical symmetry in each equivalence class? Start from the boundary condition (P_0, P_1). In general there appear several AB phases θ_j. The values of the AB phases $\theta_j^{\rm min}$ are determined from the location of the global minimum of the effective potential $V_{\rm eff}(\theta_j)$ at the quantum level. Let $\bar{A}_y$ be $(\pi R)^{-1/2} A_y^{(0)}$ corresponding to $\theta_j^{\rm min}$. By making a large gauge transformation $\Omega = \exp\{ig\bar{A}_y y\}$, we have $A_y^{(0)} \to A_y^{\prime(0)} = 0$ as in (5.38), and the effective potential in the new gauge $V_{\rm eff}(\theta'_j)$ is minimized at $\theta'_j = 0$. The boundary condition is transformed to $(P'_0, P'_1) = (P_0^{\rm sym}, P_1^{\rm sym})$. Among the generators of the group $\mathcal{G}$, let $\{T^\alpha\}$ be the generators commuting with $P_0^{\rm sym}$ and $P_1^{\rm sym}$. The **physical gauge symmetry** is the subgroup $\mathcal{H}^{\rm phys}$ whose generators are given by $\{T^\alpha\}$. [34]

Given the gauge group, the number of the equivalence classes of boundary conditions is finite. For instance, in $SU(N)$ gauge theory on $M^4 \times (S^1/Z_2)$ the number of the equivalence classes is $(N+1)^2$. [35, 36] In one equivalence class physics is the same as a result of dynamics of AB phases. In a different equivalence class physics is different. At the moment a mechanism for selecting an equivalence class is not known.

5.8 Model of universal extra dimensions (UED)

A simple extension of the SM to extra-dimensions is the UED (Universal Extra Dimensions) model. [37]. The model is arranged such that all fields and particles in the SM appear as zero modes on $M^4 \times (S^1/Z_2)$. The gauge group is the SM gauge group $\mathcal{G}_{\rm SM} = SU(3)_C \times SU(2)_L \times U(1)_Y$. As in Section 2.5, the respected gauge fields are written as G_M, A_M and B_M. The orbifold boundary conditions are given by

$$\begin{pmatrix} G_\mu, G_y \\ A_\mu, A_y \\ B_\mu, B_y \end{pmatrix} (x, y_j - y) = \begin{pmatrix} G_\mu, -G_y \\ A_\mu, -A_y \\ B_\mu, -B_y \end{pmatrix} (x, y_j + y) . \tag{5.40}$$

G_y, A_y and B_y do not have a zero mode. The KK expansion is the one in case (i) $(P_0, P_1) = (+,+)$ in Eq. (5.23). The Higgs field Φ satisfies

$$\Phi(x, y_j - y) = \Phi(x, y_j + y) . \tag{5.41}$$

5D fermion multiplets need to be prepared for left-handed and right-handed quarks separately. 5D fields $Q(x,y)$, $U(x,y)$ and $D(x,y)$, which contain $SU(2)_L$ doublet left-handed quarks $q_L = (u_L, d_L)^t$ and singlet right-handed quarks u_R, d_R, satisfy boundary conditions

$$\begin{aligned} Q(x, y_j - y) &= -\Gamma^5 Q(x, y_j + y) , \\ U(x, y_j - y) &= +\Gamma^5 U(x, y_j + y) , \\ D(x, y_j - y) &= +\Gamma^5 D(x, y_j + y) . \end{aligned} \tag{5.42}$$

Here Q, U, D are four-component fermion fields, each of which has both 4D left-handed and right-handed components. The KK expansions are given by

$$Q(x,y) = \begin{pmatrix} \hat{U}(x,y) \\ \hat{D}(x,y) \end{pmatrix},$$

$$\begin{pmatrix} Q_R \\ Q_L \end{pmatrix} = \frac{1}{\sqrt{\pi R}} \begin{pmatrix} 0 \\ q_L^{(0)}(x) \end{pmatrix} + \sqrt{\frac{2}{\pi R}} \sum_{n=1}^{\infty} \begin{pmatrix} q_R'^{(n)}(x) \sin \frac{ny}{R} \\ q_L^{(n)}(x) \cos \frac{ny}{R} \end{pmatrix},$$

$$q_L^{(n)} = \begin{pmatrix} \hat{u}_L^{(n)} \\ \hat{d}_L^{(n)} \end{pmatrix}, \quad q_R'^{(n)} = \begin{pmatrix} \hat{u}_R'^{(n)} \\ \hat{d}_R'^{(n)} \end{pmatrix},$$

$$U(x,y) = \frac{1}{\sqrt{\pi R}} \begin{pmatrix} u_R^{(0)}(x) \\ 0 \end{pmatrix} + \sqrt{\frac{2}{\pi R}} \sum_{n=1}^{\infty} \begin{pmatrix} u_R^{(n)}(x) \cos \frac{ny}{R} \\ u_L'^{(n)}(x) \sin \frac{ny}{R} \end{pmatrix},$$

$$D(x,y) = \frac{1}{\sqrt{\pi R}} \begin{pmatrix} d_R^{(0)}(x) \\ 0 \end{pmatrix} + \sqrt{\frac{2}{\pi R}} \sum_{n=1}^{\infty} \begin{pmatrix} d_R^{(n)}(x) \cos \frac{ny}{R} \\ d_L'^{(n)}(x) \sin \frac{ny}{R} \end{pmatrix}. \tag{5.43}$$

Lepton multiplets are introduced in a similar manner. For the gauge and Higgs fields one 5D multiplet field corresponds to one 4D field, whereas for quarks and leptons two 5D multiplet fields correspond to one 4D field. The number of degrees of freedom is doubled.

Gauge interactions of the matter fields are introduced, as in the SM, by replacing the derivative ∂_M by the covariant derivative $D_M = \partial_M - ig_s G_M - igA_M - iYg'B_M$. Gauge coupling constants are related to the 4D coupling constants by $g_{4d} = g/\sqrt{\pi R}$ etc.

In the UED model gauge symmetry is spontaneously broken by the Higgs mechanism. The action for the 5D Higgs field $\Phi(x,y)$ is given in the same form as that in Eq. (2.36). With $m^2 < 0$ the zero mode $\Phi^{(0)}(x)$ develops a non-vanishing vacuum expectation value $\langle \Phi^{(0)}(x) \rangle = (0, v/\sqrt{2})^t$ as in Eq. (2.39). There are Yukawa interactions among the Higgs field and fermion fields in the form of Eq. (2.37). For instance,

$$\mathcal{L}_{5d}^{\text{Yukawa}} = -y_{5d} \overline{D} \Phi^\dagger Q - \text{H.c.}. \tag{5.44}$$

with the vacuum expectation value of Φ gives rise to a Yukawa mass term between D and $\hat{D}$ of $Q = (\hat{U}, \hat{D})^t$. 4D Yukawa coupling constant is given by $y_{4d} = y_{5d}/\sqrt{\pi R}$. Terms bilinear in $D, \hat{D}$ are, with $\tilde{m} = y_{4d} v/\sqrt{2}$,

$$\mathcal{L} = \overline{D} \Gamma^M \partial_M D + \overline{\hat{D}} \Gamma^M \partial_M \hat{D} - \tilde{m}(\overline{D}\hat{D} + \overline{\hat{D}} D) . \tag{5.45}$$

The equations of motion are

$$\Gamma^M \partial_M D - \tilde{m}\hat{D} = 0 \ , \quad \Gamma^M \partial_M \hat{D} - \tilde{m} D = 0 \ , \tag{5.46}$$

and D and $\hat{D}$ are intertwined. In the representation of Eq. (5.7) the equations become

$$\begin{aligned} \sigma^\mu \partial_\mu \begin{pmatrix} D_L \\ \hat{D}_L \end{pmatrix} + \partial_y \begin{pmatrix} D_R \\ \hat{D}_R \end{pmatrix} - \tilde{m} \begin{pmatrix} \hat{D}_R \\ D_R \end{pmatrix} = 0 \ , \\ \bar{\sigma}^\mu \partial_\mu \begin{pmatrix} D_R \\ \hat{D}_R \end{pmatrix} - \partial_y \begin{pmatrix} D_L \\ \hat{D}_L \end{pmatrix} - \tilde{m} \begin{pmatrix} \hat{D}_L \\ D_L \end{pmatrix} = 0 \ . \end{aligned} \tag{5.47}$$

Inserting the KK expansion (5.43) into (5.47), one finds that

$$\begin{aligned} \sigma^\mu \partial_\mu \left(d_L'^{(n)} \pm i \hat{d}_L^{(n)} \right) = \left(\tilde{m} \mp i \frac{n}{R} \right) \left(\hat{d}_R'^{(n)} \pm i d_R^{(n)} \right) \ , \\ \bar{\sigma}^\mu \partial_\mu \left(\hat{d}_R'^{(n)} \pm i d_R^{(n)} \right) = \left(\tilde{m} \pm i \frac{n}{R} \right) \left(d_L'^{(n)} \pm i \hat{d}_L^{(n)} \right) \ . \end{aligned} \tag{5.48}$$

Combining these two equations, one finds that the mass of the n-th KK mode is given by

$$m_n = \sqrt{\tilde{m}^2 + \frac{n^2}{R^2}} \ . \tag{5.49}$$

The mass of the zero mode comes solely from the Yukawa interaction.

In the UED model all fields and particles in the SM appear as zero modes of the 5D gauge, Higgs and fermion fields, and the interactions among them become the same as in the SM. There are selection rules in the interactions among the KK excited modes. The cubic couplings $\phi_1^{(n_1)} \phi_2^{(n_2)} \phi_3^{(n_3)}$ are allowed for $(n_1, n_2, n_3) = (0,0,0), (1,1,0), (1,1,2)$ etc., but are forbidden for $(0,0,1), (1,1,1), (0,0,2)$ etc. In the pair creation process of KK excited modes in the collision of two SM particles, the first excited ($n = 1$) modes can be produced only in a pair. A single $n = 2$ mode can be produced through an internal triangular loop diagram in which $n = 1$ modes run along the loop. There is the KK parity P_K. $A_\mu^{(n)}, q_L^{(n)}, u_R^{(n)}, d_R^{(n)}, \Phi^{(n)}$ have $P_K = (-1)^n$, whereas $A_y^{(n)}, q_R'^{(n)}, u_L'^{(n)}, d_L'^{(n)}$ have $P_K = (-1)^{n+1}$. The lightest particle among the $n = 1$ particles becomes stable, serving as a candidate for dark matter. $B_\mu^{(1)}$ is such a candidate.

The UED model is the simplest extension of the SM to extra dimensions. Unfortunately it does not solve the gauge hierarchy problem.

5.9 Kawamura model of $SU(5)$ grand unification

Chiral fermions are naturally incorporated in theory on $M^4 \times (S^1/Z_2)$, and gauge symmetry is partially broken by orbifold boundary conditions. In the $SU(5)$ grand unified theory (GUT) [32] there is the doublet-triplet splitting problem of the Higgs boson. It is interesting that this problem is brilliantly solved by the orbifold boundary conditions on $M^4 \times (S^1/Z_2)$.

In $SU(5)$ GUT in four dimensions one needs to introduce at least two kinds of Higgs scalar fields. One of them is a scalar field H in the adjoint representation (**24**) of $SU(5)$, which is expressed as a 5×5 hermitian matrix ($H^\dagger = H$) satisfying $\mathrm{Tr}\, H = 0$. The other is a scalar field ϕ in the fundamental representation. By the Higgs mechanism the two scalar fields have non-vanishing vacuum expectation values $\langle H \rangle \neq 0$, $\langle \phi \rangle \neq 0$, and $SU(5)$ gauge symmetry spontaneously breaks in two steps as

$$\begin{aligned} SU(5) &\to SU(3)_C \times SU(2)_L \times U(1)_Y \quad \text{by } \langle H \rangle \neq 0 \\ &\to SU(3)_C \times U(1)_{\mathrm{EM}} \quad \text{by } \langle \phi \rangle \neq 0 \,. \end{aligned} \tag{5.50}$$

Concretely we take a potential

$$\begin{aligned} V(H, \phi) = &- \mu_1^2 \,\mathrm{Tr}\, H^2 + \lambda_1 (\mathrm{Tr}\, H^2)^2 + \lambda_2 \,\mathrm{Tr}\, H^4 \\ &- \mu_2^2 \,\phi^\dagger \phi + \lambda_3 (\phi^\dagger \phi)^2 \\ &+ \lambda_4 (\mathrm{Tr}\, H^2)\, \phi^\dagger \phi + \lambda_5 \,\phi^\dagger H^2 \phi \,. \end{aligned} \tag{5.51}$$

We assume that $|\langle H \rangle| \gg |\langle \phi \rangle|$, and first look for the minimum of $V(H, 0)$. For $\lambda_2 > 0$ and $30\lambda_1 + 7\lambda_2 > 0$ the minimum is found at

$$\langle H \rangle = v_1 \begin{pmatrix} 2 & 0 & 0 & 0 & 0 \\ 0 & 2 & 0 & 0 & 0 \\ 0 & 0 & 2 & 0 & 0 \\ 0 & 0 & 0 & -3 & 0 \\ 0 & 0 & 0 & 0 & -3 \end{pmatrix},$$

$$v_1 = \frac{\mu_1}{\sqrt{2(30\lambda_1 + 7\lambda_2)}} \,. \tag{5.52}$$

With this $\langle H \rangle$, $SU(5)$ symmetry breaks down to $SU(3)_C \times SU(2)_L \times U(1)_Y$. v_1 represents the energy scale characterizing the grand unification, being about 10^{15} GeV. We decompose ϕ into $SU(3)_C$ triplet and $SU(2)_L$ doublet:

$$\phi = \begin{pmatrix} \boldsymbol{\phi_3} \\ \boldsymbol{\phi_2} \end{pmatrix}, \quad \boldsymbol{\phi_3} = \begin{pmatrix} \phi_1 \\ \phi_2 \\ \phi_3 \end{pmatrix}, \quad \boldsymbol{\phi_2} = \begin{pmatrix} \phi_4 \\ \phi_5 \end{pmatrix}. \tag{5.53}$$

Insertion of $\langle H \rangle$ in (5.52) into the third line of (5.51) yields the potential for $\phi_{\mathbf{3}}$ and $\phi_{\mathbf{2}}$ given by

$$\begin{aligned} V(H,\phi) \Rightarrow & \big\{ -\mu_2^2 + (30\lambda_4 + 4\lambda_5)v_1^2 \big\} \phi_{\mathbf{3}}^\dagger \phi_{\mathbf{3}} \\ & + \big\{ -\mu_2^2 + (30\lambda_4 + 9\lambda_5)v_1^2 \big\} \phi_{\mathbf{2}}^\dagger \phi_{\mathbf{2}} \\ & + \lambda_3 (\phi_{\mathbf{3}}^\dagger \phi_{\mathbf{3}} + \phi_{\mathbf{2}}^\dagger \phi_{\mathbf{2}})^2 \ . \end{aligned} \tag{5.54}$$

To reproduce the spontaneous breaking of the electroweak symmetry in the SM, the conditions

$$\begin{aligned} m_{\mathbf{3}}^2 &= -\mu_2^2 + (30\lambda_4 + 4\lambda_5)v_1^2 > 0 \ , \\ m_{\mathbf{2}}^2 &= -\mu_2^2 + (30\lambda_4 + 9\lambda_5)v_1^2 < 0 \ , \\ v_2 &= \sqrt{\frac{-m_{\mathbf{2}}^2}{\lambda_3}} = 246 \, \mathrm{GeV} \end{aligned} \tag{5.55}$$

must be satisfied. Here the relation (2.39) in the SM has been made use of. It follows from the third relation in (5.55) that $|m_{\mathbf{2}}| = O(100)\,\mathrm{GeV}$ for $\lambda_3 = O(1)$. On the other hand $v_1 = O(10^{15})\,\mathrm{GeV}$ so that in order to satisfy the second relation the value of $\mu_2 = O(v_1)$ must be fine-tuned with 13 digits accuracy. This is the doublet-triplet splitting problem at the classical level. At the quantum level one encounters the gauge hierarchy problem further.

It was shown by Kawamura that the doublet-triplet splitting problem is naturally solved by the appropriate choice of the orbifold boundary conditions. [38] As seen in Section 5.6, orbifold boundary conditions (5.25) are imposed in $SU(5)$ gauge theory on $M^4 \times (S^1/Z_2)$. Let us take

$$P_0 = \begin{pmatrix} 1 & 0 & 0 & 0 & 0 \\ 0 & 1 & 0 & 0 & 0 \\ 0 & 0 & 1 & 0 & 0 \\ 0 & 0 & 0 & 1 & 0 \\ 0 & 0 & 0 & 0 & 1 \end{pmatrix},$$

$$P_1 = \begin{pmatrix} -1 & 0 & 0 & 0 & 0 \\ 0 & -1 & 0 & 0 & 0 \\ 0 & 0 & -1 & 0 & 0 \\ 0 & 0 & 0 & 1 & 0 \\ 0 & 0 & 0 & 0 & 1 \end{pmatrix}. \tag{5.56}$$

Then $SU(5)$ gauge symmetry breaks down to $SU(3)_C \times SU(2)_L \times U(1)_Y$ gauge symmetry by the boundary condition. We introduce a 5D scalar field $\Phi(x,y)$ in the fundamental representation (**5**) which satisfies

$$\Phi(x, y_j - y) = P_j \, \Phi(x, y_j + y) \, . \tag{5.57}$$

As in (5.53), we decompose $\Phi(x,y)$ into

$$\Phi(x,y) = \begin{pmatrix} \boldsymbol{\Phi_3} \\ \boldsymbol{\Phi_2} \end{pmatrix} (x,y) \, . \tag{5.58}$$

Then the boundary condition (5.57) implies that

$$\begin{aligned} &\boldsymbol{\Phi_3}(x,-y) = \boldsymbol{\Phi_3}(x,y), \quad \boldsymbol{\Phi_3}(x,\pi R - y) = -\boldsymbol{\Phi_3}(x,\pi R + y), \\ &\boldsymbol{\Phi_2}(x,-y) = \boldsymbol{\Phi_2}(x,y), \quad \boldsymbol{\Phi_2}(x,\pi R - y) = \boldsymbol{\Phi_2}(x,\pi R + y), \end{aligned} \tag{5.59}$$

$\boldsymbol{\Phi_2}$ has a zero mode, but $\boldsymbol{\Phi_3}$ does not. The KK expansions are given by

$$\begin{aligned} \boldsymbol{\Phi_3}(x,y) &= \sqrt{\frac{2}{\pi R}} \sum_{n=1}^{\infty} \phi_{\boldsymbol{3}}^{(n)}(x) \cos \frac{(n-\frac{1}{2})y}{R} \, , \\ \boldsymbol{\Phi_2}(x,y) &= \frac{1}{\sqrt{\pi R}} \, \phi_{\boldsymbol{2}}(x) + \sqrt{\frac{2}{\pi R}} \sum_{n=1}^{\infty} \phi_{\boldsymbol{2}}^{(n)}(x) \cos \frac{ny}{R} \, . \end{aligned} \tag{5.60}$$

The lightest KK mode of $\boldsymbol{\Phi_3}$ is $\phi_{\boldsymbol{3}}^{(1)}(x)$, which has a mass $1/(2R) \sim 10^{15}$ GeV. $\phi_{\boldsymbol{2}}(x)$ is the Higgs field appearing in the SM. With 5D potential $V(\Phi) = m^2 \Phi^\dagger \Phi + \lambda \pi R (\Phi^\dagger \Phi)^2$ ($m^2 < 0$), we set $-m^2 = \lambda v_2^2$ ($v_2 = 246$ GeV). The doublet-triplet splitting problem is solved in this manner. The scenario can be extended to supersymmetric theory.

It is difficult to incorporate quark and lepton multiplets as 5D fields, however. As explained in Section 5.1, in $SU(5)$ GUT quarks and leptons are introduced in the fundamental representation (**5**) and in the antisymmetric representation (**10**). Similarly to the case of Φ in the **5** representation, with the boundary condition (P_0, P_1) in (5.56) only a part of quarks and leptons appear as zero modes. For this reason quark and lepton fields are assumed to exist only on the $y = 0$ brane.

5.10 Electroweak unification – $SU(3)$ gauge-Higgs unification

A simple and typical model of GHU in which the 4D Higgs field appears as a part of the fifth-dimensional component of gauge fields is the $SU(3)$ model on $M^4 \times (S^1/Z_2)$ introduced by Kubo, Lim and Yamashita. [39] To have the SM symmetry $\mathcal{G}_{\rm SM} = SU(2)_L \times U(1)_Y$, we take the boundary condition (5.27), with which $A_M(x, y + 2\pi R) = A_M(x, y)$. We identify $\mathcal{H}^{\rm BC} = SU(2) \times U(1)$ with $\mathcal{G}_{\rm SM}$. As seen from (5.28) the $SU(2)_L$ doublet 4D Higgs field appears as a zero mode of the $\mathcal{G}/\mathcal{H}^{\rm BC}$ part of A_y.

In evaluating the effective potential for A_y, one may take, with the $SU(2)_L$ invariance in mind,

$$\langle A_y \rangle = \frac{1}{\sqrt{\pi R}} \langle A_y^{(0)} \rangle = \frac{1}{\sqrt{\pi R}} \begin{pmatrix} 0 & 0 & 0 \\ 0 & 0 & v \\ 0 & v & 0 \end{pmatrix} . \tag{5.61}$$

The eigenvalues of $P \exp\big\{ -ig \int_0^{2\pi R} dy \langle A_y \rangle \big\}$ are $\{1, e^{\pm i\theta_H}\}$, and the AB phase is given by

$$\theta_H = 2gv\sqrt{\pi R} = 2\pi R\, g_4\, v \ . \tag{5.62}$$

$g_4 = g/\sqrt{\pi R}$ is the $SU(2)_L$ gauge coupling constant in four dimensions. Suppose that $V_{\rm eff}(\theta_H)$ is minimized at $\theta_H = \theta_H^{\rm min}$. We apply a large gauge transformation (5.37) with $\beta = \theta_H^{\rm min}$. Then $V_{\rm eff}(\theta'_H)$ in the new gauge is minimized at $\theta'_H = 0$. The boundary condition matrices in the new gauge are given by (5.38). Dropping the superscript $^{\rm min}$, we have

$$P'_0 = \begin{pmatrix} -1 & 0 & 0 \\ 0 & -1 & 0 \\ 0 & 0 & 1 \end{pmatrix} , \quad P'_1 = \begin{pmatrix} -1 & 0 & 0 \\ 0 & -\cos\theta_H & i\sin\theta_H \\ 0 & -i\sin\theta_H & \cos\theta_H \end{pmatrix} . \tag{5.63}$$

For $\theta_H \neq 0, \pi$ (mod 2π) the unbroken physical symmetry is $U(1)$, which is identified with the electromagnetic symmetry $U(1)_{\rm EM}$. The electromagnetic field $A_\mu^{\rm EM}$ and Z boson field Z_μ are given by

$$\begin{aligned} & A_\mu^{(3)} T^3 + A_\mu^{(8)} T^8 = Z_\mu T^{3'} + A_\mu^{\rm EM} T^{8'} , \\ & \begin{pmatrix} T^{3'} \\ T^{8'} \end{pmatrix} = \begin{pmatrix} \frac{1}{2} & -\frac{\sqrt{3}}{2} \\ \frac{\sqrt{3}}{2} & \frac{1}{2} \end{pmatrix} \begin{pmatrix} T^3 \\ T^8 \end{pmatrix} , \\ & T^{3'} = \tfrac{1}{2} \begin{pmatrix} 0 & 0 & 0 \\ 0 & -1 & 0 \\ 0 & 0 & 1 \end{pmatrix} , \ T^{8'} = \frac{1}{2\sqrt{3}} \begin{pmatrix} 2 & 0 & 0 \\ 0 & -1 & 0 \\ 0 & 0 & -1 \end{pmatrix} . \end{aligned} \tag{5.64}$$

In other words

$$\begin{pmatrix} Z_\mu \\ A_\mu^{\rm EM} \end{pmatrix} = \begin{pmatrix} \frac{1}{2} & -\frac{\sqrt{3}}{2} \\ \frac{\sqrt{3}}{2} & \frac{1}{2} \end{pmatrix} \begin{pmatrix} A_\mu^{(3)} \\ A_\mu^{(8)} \end{pmatrix} \tag{5.65}$$

and therefore $\sin^2\theta_W = \frac{3}{4}$ which substantially differs from the observed value $\sin^2\theta_W \simeq 0.2312$. The electromagnetic coupling is $e = g_4 \sin\theta_W = \frac{\sqrt{3}}{2} g_4$ and the electric charge $eQ_{\rm EM}$ is

$$eQ_{\rm EM} = g_4 T^{8'} = e \begin{pmatrix} \frac{2}{3} & 0 & 0 \\ 0 & -\frac{1}{3} & 0 \\ 0 & 0 & -\frac{1}{3} \end{pmatrix} . \tag{5.66}$$

The electric charge of the $W^\pm$ boson, which corresponds to $\frac{1}{\sqrt{2}}(T^1 \pm iT^2)$, is $\pm e$. For $\theta_H = \theta_H^{\rm min} = \pi$ the unbroken physical symmetry is $U(1) \times U(1)$, and not only the photon but also the Z boson becomes massless.

It is natural to introduce a quark multiplet in the fundamental representation of $SU(3)$.

$$\Psi_q = \begin{pmatrix} u \\ d \\ d' \end{pmatrix} , \quad \Psi_q(x, y_j - y) = P_j \Gamma^5 \Psi_q(x, y_j + y) . \tag{5.67}$$

With the above boundary condition there appear zero modes in the u_L, d_L, d'_R components. The zero modes of (u_L, d_L) couple with the W boson as an $SU(2)_L$ doublet. There is no zero mode to be identified with the right-handed up-quark. When $\theta_H \neq 0$, d and d' couple with each other, generating the mass of the down quark. However, the up quark remains massless. A zero mode corresponding to the right-handed up quark can be incorporated in a multiplet in higher dimensional representation of $SU(3)$, but it necessarily gives rise to extra exotic zero modes. It is also difficult to have lepton multiplets with correct charges. [40–46]

Although it is difficult to have a realistic model in the scheme of $SU(3)$ GHU, it is useful to learn how the W boson, Z boson, Higgs field and quarks acquire masses in this model.

The mass spectrum of gauge bosons and quarks for $\theta_H \neq 0$ is determined in the following manner. We decompose A_M as $A_M = A_M^{\rm c} + A_M^{\rm q}$ where $A_\mu^{\rm c} = 0$ and

$$A_y^{\rm c} = \frac{\theta_H}{2g\pi R} \begin{pmatrix} 0 & 0 & 0 \\ 0 & 0 & 1 \\ 0 & 1 & 0 \end{pmatrix} = \frac{\theta_H}{g\pi R} T^6 . \tag{5.68}$$

Following the discussion in Section 3.4, we take a gauge-fixing term

$$\mathcal{L}_{\rm gf} = -\frac{1}{\xi} {\rm Tr} \left\{ \partial_\mu A^\mu + \xi(\partial_y A_y - ig[A_y^{\rm c}, A_y]) \right\}^2 . \tag{5.69}$$

The bilinear terms in the Lagrangian density take the form of (3.29). The term containing $\mathcal{D}_y^{\rm c}$ can be rewritten as $\mathrm{Tr}\, B\mathcal{D}_y^{\rm c}\mathcal{D}_y^{\rm c}B \sim -\mathrm{Tr}\,\{\mathcal{D}_y^{\rm c}B\}^2$ where B is $A_M^{\rm q}$ etc. By using $T^{3'}, T^{8'}$ in (5.64) instead of T^3, T^8 as a basis of the generators of $SU(3)$, one finds

$$
\begin{aligned}
\mathrm{Tr}\left\{\mathcal{D}_y^{\rm c}B\right\}^2 &= \frac{1}{2}(\partial_y B^1 - \alpha B^5)^2 + \frac{1}{2}(\partial_y B^5 + \alpha B^1)^2 \\
&\quad + \frac{1}{2}(\partial_y B^2 + \alpha B^4)^2 + \frac{1}{2}(\partial_y B^4 - \alpha B^2)^2 \\
&\quad + \frac{1}{2}(\partial_y B^{3'} + 2\alpha B^7)^2 + \frac{1}{2}(\partial_y B^7 - 2\alpha B^{3'})^2 \\
&\quad + \frac{1}{2}(\partial_y B^{8'})^2 + \frac{1}{2}(\partial_y B^6)^2 ~, \\
\alpha &= \frac{\theta_H}{2\pi R} ~.
\end{aligned}
\tag{5.70}
$$

The parity of the gauge fields and ghost fields with the orbifold boundary condition (P_0, P_1) is

a	$A_\mu^a, c^a, \bar{c}^a$	A_y^a
$1, 2, 3', 8'$	$(+,+)$	$(-,-)$
$4, 5, 6, 7$	$(-,-)$	$(+,+)$

. (5.71)

It is seen in (5.70) that fields with parity $(+,+)$ and $(-,-)$ form a pair to intertwine through the AB phase θ_H. A field with parity $(+,+)$ has a zero mode for $\theta_H = 0$, acquiring a mass at $\theta_H \neq 0$. Let ϕ_1 and ϕ_2 be fields with parity $(+,+)$ and $(-,-)$, respectively, and we consider

$$
\mathcal{L}_{\rm pair} = \frac{1}{2}\Big(\partial_y\phi_1 - \frac{\gamma}{R}\phi_2\Big)^2 + \frac{1}{2}\Big(\partial_y\phi_2 + \frac{\gamma}{R}\phi_1\Big)^2 ~. \tag{5.72}
$$

Inserting the KK expansion

$$
\begin{aligned}
\phi_1(x,y) &= \frac{1}{\sqrt{\pi R}}\phi_1^{(0)}(x) + \sqrt{\frac{2}{\pi R}}\sum_{n=1}^{\infty}\phi_1^{(n)}(x)\cos\frac{ny}{R} ~, \\
\phi_2(x,y) &= \qquad\qquad\qquad\quad \sqrt{\frac{2}{\pi R}}\sum_{n=1}^{\infty}\phi_2^{(n)}(x)\sin\frac{ny}{R} ~,
\end{aligned}
\tag{5.73}
$$

into (5.72), one obtains

$$\int_0^{\pi R} dy\, \mathcal{L}_{\rm pair} = \frac{\gamma^2}{2R^2}(\phi_1^{(0)})^2 + \frac{1}{2R^2}\sum_{n=1}^{\infty}\left(\phi_1^{(n)}, \phi_2^{(n)}\right)\begin{pmatrix} n^2+\gamma^2 & 2n\gamma \\ 2n\gamma & n^2+\gamma^2 \end{pmatrix}\begin{pmatrix} \phi_1^{(n)} \\ \phi_2^{(n)} \end{pmatrix} . \tag{5.74}$$

Making use of eigenvectors of the matrices, one rewrites the KK expansion as

$$\begin{pmatrix} \phi_1(x,y) \\ \phi_2(x,y) \end{pmatrix} = \frac{1}{\sqrt{\pi R}}\sum_{n=-\infty}^{\infty} \phi_n(x)\begin{pmatrix} \cos\frac{ny}{R} \\ \sin\frac{ny}{R} \end{pmatrix} . \tag{5.75}$$

Then (5.74) becomes

$$\int_0^{\pi R} dy\, \mathcal{L}_{\rm pair} = \frac{1}{2}\sum_{n=-\infty}^{\infty}\frac{(n+\gamma)^2}{R^2}\,\phi_n(x)^2. \tag{5.76}$$

The mass spectrum $\{m_n^2\}$ is given by

$$\left\{ m_n^2(\gamma) = \frac{(n+\gamma)^2}{R^2} \ , \ n \in \mathcal{Z}\ (n = 0, \pm 1, \pm 2, \cdots) \right\} . \tag{5.77}$$

The spectrum of the gauge fields is, in the general ξ gauge, given by

$$\begin{aligned}
&\begin{matrix}(A_\mu^1, A_\mu^5)\\(A_\mu^2, A_\mu^4)\end{matrix} : 3\left\{ m_n^2\left(\frac{\theta_H}{2\pi}\right)\right\},\ \left\{\xi m_n^2\left(\frac{\theta_H}{2\pi}\right)\right\} \quad (n \in \mathcal{Z}) \ , \\
&(A_\mu^{3'}, A_\mu^7) : 3\left\{ m_n^2\left(\frac{\theta_H}{\pi}\right)\right\},\ \left\{\xi m_n^2\left(\frac{\theta_H}{\pi}\right)\right\} \quad (n \in \mathcal{Z}) \ , \\
&A_\mu^{8'} \ : 3\left\{ m_n^2(0)\right\},\ \left\{\xi m_n^2(0)\right\} \quad (n \geq 0) \ , \\
&A_\mu^{6} \ : 3\left\{ m_n^2(0)\right\},\ \left\{\xi m_n^2(0)\right\} \quad (n \geq 1) \ .
\end{aligned} \tag{5.78}$$

In particular, the masses of the W and Z bosons are, for $|\theta_H| \leq \frac{1}{2}\pi$,

$$m_W = \frac{|\theta_H|}{2\pi R} \ , \quad m_Z = \frac{|\theta_H|}{\pi R} \ . \tag{5.79}$$

Similarly the spectrum for A_y and ghosts is given by

$$
\begin{aligned}
&\begin{matrix}(A_y^5, A_y^1)\ (c^1, c^5) \\ (A_y^4, A_y^2)\ (c^2, c^4)\end{matrix} : \left\{ \xi m_n^2\left(\frac{\theta_H}{2\pi}\right) \right\} \quad (n \in \mathcal{Z}) \,, \\
&(A_y^7, A_y^{3'})\ (c^{3'}, c^7) : \left\{ \xi m_n^2\left(\frac{\theta_H}{\pi}\right) \right\} \quad (n \in \mathcal{Z}) \,, \\
&A_y^{8'} \qquad c^{8'} \quad : \left\{ \xi m_n^2(0) \right\} \quad (n \ge 0) \,, \\
&A_\mu^6 \qquad c^6 \quad : \left\{ \xi m_n^2(0) \right\} \quad (n \ge 1) \,.
\end{aligned} \tag{5.80}
$$

Note that the mass spectrum in (5.78) and (5.80) is periodic in θ_H with a period 2π.

The mass spectrum of the quark multiplet Ψ_q in (5.67) is also found easily. It follows from (5.66) that the electric charge of (u, d, d') is $(\frac{2}{3}e, -\frac{1}{3}e, -\frac{1}{3}e)$. The fractional charge naturally emerges. The u field does not couple with θ_H, and its KK expansion is

$$
\begin{pmatrix} u_L \\ u_R \end{pmatrix} = \frac{1}{\sqrt{\pi R}} \begin{pmatrix} u_L^{(0)}(x) \\ 0 \end{pmatrix} + \sqrt{\frac{2}{\pi R}} \sum_{n=1}^{\infty} \begin{pmatrix} u_L^{(n)}(x) \cos \frac{ny}{R} \\ u_R^{(n)}(x) \sin \frac{ny}{R} \end{pmatrix} . \tag{5.81}
$$

The mass spectrum is

$$
u \ : \ \left\{ m_n^2(0) \right\} \quad (n \ge 0) \,. \tag{5.82}
$$

The (d, d') fields intertwine through θ_H. The KK expansion is

$$
\begin{aligned}
\begin{pmatrix} d_L \\ d_R \end{pmatrix} &= \frac{1}{\sqrt{\pi R}} \begin{pmatrix} d_L^{(0)}(x) \\ 0 \end{pmatrix} + \sqrt{\frac{2}{\pi R}} \sum_{n=1}^{\infty} \begin{pmatrix} d_L^{(n)}(x) \cos \frac{ny}{R} \\ d_R^{(n)}(x) \sin \frac{ny}{R} \end{pmatrix} , \\
\begin{pmatrix} d'_L \\ d'_R \end{pmatrix} &= \frac{1}{\sqrt{\pi R}} \begin{pmatrix} 0 \\ d_R'^{(0)}(x) \end{pmatrix} + \sqrt{\frac{2}{\pi R}} \sum_{n=1}^{\infty} \begin{pmatrix} d_L'^{(n)}(x) \sin \frac{ny}{R} \\ d_R'^{(n)}(x) \cos \frac{ny}{R} \end{pmatrix} ,
\end{aligned} \tag{5.83}
$$

and the equations of motion are

$$
\begin{aligned}
&\sigma^\mu \partial_\mu \begin{pmatrix} d_L \\ d'_L \end{pmatrix} + \left\{ \partial_y - i \frac{\theta_H}{2\pi R} \begin{pmatrix} 0 & 1 \\ 1 & 0 \end{pmatrix} \right\} \begin{pmatrix} d_R \\ d'_R \end{pmatrix} = 0 \,, \\
&\bar{\sigma}^\mu \partial_\mu \begin{pmatrix} d_R \\ d'_R \end{pmatrix} - \left\{ \partial_y - i \frac{\theta_H}{2\pi R} \begin{pmatrix} 0 & 1 \\ 1 & 0 \end{pmatrix} \right\} \begin{pmatrix} d_L \\ d'_L \end{pmatrix} = 0 \,.
\end{aligned} \tag{5.84}
$$

Inserting (5.83) into (5.84), one finds that

$$\begin{aligned}
&\sigma^\mu \partial_\mu d_L^{(0)} = i\frac{\theta_H}{2\pi R} d_R'^{(0)} ~,~~ \bar\sigma^\mu \partial_\mu d_R'^{(0)} = -i\frac{\theta_H}{2\pi R} d_L^{(0)} ~, \\
&\sigma^\mu \partial_\mu \begin{pmatrix} d_L^{(n)} \\ d_L'^{(n)} \end{pmatrix} = \frac{1}{R}\begin{pmatrix} -n & i\theta_H/2\pi \\ i\theta_H/2\pi & n \end{pmatrix} \begin{pmatrix} d_R^{(n)} \\ d_R'^{(n)} \end{pmatrix}, \\
&\bar\sigma^\mu \partial_\mu \begin{pmatrix} d_R^{(n)} \\ d_R'^{(n)} \end{pmatrix} = \frac{1}{R}\begin{pmatrix} -n & -i\theta_H/2\pi \\ -i\theta_H/2\pi & n \end{pmatrix} \begin{pmatrix} d_L^{(n)} \\ d_L'^{(n)} \end{pmatrix}.
\end{aligned} \tag{5.85}$$

For the $n \geq 1$ modes the resultant second-order differential equations are

$$\partial^2 \begin{pmatrix} d_L^{(n)} \\ d_L'^{(n)} \end{pmatrix} = \frac{1}{R^2}\begin{pmatrix} n^2 + (\theta_H/2\pi)^2 & 2in\theta_H/2\pi \\ -2in\theta_H/2\pi & n^2 + (\theta_H/2\pi)^2 \end{pmatrix} \begin{pmatrix} d_L^{(n)} \\ d_L'^{(n)} \end{pmatrix}$$

so that the mass spectrum is given by

$$(d, d') ~:~ \Big\{ m_n^2\Big(\frac{\theta_H}{2\pi}\Big) \Big\} ~~ (n \in \mathcal{Z}) ~. \tag{5.86}$$

The effective potential $V_{\rm eff}(\theta_H)$ at the one loop level is evaluated along the arguments in Sections 4.3 and 4.4. With (5.78), (5.80) and (5.86), the θ_H-dependent part of $V_{\rm eff}(\theta_H)$ is found to be

$$V_{\rm eff}(\theta_H) = \frac{3}{64\pi^6 R^4}\Big\{ -6h(\theta_H) - 3h(2\theta_H) + 4N_{\rm f} h(\theta_H) \Big\} \tag{5.87}$$

where $N_{\rm f}$ is the number of the fermion generation. The contributions which depend on the gauge parameter ξ are cancelled among those from A_μ, A_y, c and $\bar c$.

The location of the minimum of $V_{\rm eff}(\theta_H)$ depends on $N_{\rm f}$. $-h(2\theta_H)$ has a period π and is minimized at $\theta_H = 0$ and π. For $N_{\rm f} = 0$ and 1, $V_{\rm eff}(\theta_H)$ is minimized at $\theta_H = 0$ and $SU(2) \times U(1)$ symmetry remains unbroken. On the other hand for $N_{\rm f} \geq 2$, $V_{\rm eff}(\theta_H)$ is minimized at $\theta_H = \pi$ as $h(\theta_H)$ is minimized at $\theta_H = \pi$. In this case $SU(2) \times U(1)$ symmetry breaks down to $U(1) \times U(1)$ symmetry. The Z boson remains massless, whereas the W boson and down quark acquire a mass $1/2R$.

In any case the $SU(3)$ GHU model cannot be realistic.

5.11 Twisted gauge in flat space

There is an alternative powerful method to find mass spectra, wave functions, and couplings in GHU at $\theta_H \neq 0$. As shown in Section 5.7, physics is the same within an equivalence class of boundary conditions. Therefore one may perform a large gauge transformation such that the AB phase in the new gauge vanishes. This gauge is called the **twisted gauge**. [47, 48] Equations of motion in five-dimensional spacetime become simpler in the twisted gauge. On the other hand, the boundary conditions are changed. The method of the twisted gauge turns out indispensable for the analysis of GHU in the Randal-Sundrum warped space, or in the curved spacetime in general, which will be discussed in the following chapters.

Let us take the $SU(3)$ GHU on $M^4 \times (S^1/Z_2)$ in the previous section as an example. With the boundary condition (P_0, P_1) in Eq. (5.27) the AB phase θ_H is given by Eq. (5.35). With the comparison to GHU in the Randal-Sundrum warped space in mind, instead of Eq. (5.37), we take

$$\begin{aligned} &\Omega(y) = \exp\Big\{ i\theta(y) T^6 \Big\} ~, ~~ \theta(y) = \theta_H \Big(\frac{y}{\pi R} - 1 \Big) \\ &\Omega(0) = e^{-i\theta_H T^6} ~, ~~ \Omega(\pi R) = 1 ~. \end{aligned} \tag{5.88}$$

Let us denote quantities (M) in the twisted gauge by putting the tilde sign on top $(\tilde{M})$. Then the AB phase in the twisted gauge is $\tilde{\theta}_H = 0$, and

$$\begin{aligned} &\tilde{P}_0 = \Omega(-y) P_0 \Omega^{-1}(y) = \begin{pmatrix} -1 & 0 & 0 \\ 0 & -\cos\theta_H & -i\sin\theta_H \\ 0 & i\sin\theta_H & \cos\theta_H \end{pmatrix} ~, \\ &\tilde{P}_1 = \Omega(\pi R - y) P_1 \Omega^{-1}(\pi R + y) = P_1 ~. \end{aligned} \tag{5.89}$$

The boundary condition at $y = \pi R$ remains unchanged, but the boundary condition at $y = 0$ is twisted by the amount θ_H. With the decomposition $A_M = A_M^{\rm c} + A_M^{\rm q}$, we have $\tilde{A}_M^{\rm q} = \Omega(y) A_M^{\rm q} \Omega^{-1}(y)$. Since

$$\begin{aligned} &\Omega(y) \begin{pmatrix} T^1 \\ T^5 \end{pmatrix} \Omega^{-1}(y) = V\big[\tfrac{1}{2}\theta(y)\big] \begin{pmatrix} T^1 \\ T^5 \end{pmatrix} , \\ &\Omega(y) \begin{pmatrix} T^2 \\ T^4 \end{pmatrix} \Omega^{-1}(y) = V\big[-\tfrac{1}{2}\theta(y)\big] \begin{pmatrix} T^2 \\ T^4 \end{pmatrix} , \\ &\Omega(y) \begin{pmatrix} T^{3'} \\ T^7 \end{pmatrix} \Omega^{-1}(y) = V\big[-\theta(y)\big] \begin{pmatrix} T^{3'} \\ T^7 \end{pmatrix} , \\ &\Omega(y) T^{8'} \Omega^{-1}(y) = T^{8'} ~, ~~ \Omega(y) T^6 \Omega^{-1}(y) = T^6 ~, \\ &V[\alpha] = \begin{pmatrix} \cos\alpha & \sin\alpha \\ -\sin\alpha & \cos\alpha \end{pmatrix} , \end{aligned} \tag{5.90}$$

one finds that

$$\begin{pmatrix} A_M^{q\,1} \\ A_M^{q\,5} \end{pmatrix} = V\big[\tfrac{1}{2}\theta(y)\big] \begin{pmatrix} \tilde{A}_M^{q\,1} \\ \tilde{A}_M^{q\,5} \end{pmatrix}, \quad \begin{pmatrix} A_M^{q\,2} \\ A_M^{q\,4} \end{pmatrix} = V\big[-\tfrac{1}{2}\theta(y)\big] \begin{pmatrix} \tilde{A}_M^{q\,2} \\ \tilde{A}_M^{q\,4} \end{pmatrix},$$

$$\begin{pmatrix} A_M^{q\,3'} \\ A_M^{q\,7} \end{pmatrix} = V\big[-\theta(y)\big] \begin{pmatrix} \tilde{A}_M^{q\,3'} \\ \tilde{A}_M^{q\,7} \end{pmatrix}, \quad \begin{pmatrix} A_M^{q\,8'} \\ A_M^{q\,6} \end{pmatrix} = \begin{pmatrix} \tilde{A}_M^{q\,8'} \\ \tilde{A}_M^{q\,6} \end{pmatrix}. \tag{5.91}$$

In writing down KK expansions, one needs to solve free equations of motion as the background field vanishes (i.e. $\tilde{\theta}_H = 0$). At $y = \pi R$, $\tilde{A}_M^q = A_M^q$ and $\tilde{P}_1 = P_1$ so that eigenmodes satisfy

$$-\frac{d^2}{dy^2} f_\lambda(y) = \lambda^2 f_\lambda(y) , \quad f_\lambda(\pi R - y) = \pm f_\lambda(\pi R + y) . \tag{5.92}$$

For gauge fields the 4D mass m of the eigenmodes are $m^2 = \lambda^2$ or $\xi\lambda^2$. Depending on the parity, $f_\lambda(y)$ is given by

$$\begin{aligned} P_1 = + &: C_\lambda(y) = \cos\lambda(y - \pi R) , \quad C'_\lambda(\pi R) = 0 , \\ P_1 = - &: S_\lambda(y) = \sin\lambda(y - \pi R) , \quad S_\lambda(\pi R) = 0 , \\ & C'_\lambda = -\lambda S_\lambda , \quad S'_\lambda = \lambda C_\lambda . \end{aligned} \tag{5.93}$$

The eigenvalue λ is determined by the boundary condition at $y = 0$.

To find λ it is convenient to use the relations in Eq. (5.91) and the parity condition at $y = 0$ in the original gauge. For the pair $(\tilde{A}_\mu^{q\,1}, \tilde{A}_\mu^{q\,5})$, for instance, $\tilde{P}_1$ is $(+,-)$ and eigenmodes can be written as

$$\begin{pmatrix} \tilde{A}_\mu^{q\,1} \\ \tilde{A}_\mu^{q\,5} \end{pmatrix} = \begin{pmatrix} \alpha C_\lambda(y) \\ \beta S_\lambda(y) \end{pmatrix} B_{\lambda,\mu}(x) . \tag{5.94}$$

P_0 for $(A_\mu^{q\,1}, A_\mu^{q\,5})$ is $(+,-)$. It follows from $A_\mu^{q\,5}(x,0) = 0$ and $\theta(0) = -\theta_H$ that

$$\alpha \sin\tfrac{1}{2}\theta_H\, C_\lambda(0) + \beta \cos\tfrac{1}{2}\theta_H S_\lambda(0) = 0 . \tag{5.95}$$

On the other hand $A_\mu^{q\,1}$ satisfies the Neumann condition $\partial A_\mu^{q\,1}/\partial y|_{y=0} = 0$ so that

$$\begin{aligned} &\alpha \cos\tfrac{1}{2}\theta_H\, C'_\lambda(0) - \beta \sin\tfrac{1}{2}\theta_H S'_\lambda(0) \\ &+ \tfrac{1}{2}\theta'(0)\big\{\alpha \sin\tfrac{1}{2}\theta_H\, C_\lambda(0) + \beta \cos\tfrac{1}{2}\theta_H S_\lambda(0)\big\} = 0 . \end{aligned} \tag{5.96}$$

Combining (5.95) and (5.96), one finds

$$\begin{pmatrix} \sin\tfrac{1}{2}\theta_H\, C_\lambda & \cos\tfrac{1}{2}\theta_H S_\lambda \\ \cos\tfrac{1}{2}\theta_H\, \lambda S_\lambda & \sin\tfrac{1}{2}\theta_H\, \lambda C_\lambda \end{pmatrix} \begin{pmatrix} \alpha \\ \beta \end{pmatrix} = 0 \tag{5.97}$$

where $C_\lambda = C_\lambda(0)$ and $S_\lambda = S_\lambda(0)$. For nontrivial solutions to exist, we have

$$(A_\mu^{q\,1}, A_\mu^{q\,5}) : \ -S_\lambda^2 + \sin^2 \tfrac{1}{2}\theta_H = -\sin^2 \pi R\lambda + \sin^2 \tfrac{1}{2}\theta_H = 0 \ , \qquad (5.98)$$

which yields the same mass spectrum as in (5.78). For $\lambda R = n + (\theta_H/2\pi)$, $C_\lambda(0) = (-1)^n \cos \frac{1}{2}\theta_H$ and $S_\lambda(0) = -(-1)^n \sin \frac{1}{2}\theta_H$ so that $\alpha = \beta$ must hold. Consequently the eigenfunction in the original gauge takes the same form as in Eq. (5.75) apart from an overall factor $(-1)^n$. Similarly, for $(A_\mu^{q\,3'}, A_\mu^{q\,7})$ one finds

$$(A_\mu^{q\,3'}, A_\mu^{q\,7}) : \ -S_\lambda^2 + \sin^2 \theta_H = 0 \ . \qquad (5.99)$$

For the quark multiplet Ψ_q, $\tilde{\Psi}_q = \Omega(y)\Psi_q$ and

$$u = \tilde{u} \ , \quad \begin{pmatrix} d \\ d' \end{pmatrix} = \Big\{ \cos \tfrac{1}{2}\theta(y) - i \sin \tfrac{1}{2}\theta(y)\tau_1 \Big\} \begin{pmatrix} \tilde{d} \\ \tilde{d}' \end{pmatrix} . \qquad (5.100)$$

In particular, $\tilde{P}_1 = +1$ for $(\tilde{d}_L, \tilde{d}'_R)$ and $\tilde{P}_1 = -1$ for $(\tilde{d}_R, \tilde{d}'_L)$. Eigenmodes can be written as

$$\begin{pmatrix} \tilde{d}_R \\ \tilde{d}'_R \end{pmatrix} = \begin{pmatrix} \alpha_R S_\lambda(y) \\ \beta_R C_\lambda(y) \end{pmatrix} f_{\lambda,R}(x) \ ,$$

$$\begin{pmatrix} \tilde{d}_L \\ \tilde{d}'_L \end{pmatrix} = \begin{pmatrix} -\alpha_L C_\lambda(y) \\ \beta_L S_\lambda(y) \end{pmatrix} f_{\lambda,L}(x) \ , \qquad (5.101)$$

where

$$\bar{\sigma}^\mu \partial_\mu f_{\lambda,R}(x) = \lambda f_{\lambda,L}(x) \ ,$$

$$\sigma^\mu \partial_\mu f_{\lambda,L}(x) = \lambda f_{\lambda,R}(x) \ . \qquad (5.102)$$

It follows from the equations of motion that $(\alpha_R, \beta_R) = (\alpha_L, \beta_L)$. At $y = 0$ $P_0 = +1$ for (d_L, d'_R) and $P_0 = -1$ for (d_R, d'_L). From the boundary conditions for (d_R, d'_R) one finds that

$$\begin{pmatrix} \cos \frac{1}{2}\theta_H S_\lambda & i \sin \frac{1}{2}\theta_H C_\lambda \\ i \sin \frac{1}{2}\theta_H \, \lambda C_\lambda & -\cos \frac{1}{2}\theta_H \, \lambda S_\lambda \end{pmatrix} \begin{pmatrix} \alpha_R \\ \beta_R \end{pmatrix} = 0 \ . \qquad (5.103)$$

From the boundary conditions for (d_L, d'_L) one obtains a condition equivalent to (5.103). It follows that

$$(d, d') : \ -S_\lambda^2 + \sin^2 \tfrac{1}{2}\theta_H = 0 \ , \qquad (5.104)$$

which reproduces the mass spectrum in Eq. (5.86).

In summary, in the twisted gauge the equations of motion become those in the absence of background fields and the effects of the AB phases are reflected in the boundary conditions. The mass spectrum of each KK tower is determined by the zeros of a certain function of λ as in (5.98), (5.99) and (5.104). Once the mass spectrum is determined, the corresponding wave functions are also fixed and various physical quantities can be evaluated.

Chapter 6

Randall-Sundrum warped space

Realistic models of gauge-Higgs unification are constructed in curved spacetime. In this chapter we review the Einstein equations in the tetrad formalism, and derive the Randall-Sundrum (RS) warped-space solution. We examine scalar fields, gauge fields and fermion fields in the RS space, and see that their mode functions are expressed in terms of Bessel functions.

6.1 Einstein equations

The Randall-Sundrum (RS) warped space becomes crucially important in constructing realistic models of gauge-Higgs unification. The RS space is a curved space. It is a solution to the Einstein equations which describe the gravity. In this section we explain the Einstein equations and one convenient way to solve them.

The world interval ds in a curved space is given by

$$-ds^2 = g_{\mu\nu}(x)dx^\mu dx^\nu \ , \tag{6.1}$$

and is invariant under general coordinate transformations $x^\mu \to x'^\mu = x'^\mu(x)$. Here $g_{\mu\nu}$ is the metric of spacetime, its inverse matrix $g^{\nu\rho}$ being defined by $g_{\mu\nu}g^{\nu\rho} = \delta_\mu{}^\rho$. At each spacetime point an orthonormal frame, a tetrad $e^a{}_\mu(x)$, is introduced by

$$\begin{aligned}
&g_{\mu\nu} = \eta_{ab}e^a{}_\mu e^b{}_\nu = e^a{}_\mu e_{a\nu} \ , \\
&e_a{}^\mu e_{b\mu} = g^{\mu\nu}e_{a\mu}e_{b\nu} = \eta_{ab} \ , \quad e_a{}^\mu e^b{}_\mu = \delta_a{}^b \ , \\
&\eta_{ab} = \text{diag}\,(-1,1,1,1) = \eta^{ab} \ .
\end{aligned} \tag{6.2}$$

η_{ab} is the metric of flat Minkowski spacetime. Subscripts/superscripts a, b are raised and lowered by η_{ab}, whereas μ, ν are raised and lowered by $g_{\mu\nu}$.

$e^a{}_\mu$ transforms under local Lorentz transformation $\Lambda^a{}_b(x)$ and general coordinate transformation as

$$e^a{}_\mu(x) \to e'^a{}_\mu(x') = \Lambda^a{}_b \frac{\partial x^\nu}{\partial x'^\mu} e^b{}_\nu(x) ,$$

$$\eta_{ab}\Lambda^a{}_c\Lambda^b{}_d = \eta_{cd} , \quad (\Lambda^{-1})^a{}_b = \Lambda_b{}^a . \tag{6.3}$$

The tetrad 1-form e_a is defined by

$$e_a = e_{a\mu}dx^\mu . \tag{6.4}$$

The spin connection 1-form ω_{ab} is introduced by

$$\omega_{ab} = -\omega_{ba} = \omega_{ab\mu}dx^\mu , \tag{6.5}$$

$$de_a + \omega_a{}^b \wedge e_b = 0 . \tag{6.6}$$

The antisymmetric property (6.5) represents general relativity without torsion. The curvature 2-form $\mathcal{R}_{ab}$ is given, in terms of ω_{ab}, by

$$\mathcal{R}_{ab} = d\omega_{ab} + \omega_{ac} \wedge \omega^c{}_b = \frac{1}{2} R_{abcd}\, e^c \wedge e^d ,$$

$$R_{abcd} = e_a{}^\mu e_b{}^\nu e_c{}^\rho e_d{}^\sigma R_{\mu\nu\rho\sigma} ,$$

$$R_{\mu\nu\rho\sigma} = e^a{}_\mu e^b{}_\nu e^c{}_\rho e^d{}_\sigma R_{abcd} . \tag{6.7}$$

$\mathcal{R}_{ab}$ is antisymmetric, $\mathcal{R}_{ab} = -\mathcal{R}_{ba}$, and $R_{\mu\nu\rho\sigma}$ is the Riemann curvature tensor. The Ricci tensor $(\mathcal{R}icci)_{ab}$ and scalar curvature R are given by

$$(\mathcal{R}icci)_{ab} = R_{acb}{}^c = \eta^{cd}R_{acbd} ,$$

$$R = (\mathcal{R}icci)_a{}^a = \eta^{ab}(\mathcal{R}icci)_{ab} . \tag{6.8}$$

The gravity part of the action in general relativity is given by

$$-\frac{c^3}{16\pi G}\int d^4x\sqrt{-g}\,(R + 2\Lambda) \tag{6.9}$$

where $g = \det g_{\mu\nu}$. $G = 6.67 \times 10^{-11}\, m^3kg^{-1}s^{-2}$ is the gravitational constant and Λ is the cosmological constant. The Einstein equations in the natural units in which $c = 1$ are

$$(\mathcal{R}icci)_{ab} - \frac{1}{2}R\,\eta_{ab} = 8\pi G\, T_{ab} - \Lambda\,\eta_{ab} . \tag{6.10}$$

T_{ab} is the energy-momentum tensor of matter fields.

Given the metric $g_{\mu\nu}$, the tetrad satisfying (6.2) is determined. The next step of finding the spin connection 1-form ω_{ab} satisfying (6.6) is, in general, nontrivial. In many cases such as the RS warped space, the Robertson-Walker spacetime representing the expanding universe and the Schwarzshild solution, ω_{ab} can be found in a simple manner. This method is much easier and requires less amount of labor than first evaluating Christoffel symbols $\Gamma^\mu_{\nu\rho}$ and then evaluating $R_{\mu\nu\rho\sigma}$. Further ω_{ab} becomes absolutely necessary to describe fermion fields in curved space.

6.2 Randall-Sundrum solution

Let us consider five-dimensional spacetime with coordinates $x^M = (x^\mu, y)$ whose world interval is given by

$$-ds^2 = e^{-2\sigma(y)}\eta^{(4)}_{\mu\nu}dx^\mu dx^\nu + dy^2 \ ,$$
$$\sigma(y) = \sigma(y+2L) = \sigma(-y) \ . \tag{6.11}$$

Here $\eta^{(4)}_{\mu\nu} = \mathrm{diag}\,(-1,1,1,1)$ is the metric of four-dimensional Minkowski spacetime. $\sigma(y)$ is an even function with a period $2L$. Spacetime points $(x^\mu, y+2L)$ and $(x^\mu, -y)$ are identified with (x^μ, y).

$$(x^\mu, y) \sim (x^\mu, y+2L) \sim (x^\mu, -y) \ . \tag{6.12}$$

The topology of this spacetime is the same as the orbifold $M^4 \times (S^1/Z_2)$ introduced in Section 5.3. There are four-dimensional branes at $y=0$ and $y=L$. The fundamental region may be taken to be $0 \le y \le L$. $\sigma(y)$ is determined to satisfy the Einstein equations.

The metric g_{MN} determined by (6.11) is diagonal, and the tetrad 1-form e_A $(A = 0,1,2,3,5)$ is given by

$$g_{\mu\nu} = e^{-2\sigma}\eta^{(4)}_{\mu\nu} \ , \quad g_{yy} = 1 \ ,$$
$$e_a = e^{-\sigma}dx^a \quad (a = 0,1,2,3) \ , \quad e_5 = dy \ . \tag{6.13}$$

Since $de_a = \sigma' e_a \wedge e_5$ and $de_5 = 0$, the spin connection is found from (6.6) to be

$$\omega_{ab} = 0 \ , \quad \omega_{a5} = -\sigma' e_a \quad (a, b = 0 \sim 3). \tag{6.14}$$

Noting $d\omega_{a5} = (\sigma'' - \sigma'^{\,2})\, e_a \wedge e_5$, one finds from (6.7) that the curvature 2-form is

$$\mathcal{R}_{ab} = -(\sigma')^2\, e_a \wedge e_b \ ,$$
$$\mathcal{R}_{a5} = \big\{\sigma'' - (\sigma')^2\big\}\, e_a \wedge e_5 \ . \tag{6.15}$$

Nonvanishing components of the Riemann tensor ${R_{AB}}^{CD}$ are

$${R_{ab}}^{ab} = -{R_{ba}}^{ab} = -(\sigma')^2 \ ,$$
$${R_{a5}}^{a5} = -{R_{5a}}^{a5} = -{R_{a5}}^{5a} = {R_{5a}}^{5a} = \sigma'' - (\sigma')^2 \ . \tag{6.16}$$

The Ricci tensor and scalar curvature are found to be

$$(\mathcal{R}icci)_a{}^b = \big\{\sigma'' - 4(\sigma')^2\big\}\, \delta_a{}^b \ ,$$
$$(\mathcal{R}icci)_5{}^5 = 4\sigma'' - 4(\sigma')^2 \ ,$$
$$(\mathcal{R}icci)_a{}^5 = (\mathcal{R}icci)_5{}^a = 0 \ ,$$
$$R = 4\big\{2\sigma'' - 5(\sigma')^2\big\} \ . \tag{6.17}$$

The Einstein equations (6.10) become

$$\begin{aligned} \{ -3\sigma'' + 6(\sigma')^2 \} \, \delta_a{}^b &= 8\pi G \, T_a{}^b - \Lambda \, \delta_a{}^b \, , \\ 6(\sigma')^2 &= 8\pi G \, T_5{}^5 - \Lambda \, , \\ 0 &= 8\pi G \, T_a{}^5 \, . \end{aligned} \tag{6.18}$$

From the first equation it results that $T_a{}^b = T_0{}^0 \, \delta_a{}^b$. Assuming that $T_{AB} = 0$ in the region $0 < y < L$, one finds that $\sigma'' = 0$ and $\Lambda = -6(\sigma')^2 \leq 0$ there. If $\Lambda = 0$, σ becomes constant, leading to the flat Minkowski spacetime. For $\Lambda < 0$

$$\sigma' = \pm k \ , \quad \Lambda = -6k^2 \tag{6.19}$$

so that the general solution in the region $|y| \leq L$ is $\sigma = \alpha \pm k|y|$. α can be absorbed in the redefinition of L. It turns out later that the sign $\pm$ corresponds to interchanging the branes at $y = 0$ and $y = L$. Hence without loss of generality one can take

$$\begin{aligned} &\sigma(y) = k|y| \quad \text{for} \ -L \leq y \leq +L \, , \\ &\sigma(y + 2L) = \sigma(y) \, . \end{aligned} \tag{6.20}$$

It follows from (6.20) that

$$\sigma'(y) = k \, \epsilon_{2L}(y) \ , \quad \sigma''(y) = 2k \left\{ \delta_{2L}(y) - \delta_{2L}(y - L) \right\} \tag{6.21}$$

where $\epsilon_T(y)$ and $\delta_T(y)$ are sign and delta functions with a period T, respectively. $\epsilon_T(y) = +1 \, (-1)$ for $0 < y < \frac{1}{2}T$ $(-\frac{1}{2}T < y < 0)$. In order for the metric to be a solution to the Einstein equations, the energy-momentum tensor of the matter fields must be

$$\begin{aligned} &T_{ab} = -\frac{3k}{4\pi G} \left\{ \delta_{2L}(y) - \delta_{2L}(y - L) \right\} \eta_{ab} \ , \\ &T_{55} = T_{a5} = T_{5a} = 0 \ . \end{aligned} \tag{6.22}$$

The spacetime specified with (6.11) and (6.20) is called as the Randall-Sundrum warped space (RS space). [49] The region $0 < y < L$ is anti-de Sitter (AdS) space with $\Lambda < 0$. As is seen from (6.22), there exist physical four-dimensional branes at $y = 0$ and $y = L$ whose tensions are positive and negative, respectively. The RS space is the AdS space sandwiched by UV brane at $y = 0$ and IR brane at $y = L$. At the moment it is not clear how this kind of spacetime is dynamically generated.

The metric of the RS space is given, in the region $0 \leq y \leq L$, by

$$-ds^2 = e^{-2ky} \eta_{\mu\nu}^{(4)} dx^\mu dx^\nu + dy^2 \ . \tag{6.23}$$

The factor e^{-2ky} implies that the unit of the length becomes smaller by a factor e^{-kL} as y changes from 0 to L. $z_L = e^{kL}$ is called the warp factor. In gauge-Higgs unification we consider the case with $z_L \gg 1$. Originally Randall and Sundrum proposed the RS space with $z_L \sim 10^{17}$ as spacetime in which the Planck scale (10^{19} GeV) characterizing quantum theory of gravity is naturally connected to the weak scale (100 GeV).

In analyzing field theory in the RS space the usage of the conformal coordinate $z = e^{ky}$ in place of the coordinate y becomes useful. [50] In terms of z we have

$$-ds^2 = \frac{1}{z^2}\left\{\eta_{\mu\nu}^{(4)} dx^\mu dx^\nu + \frac{dz^2}{k^2}\right\} \qquad (1 \le z \le z_L) \ . \tag{6.24}$$

It is seen that the RS space is conformally flat.

6.3 Scalar fields in the RS space

The action and Euler equation for a real scalar field ϕ in the RS space are given by

$$\begin{aligned} I &= \int d^5x \sqrt{-g} \Big\{ -\tfrac{1}{2} g^{MN} \partial_M \phi \partial_N \phi - \tfrac{1}{2} \mu_0^2 \phi^2 \Big\} \ , \\ & \frac{1}{\sqrt{-g}} \partial_M \Big\{ \sqrt{-g} g^{MN} \partial_N \phi \Big\} - \mu_0^2 \phi = 0 \ . \end{aligned} \tag{6.25}$$

In the conformal coordinate system $x^M = (x^\mu, z)$, $\sqrt{-g} = 1/kz^5$ and the equation becomes

$$\left\{ \Box_4 + k^2 z^3 \frac{\partial}{\partial z} \frac{1}{z^3} \frac{\partial}{\partial z} - \frac{\mu_0^2}{z^2} \right\} \phi(x, z) = 0 \ . \tag{6.26}$$

With boundary conditions for ϕ at $z = 1, z_L$ given, $\phi(x, z)$ is expanded as

$$\phi(x, z) = \sum_n \phi_n(x)\, f_n(z) \tag{6.27}$$

where $\{f_n(z)\}$ is the complete set of the solutions to

$$\begin{aligned} & \left\{ -k^2 z^3 \frac{d}{dz} \frac{1}{z^3} \frac{d}{dz} + \frac{\mu_0^2}{z^2} \right\} f_n(z) = m_n^2 f_n(z) \ , \\ & \int_1^{z_L} \frac{dz}{kz^3}\, f_n(z)^* f_\ell(z) = \delta_{n\ell} \ . \end{aligned} \tag{6.28}$$

$\phi_n(x)$ in (6.27) transforms as a scalar field with mass m_n in four dimensions.

The boundary conditions at $z = 1, z_L$ are determined by the parity P at $y = 0, L$ in the y-coordinate. With the action (6.25), $\phi(x, z)$ satisfies the

Neumann (N) condition $\partial_z \phi(x,z) = 0$ for even parity ($P = +$), and the Dirichlet (D) condition $\phi(x,z) = 0$ for odd parity ($P = -$).

The equation (6.28) can be written in the form

$$\left\{ \frac{d^2}{dz^2} + \frac{1}{z}\frac{d}{dz} + \lambda_n^2 - \frac{\nu^2}{z^2} \right\} \frac{f_n(z)}{z^2} = 0 \ ,$$

$$\lambda_n = \frac{m_n}{k} \ , \quad \nu = \sqrt{4 + \frac{\mu_0^2}{k^2}} \ , \tag{6.29}$$

For $\lambda_0 = 0$, $f_0(z) = c_1 z^{2+\nu} + c_2 z^{2-\nu}$. The zero mode solution exists only if $\mu_0 = 0$, $\nu = 2$ and the boundary condition at $z = (1, z_L)$ is (N, N). The zero mode solution is given by $f_0(z) = c_2$ (constant).

For $\lambda_n > 0$, we make a change of variable from z to $w = \lambda_n z$. Then for $u_n(w) = z^{-2} f_n(z)$ Eq. (6.29) reduces to the Bessel equation

$$\left\{ \frac{d^2}{dw^2} + \frac{1}{w}\frac{d}{dw} + 1 - \frac{\nu^2}{w^2} \right\} u_n(w) = 0 \ . \tag{6.30}$$

Two independent solutions are given by $J_\nu(w)$ and $Y_\nu(w)$. Hence the solution to (6.29) is expressed as $z^{-2} f_n(z) = \alpha_n J_\nu(\lambda_n z) + \beta_n Y_\nu(\lambda_n z)$. The eigenvalues $\{\lambda_n\}$ are determined by the boundary conditions.

We note that in the RS space, or more generally in the AdS space, a theory of scalar fields with negative (mass)^2 can be defined. Indeed, if $\mu_0^2 \geq -4k^2$, ν in (6.29) is a real number, and physical solutions exist. To formulate supersymmetry in the RS space one needs a scalar field with $\mu_0^2 < 0$. In the following we restrict ourselves to the simplest case $\mu_0 = 0$, and therefore $\nu = 2$.

(i) Boundary condition (N, N)

The zero mode $\lambda_0 = 0$, $f_0(z) =$ constant exists. The wave function of the $\lambda_n > 0$ mode is $f_n(z) = z^2 \{\alpha_n J_2(\lambda_n z) + \beta_n Y_2(\lambda_n z)\}$. By using a relation for the Bessel function $Z_\nu (= J_\nu, Y_\nu)$

$$\frac{d}{dz} z^{\pm\nu} Z_\nu(\lambda z) = \pm \lambda z^{\pm\nu} Z_{\nu\mp1}(\lambda z) \ , \tag{6.31}$$

the boundary condition at $z = 1, z_L$ becomes

$$\begin{aligned} &\alpha_n J_1(\lambda_n) + \beta_n Y_1(\lambda_n) = 0 \ , \\ &\alpha_n J_1(\lambda_n z_L) + \beta_n Y_1(\lambda_n z_L) = 0 \ . \end{aligned} \tag{6.32}$$

We define

$$F_{\alpha,\beta}(u,v) = J_\alpha(u) Y_\beta(v) - Y_\alpha(u) J_\beta(v) \ . \tag{6.33}$$

(Various useful relations concerning the Bessel functions are summarized in Appendix C.) The eigenvalues λ_n and mode functions $f_n(z)$ are determined by

$$\begin{aligned} &F_{1,1}(\lambda_n, \lambda_n z_L) = 0 \ , \\ &f_n(z) = N_n z^2 \, F_{2,1}(\lambda_n z, \lambda_n z_L) = N_n' z^2 \, F_{2,1}(\lambda_n z, \lambda_n) \ . \end{aligned} \tag{6.34}$$

Here N_n and N_n' are normalization constants.

(ii) Boundary condition (D, D)

There is no zero mode. The boundary condition is

$$\begin{aligned} &\alpha_n J_2(\lambda_n) + \beta_n Y_2(\lambda_n) = 0 \ , \\ &\alpha_n J_2(\lambda_n z_L) + \beta_n Y_2(\lambda_n z_L) = 0 \ . \end{aligned} \tag{6.35}$$

λ_n and $f_n(z)$ are determined by

$$\begin{aligned} &F_{2,2}(\lambda_n, \lambda_n z_L) = 0 \ , \\ &f_n(z) = N_n z^2 \, F_{2,2}(\lambda_n z, \lambda_n z_L) = N_n' z^2 \, F_{2,2}(\lambda_n z, \lambda_n) \ . \end{aligned} \tag{6.36}$$

(iii) Boundary condition (N, D)

There is no zero mode. The boundary condition is

$$\begin{aligned} &\alpha_n J_1(\lambda_n) + \beta_n Y_1(\lambda_n) = 0 \ , \\ &\alpha_n J_2(\lambda_n z_L) + \beta_n Y_2(\lambda_n z_L) = 0 \ . \end{aligned} \tag{6.37}$$

λ_n and $f_n(z)$ are determined by

$$\begin{aligned} &F_{1,2}(\lambda_n, \lambda_n z_L) = 0 \ , \\ &f_n(z) = N_n z^2 \, F_{2,2}(\lambda_n z, \lambda_n z_L) = N_n' z^2 \, F_{2,2}(\lambda_n z, \lambda_n) \ . \end{aligned} \tag{6.38}$$

(iv) Boundary condition (D, N)

There is no zero mode. The boundary condition is

$$\begin{aligned} &\alpha_n J_2(\lambda_n) + \beta_n Y_2(\lambda_n) = 0 \ , \\ &\alpha_n J_1(\lambda_n z_L) + \beta_n Y_1(\lambda_n z_L) = 0 \ . \end{aligned} \tag{6.39}$$

λ_n and $f_n(z)$ are determined by

$$\begin{aligned} &F_{2,1}(\lambda_n, \lambda_n z_L) = 0 \ , \\ &f_n(z) = N_n z^2 \, F_{2,1}(\lambda_n z, \lambda_n z_L) = N_n' z^2 \, F_{2,1}(\lambda_n z, \lambda_n) \end{aligned} \tag{6.40}$$

In Table 6.1 the equation determining the spectrum and the wave functions for each boundary condition are summarized. The mass eigenvalues $\{m_n = k\lambda_n\}$ are determined by $F_{\alpha,\beta}(\lambda_n, \lambda_n z_L) = 0$ in all cases. By using the asymptotic form of the Bessel function (C.5), one recognizes that

Table 6.1 The spectrum and wave functions of a scalar field ($\mu_0 = 0$) in the RS space. A zero mode ($\lambda_0 = 0$) appears only when the boundary condition (BC) is (N, N).

	BC ϕ	Spectrum $\{m_n = k\lambda_n\}$	Wave functions $(n \geq 1)$ $f_n(z)$
(i)	(N, N)	$F_{1,1}(\lambda_n, \lambda_n z_L) = 0$	$z^2 F_{2,1}(\lambda_n z, \lambda_n z_L)$
(ii)	(D, D)	$F_{2,2}(\lambda_n, \lambda_n z_L) = 0$	$z^2 F_{2,2}(\lambda_n z, \lambda_n z_L)$
(iii)	(N, D)	$F_{1,2}(\lambda_n, \lambda_n z_L) = 0$	$z^2 F_{2,2}(\lambda_n z, \lambda_n z_L)$
(iv)	(D, N)	$F_{2,1}(\lambda_n, \lambda_n z_L) = 0$	$z^2 F_{2,1}(\lambda_n z, \lambda_n z_L)$

$m_n \sim m_{\rm KK}\, n$ ($n \gg 1$) for sufficiently large eigenvalues, where the KK mass scale $m_{\rm KK}$ is given by

$$m_{\rm KK} = \frac{\pi k}{z_L - 1} = \begin{cases} \pi k\, e^{-kL} & \text{for } z_L = e^{kL} \gg 1 \\ \pi/L & \text{for } kL \ll 1 \end{cases} . \tag{6.41}$$

In the $k \to 0$ limit the RS space becomes flat $M^4 \times (S^1/Z_2)$ $(L = \pi R)$, and $m_{\rm KK} \to 1/R$.

The mass spectra of scalar fields with the boundary conditions in Table 6.1 are plotted in Fig. 6.1 in the unit of $m_{\rm KK}$. In the warp factor $z_L \to 1$ limit the spectra become those in the $M^4 \times (S^1/Z_2)$ spacetime. The spectrum of the KK excited modes $(n \geq 1)$ is mostly determined by the boundary condition at the IR brane for $z_L \geq 10^3$.

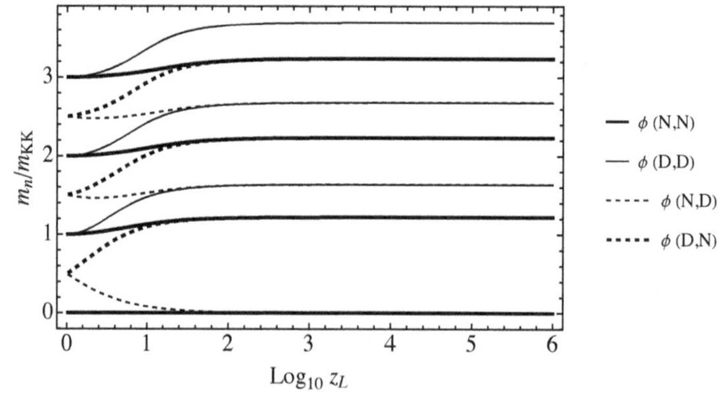

Fig. 6.1 The mass spectra of scalar fields ϕ ($\mu_0 = 0$) with four types of boundary conditions in Table 6.1 are plotted as functions of z_L. $m_n/m_{\rm KK} = \lambda_n(z_L - 1)/\pi$.

6.4 Gauge fields in the RS space

To quantize gauge fields in the RS space, as seen in Section 3.4, one needs to introduce a gauge-fixing function and accompanying ghost fields. There is some arbitrariness in the choice of a gauge-fixing function. In analyzing the Hosotani mechanism in curved space, it is possible to introduce a gauge fixing function invariant under general coordinate transformations, and develop general theory of spontaneous gauge symmetry breaking. [14] Anticipating the need to consider the case in the presence of background gauge fields $A^{\rm c}_M$, we write $A_M = A^{\rm c}_M + A^{\rm q}_M$ and choose a gauge fixing function $f_{\rm gf}(A) = D_M(A^{\rm c})A^M$. Then we have

$$\begin{aligned}
I = \int d^5x \sqrt{-g} \bigg\{ - \frac{1}{2} \mathrm{Tr}\, F_{MN} F^{MN} - \mathrm{Tr} \left[D_M(A^{\rm c}) A^M \right]^2 \\
- \mathrm{Tr}\, \bar{c} D_M(A^{\rm c}) D^M(A)\, c \bigg\} ,
\end{aligned}$$

$$\begin{aligned}
D_M(A) B^M &= \frac{1}{\sqrt{-g}} \partial_M \big(\sqrt{-g}\, B^M \big) + ig [A_M, B^M] , \\
D^M(A)\, c &= g^{MN} \{ \partial_N c + ig [A_N, c] \} . \qquad (6.42)
\end{aligned}$$

Retaining bilinear terms in $A^{\rm q}_M, c, \bar{c}$, we find

$$\begin{aligned}
I &\sim \int d^5x \sqrt{-g} \Big\{ - \mathrm{Tr}\, A^{{\rm q}M} \mathcal{M}^{\rm g}_{MN} A^{{\rm q}N} - \mathrm{Tr}\, \bar{c} \mathcal{M}^{\rm gh} c \Big\} , \\
\mathcal{M}^{\rm g}_{MN} &= - g_{MN} D_P(A^{\rm c}) D^P(A^{\rm c}) + R_{MN} - 4ig F^{\rm c}_{MN} , \\
\mathcal{M}^{\rm gh} &= D_P(A^{\rm c}) D^P(A^{\rm c}) , \qquad (6.43)
\end{aligned}$$

where $F^{\rm c}_{MN} = \partial_M A^{\rm c}_N - \partial_N A^{\rm c}_M - ig[A^{\rm c}_M, A^{\rm c}_N]$. One loop effective potential for $A^{\rm c}_M$ is

$$\begin{aligned}
V_{\rm eff}[A^{\rm c}]^{\rm g} &= - \frac{i}{2} \,\mathrm{tr}\, \ln \mathcal{M}^{\rm g} , \\
V_{\rm eff}[A^{\rm c}]^{\rm gh} &= + i \,\mathrm{tr}\, \ln \mathcal{M}^{\rm gh} . \qquad (6.44)
\end{aligned}$$

Here tr includes, in addition to the trace Tr in the gauge group, the sum of degrees of freedom in the phase space $\sum_n \int d^4p/(2\pi)^4$. In practice, however, the evaluation of $D_P D^P$ in the $A^{\rm q}$ space becomes highly complicated.

There is a better way of gauge-fixing in the RS space. In place of (6.42), we adopt, in the conformal coordinate system (x^μ, z),

$$I = \int d^5x \sqrt{-g}\,(-)\mathrm{Tr}\left\{\frac{1}{2}F_{MN}F^{MN} + \frac{1}{\xi}(f_{\rm gf})^2 + \mathcal{L}_{\rm gh}\right\},$$

$$f_{\rm gf} = z^2\Big\{\eta^{\mu\nu}\mathcal{D}_\mu(A^{\rm c})A^{\rm q}_\nu + \xi k^2 z\mathcal{D}_z(A^{\rm c})\Big(\frac{1}{z}A^{\rm q}_z\Big)\Big\},$$

$$\mathcal{L}_{\rm gh} = \bar{c}\Big\{\eta^{\mu\nu}\mathcal{D}_\mu(A^{\rm c})\mathcal{D}_\nu(A) + \xi k^2 z\,\mathcal{D}_z(A^{\rm c})\frac{1}{z}\,\mathcal{D}_z(A)\Big\}c,$$

$$\mathcal{D}_M(A^{\rm c})A^{\rm q}_N = \partial_M A^{\rm q}_N - ig[A^{\rm c}_M, A^{\rm q}_N]\,. \tag{6.45}$$

The gauge-fixing function $f_{\rm gf}$ has been chosen such that when bilinear terms in $A^{\rm q}_M$ are retained, intertwining terms in $A^{\rm q}_\mu$ and $A^{\rm q}_z$ cancell among themselves. $\xi \geq 0$ is an arbitrary gauge parameter. (6.45) generalizes the R_ξ-gauge in four-dimensional gauge theory. $[\mathcal{D}^{\rm c}_M, \mathcal{D}^{\rm c}_N] = 0$ when $F^{\rm c}_{MN} = 0$. In the case where $A^{\rm c}_\mu = 0$ and $A^{\rm c}_z \neq 0$, the bilinear terms in $A^{\rm q}_M, c, \bar{c}$ in the action I become

$$I = \int d^4x\,\frac{dz}{kz}\,\mathrm{tr}\left\{\frac{1}{2}A^{\rm q}_\mu\Big(\eta^{\mu\nu}(\Box + k^2\mathcal{P}_4) - (1-\frac{1}{\xi})\partial^\mu\partial^\nu\Big)A^{\rm q}_\nu\right.$$
$$\left. + \frac{1}{2}k^2 A^{\rm q}_z(\Box + \xi k^2\mathcal{P}_z)A^{\rm q}_z + \bar{c}_A(\Box + \xi k^2\mathcal{P}_4)c_A\right\},$$

$$\Box \equiv \eta^{\mu\nu}\partial_\mu\partial_\nu\,,\quad \partial^\mu = \eta^{\mu\nu}\partial_\nu\,,$$

$$\mathcal{P}_4 \equiv z\,\mathcal{D}^{\rm c}_z\,\frac{1}{z}\,\mathcal{D}^{\rm c}_z\,,\quad \mathcal{P}_z \equiv \mathcal{D}^{\rm c}_z\,z\,\mathcal{D}^{\rm c}_z\,\frac{1}{z}\,. \tag{6.46}$$

When $A^{\rm c}_z = 0$, $\mathcal{D}^{\rm c}_z = \partial_z$. With the boundary conditions at $z = 1$ and z_L given, one needs to find, for $A_\mu, c, \bar{c}$, the complete set of

$$-z\frac{d}{dz}\frac{1}{z}\frac{d}{dz}f_n(z) = \lambda_n^2 f_n(z),\quad \int_1^{z_L}\frac{dz}{z}f_n(z)f_\ell(z) = \delta_{n\ell}\,, \tag{6.47}$$

and for A_z the complete set of

$$-\frac{d}{dz}z\frac{d}{dz}\frac{1}{z}h_n(z) = \tilde{\lambda}_n^2 h_n(z),\quad \int_1^{z_L}\frac{dz}{z}h_n(z)h_\ell(z) = \delta_{n\ell}\,. \tag{6.48}$$

General solutions are given in terms of basis functions summarized in Appendix C. Solutions $f_n(z)$ to Eq. (6.47) are given by linear combinations of $C(z;\lambda)$ and $S(z;\lambda)$ in (C.14) and (C.15). Solutions $h_n(z)$ to Eq. (6.48) are given by linear combinations of $C'(z;\lambda)$ and $S'(z;\lambda)$. Here C, S are expressed in terms of $F_{\alpha,\beta}(u,v)$ in (6.33).

For each component in the gauge group, the KK expansion of $A_\mu, A_z, c, \bar{c}$ are given by

$$\begin{aligned}
A_\mu(x,z) &= \sqrt{k} \sum_n A_\mu^{(n)}(x)\, f_n(z), \\
A_z(x,z) &= \frac{1}{\sqrt{k}} \sum_n A_z^{(n)}(x)\, h_n(z), \\
c(x,z) &= \sqrt{k} \sum_n c^{(n)}(x)\, f_n(z), \\
\bar{c}(x,z) &= \sqrt{k} \sum_n \bar{c}^{(n)}(x)\, f_n(z).
\end{aligned} \tag{6.49}$$

With $m_n = k\lambda_n, \tilde{m}_n = k\tilde{\lambda}_n$, the Fourier transform of the two point function of each KK mode is given by

$$\begin{aligned}
\langle A_\mu^{(n)} A_\nu^{(n)} \rangle &: \frac{\eta_{\mu\nu}}{p^2 + m_n^2} + \frac{p_\mu p_\nu}{m_n^2} \left(\frac{1}{p^2 + m_n^2} - \frac{1}{p^2 + \xi m_n^2} \right), \\
\langle A_z^{(n)} A_z^{(n)} \rangle &: \frac{1}{p^2 + \xi \tilde{m}_n^2}, \\
\langle c^{(n)}\, \bar{c}^{(n)} \rangle &: \frac{1}{p^2 + \xi m_n^2}.
\end{aligned} \tag{6.50}$$

ξ-dependent terms correspond to unphysical degrees of freedom. In particular, only the zero mode ($\tilde{m}_0 = 0$) is physical for A_z.

To pin down the boundary conditions for gauge fields, one needs to go back to the action principle in the (x^μ, y) coordinates. The boundary conditions must be set such that in the course of deriving Euler equations in $\delta I = \delta \int d^4x \int_0^L dy\,(\cdots)$, the surface terms appearing in partial integration

$$\begin{aligned}
&- \int d^4x \Big[e^{-2ky} \eta^{\mu\nu} \mathrm{Tr}\, F_{\mu y} \delta A_\nu \Big]_{y=0}^{y=L} \\
&- \int d^4x \Big[e^{-2ky} \mathrm{Tr}\, \delta A_y^{\mathrm{q}} \Big\{ \eta^{\mu\nu} \partial_\mu A_\nu + \frac{\partial}{\partial y} (e^{-2ky} A_y^{\mathrm{q}}) \\
&\qquad\qquad - ig e^{-ky} [A_y^{\mathrm{c}}, A_y^{\mathrm{q}}] \Big\} \Big]_{y=0}^{y=L}
\end{aligned} \tag{6.51}$$

vanish. Note that the δA_y^{q} term emerges from the gauge-fixing term. In the z coordinate they become

$$\mathrm{Tr}\,(\partial_\mu A_z - \partial_z A_\mu - ig_A[A_\mu, A_z])\,\delta A_\nu = 0\ ,$$
$$\mathrm{Tr}\left[\eta^{\mu\nu}\partial_\mu A_\nu + k^2\left\{z\frac{\partial}{\partial z}\Big(\frac{1}{z}A_z^q\Big) - ig_A[A_z^c, A_z^q]\right\}\right]\delta A_z^q = 0\ ,$$
$$\text{at } z = 1, z_L\ . \tag{6.52}$$

The RS space has the same orbifold structure as S^1/Z_2, and fields satisfy the parity conditions (P_0, P_1) at $(y_0, y_1) = (0, L)$ given in (5.25). The components of odd parity ($P = -$) satisfy the Dirichlet (D) condition $A_\mu, A_y = 0$. The components of even parity ($P = +$) satisfy the Neumann (N) condition. The form of (6.52) indicates

$$\text{Neumann } (N) \text{ condition} \begin{cases} A_\mu : & \dfrac{\partial}{\partial z}A_\mu = 0\ , \\[2ex] A_z : & \dfrac{\partial}{\partial z}\dfrac{1}{z}A_z = 0\ . \end{cases} \tag{6.53}$$

To confirm the above assertion (6.53), one needs to take a close look at the $[A_M, A_N]$ term in F_{MN}. In each gauge group component A_μ and A_y have opposite parity in orbifold gauge theory. Let $\check{T}$ ($\hat{T}$) represent the generators of the gauge group for which A_μ has parity $+$ ($-$). It is confirmed that under orbifold boundary conditions

$$[\check{T}\,,\check{T}\,] \sim \check{T}\ ,\ [\hat{T}\,,\hat{T}\,] \sim \check{T}\ ,\ [\check{T}\,,\hat{T}\,] \sim \hat{T}\ . \tag{6.54}$$

When $\delta\check{A}_\mu \neq 0$, $\partial_z\check{A}_\mu = 0$ and $\check{A}_z = 0$. Further, since $[\check{T}\,,\check{T}\,] \sim \check{T}$, $[\hat{T}\,,\hat{T}\,] \sim \check{T}$, the first condition in (6.52) is satisfied. When $\delta\hat{A}_z^q \neq 0$, $\hat{A}_\mu = 0$ and $\partial_z(z^{-1}\hat{A}_z) = 0$. Further, since $[\check{T}\,,\hat{T}\,] \sim \hat{T}$, the second condition in (6.52) is satisfied.

With boundary conditions of A_μ and A_z at $z = 1$ and z_L specified, one can determine the spectra λ_n and $\tilde{\lambda}_n$ in (6.47) and (6.48) and the corresponding mode functions $f_n(z)$ and $h_n(z)$. When $A_z^c = 0$, the boundary conditions are classified in four categories. In either case it turns out that $\lambda_n = \tilde{\lambda}_n$ except for the zero mode.

(i) Boundary condition $A_\mu : (N, N)$, $A_z : (D, D)$

There is a zero mode ($\lambda_0 = 0$) for A_μ. The general solution f_0 is $f_0(z) = c_1 + c_2 z^2$. As $f_0' = 0$ at $z = 1, z_L$,

$$\lambda_0 = 0\ ,\ \ f_0(z) = \frac{1}{\sqrt{kL}}\ . \tag{6.55}$$

For $\lambda_n \neq 0$

$$-z\,\frac{d}{dz}\,\frac{1}{z}\,\frac{d}{dz}\,f_n(z) = \lambda_n^2\,f_n(z)$$

so that

$$h_n = \frac{1}{\lambda_n}\,\frac{df_n}{dz} \tag{6.56}$$

satisfies

$$-\frac{d}{dz}\,z\,\frac{d}{dz}\,\frac{1}{z}\,h_n = \lambda_n^2\,h_n \ , \quad \int_1^{z_L} \frac{dz}{z}\,h_n h_\ell = \delta_{n\ell} \ .$$

$h_n(1) = h_n(z_L) = 0$, and the boundary condition (D, D) is satisfied. One sees that $\tilde{\lambda}_n = \lambda_n$. $f_n(z)$ satisfying the Neumann condition at z_L is $C(z;\lambda_n)$. To satisfy the Neumann condition at $z = 1$, $C'(1;\lambda_n) = 0$ must hold. To summarize

$$\begin{aligned}
&\lambda_n \neq 0: \quad C'(1;\lambda_n) = 0 \ , \\
&f_n(z) = \frac{1}{\sqrt{r_n}}\,C(z;\lambda_n) \ , \quad r_n = \int_1^{z_L} \frac{dz}{z}\,C(z;\lambda_n)^2 \ , \\
&h_n(z) = \frac{1}{\lambda_n\sqrt{r_n}}\,C'(z;\lambda_n) \ .
\end{aligned} \tag{6.57}$$

(ii) Boundary condition $A_\mu : (D, D)$, $A_z : (N, N)$

There is a zero mode $(\tilde{\lambda}_0 = 0)$ for A_z. The general solution h_0 is $h_0(z) = c_1 z + c_2 z \ln z$. As $(h_0/z)' = 0$ at $z = 1$ and z_L, we have $c_2 = 0$ and

$$\tilde{\lambda}_0 = 0 \ , \quad h_0(z) = \sqrt{\frac{2}{z_L^2 - 1}}\,z \ . \tag{6.58}$$

For $\tilde{\lambda}_n \neq 0$

$$-\frac{d}{dz}\,z\,\frac{d}{dz}\,\frac{1}{z}\,h_n(z) = \tilde{\lambda}_n^2\,h_n(z)$$

so that

$$f_n = -\frac{1}{\tilde{\lambda}_n}\,z\,\frac{d}{dz}\,\frac{1}{z}\,h_n(z) \tag{6.59}$$

satisfies

$$-z\,\frac{d}{dz}\,\frac{1}{z}\,\frac{d}{dz}\,f_n = \tilde{\lambda}_n^2\,f_n \ , \quad \int_1^{z_L} \frac{dz}{z}\,f_n f_\ell = \delta_{n\ell} \ .$$

The condition $f_n(1) = f_n(z_L) = 0$ is satisfied. One sees that $\lambda_n = \tilde{\lambda}_n$. The relation (6.56) also holds. From the boundary condition at z_L, $f_n \propto$

$S(z;\lambda_n)$ and $h_n \propto S'(z;\lambda_n)$. The boundary condition at $z=1$ implies $S(1;\lambda_n)=0$. To summarize

$$\lambda_n \neq 0: \quad S(1;\lambda_n)=0 \ ,$$
$$f_n(z)=\frac{1}{\sqrt{r_n}}S(z;\lambda_n) \ , \quad r_n=\int_1^{z_L}\frac{dz}{z}\,S(z;\lambda_n)^2 \ ,$$
$$h_n(z)=\frac{1}{\lambda_n\sqrt{r_n}}S'(z;\lambda_n) \ . \tag{6.60}$$

(iii) Boundary condition $A_\mu:(N,D)$, $A_z:(D,N)$

There is no zero mode for A_μ and A_z. The relations (6.56) and (6.59) are valid in this case as well since the boundary conditions N and D are reversed for A_μ and A_z at both ends. It follows that $\lambda_n=\tilde{\lambda}_n$. One finds

$$\lambda_n \neq 0: \quad S'(1;\lambda_n)=0 \ ,$$
$$f_n(z)=\frac{1}{\sqrt{r_n}}S(z;\lambda_n) \ , \quad r_n=\int_1^{z_L}\frac{dz}{z}\,S(z;\lambda_n)^2 \ ,$$
$$h_n(z)=\frac{1}{\lambda_n\sqrt{r_n}}S'(z;\lambda_n) \ . \tag{6.61}$$

(iv) Boundary condition $A_\mu:(D,N)$, $A_z:(N,D)$

There is no zero mode for A_μ and A_z. As in the case (iii) the relations (6.56) and (6.59) are valid and $\lambda_n=\tilde{\lambda}_n$. One finds

$$\lambda_n \neq 0: \quad C(1;\lambda_n)=0 \ ,$$
$$f_n(z)=\frac{1}{\sqrt{r_n}}C(z;\lambda_n) \ , \quad r_n=\int_1^{z_L}\frac{dz}{z}\,C(z;\lambda_n)^2 \ ,$$
$$h_n(z)=\frac{1}{\lambda_n\sqrt{r_n}}C'(z;\lambda_n) \ . \tag{6.62}$$

In Table 6.2 the spectrum determining equations and wave functions $f_n(z)$ and $h_n(z)$ are tabulated. The wave function $f_0(z)$ of the zero mode of A_μ in the case (i) is constant, whereas the wave function $h_0(z)$ of the zero mode of A_z in the case (ii) is $h_0(z)\propto z$ and localized near the IR brane at $z=z_L$.

The mass spectra of gauge fields with the boundary conditions in Table 6.2 are plotted in Fig. 6.2 in the unit of $m_{\rm KK}$. In the warp factor $z_L \to 1$ limit the spectra become those in the $M^4\times(S^1/Z_2)$ spacetime. The spectrum of the KK excited modes ($n\geq 1$) is mostly determined by the boundary condition at the IR brane for $z_L \geq 10^5$.

Table 6.2 The spectrum and wave functions of gauge fields A_μ and A_z in the RS space. Zero modes ($\lambda_0 = 0$) appear only when the boundary condition (BC) is (N, N). The functions C and S are defined in Appendix C.

	BC		Spectrum	Wave functions ($n \geq 1$)	
	A_μ	A_z	$\{m_n = k\lambda_n\}$	$f_n(z)$	$h_n(z)$
(i)	(N, N)	(D, D)	$C'(1; \lambda_n) = 0$	$C(z; \lambda_n)$	$C'(z; \lambda_n)$
(ii)	(D, D)	(N, N)	$S(1; \lambda_n) = 0$	$S(z; \lambda_n)$	$S'(z; \lambda_n)$
(iii)	(N, D)	(D, N)	$S'(1; \lambda_n) = 0$	$S(z; \lambda_n)$	$S'(z; \lambda_n)$
(iv)	(D, N)	(N, D)	$C(1; \lambda_n) = 0$	$C(z; \lambda_n)$	$C'(z; \lambda_n)$

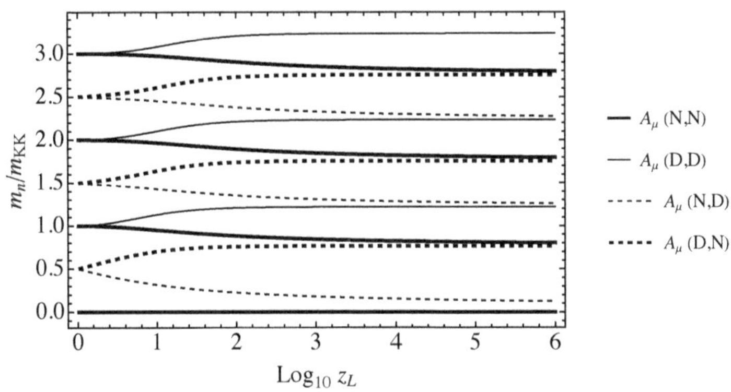

Fig. 6.2 The mass spectra of gauge fields A_μ with four types of boundary conditions in Table 6.2 are plotted as functions of z_L. $m_n/m_{\text{KK}} = \lambda_n(z_L - 1)/\pi$.

6.5 Fermion fields in curved spacetime

In flat spacetime, under Lorentz transformations boson fields transform as fields with integer spins $(0, 1, 2, \cdots)$, whereas fermion fields transform as fields with half-odd integer spins $(\frac{1}{2}, \frac{3}{2}, \frac{5}{2} \cdots)$. In curved spacetime a fermion field $\psi(x)$ with spin $\frac{1}{2}$ is defined as a spinor (a quantity with spin $\frac{1}{2}$ in four dimensions) under local Lorentz transformations $\Lambda^a{}_b(x)$ at each spacetime point. Dirac matrices $\{\gamma^a\}$ are matrices acting on spinor components of $\psi(x)$. Spacetime derivatives $\partial_\mu = \partial/\partial x^\mu$ are transformed to derivatives with spinorial subscripts by $e_a{}^\mu \partial_\mu \psi(x)$ where $e^a{}_\mu(x)$ is a tetrad introduced in Section 6.1. The action must be invariant under local Lorentz transformations as well. Spacetime derivatives appear as covariant derivatives containing spin connections.

Under local Lorentz transformations, $e^a(x) = e^a{}_\mu(x)dx^\mu$ and $\psi(x)$ transform as

$$\begin{aligned} e^a(x) &\to e'^a(x) = \Lambda^a{}_b(x)\, e^b(x) \;, \\ \psi(x) &\to \psi'(x) = S[\Lambda(x)]\, \psi(x) \;. \end{aligned} \tag{6.63}$$

Here $S(\Lambda)$ satisfies

$$\begin{aligned} S(\Lambda)^{-1}\, \gamma^a\, S(\Lambda) &= \Lambda^a{}_b \gamma^b \;, \\ S(\Lambda)\, \gamma^a\, S(\Lambda)^{-1} &= \gamma^b\, \Lambda_b{}^a \;. \end{aligned} \tag{6.64}$$

The spin connection 1-form $\omega^a{}_b$ is defined by (6.6). To maintain the relation $de'^a + \omega'^a{}_b \wedge e'^b = 0$ after the transformations,

$$\begin{aligned} \omega'^a{}_b &= \Lambda^a{}_c\, \omega^c{}_d\, (\Lambda^{-1})^d{}_b + \Lambda^a{}_c\, d(\Lambda^{-1})^c{}_b \;, \\ \omega'_{ab} &= \Lambda_a{}^c\, \Lambda_b{}^d\, \omega_{cd} + \Lambda_{ac}\, d\Lambda_b{}^c \end{aligned} \tag{6.65}$$

must hold. The first equation may be written, in short, as $\omega' = \Lambda\, \omega\, \Lambda^{-1} + \Lambda\, d\Lambda^{-1}$. The curvature 2-form $\mathcal{R} = d\omega + \omega \wedge \omega$ in (6.7) transforms as $\mathcal{R}' = \Lambda\, \mathcal{R}\, \Lambda^{-1}$.

Next we define

$$\Gamma = \frac{1}{8}\, \omega_{ab}\, [\gamma^a, \gamma^b] = \Gamma_\mu\, dx^\mu \;. \tag{6.66}$$

It follows from (6.64) and (6.65) that under local Lorentz transformations

$$\Gamma' = S\, \Gamma\, S^{-1} - \frac{n}{4}\, dS\, S^{-1} + \frac{1}{4} S\, \gamma_a S^{-1}\, dS\, \gamma^a\, S^{-1}$$

where n is the spacetime dimension and the relation $\gamma_a \gamma^a = n\, I$ has been used. For infinitesimal local Lorentz transformations $\Lambda^a{}_b = \delta^a{}_b + \epsilon^a{}_b$ ($\epsilon_{ab} = -\epsilon_{ba}$), $S(\Lambda) = I + \frac{1}{8}\epsilon_{ab}[\gamma^a, \gamma^b]$ so that one finds $\gamma_a S^{-1}\, dS\, \gamma^a = (n-4)\, S^{-1}\, dS$. Hence irrespective of the spacetime dimension we have

$$\begin{aligned} \Gamma' &= S\, \Gamma\, S^{-1} + S\, dS^{-1} \;, \\ \Gamma'_\mu &= S\, \Gamma_\mu\, S^{-1} + S\, \partial_\mu S^{-1} \;. \end{aligned} \tag{6.67}$$

The covariant derivative of the spinor field $\psi(x)$ in curved spacetime is given by

$$\begin{aligned} D\, \psi &= (d + \Gamma)\, \psi = D_\mu \psi\, dx^\mu \;, \\ D_\mu \psi &= \left\{ \partial_\mu + \frac{1}{8}\, \omega_{ab\mu}\, [\gamma^a, \gamma^b] \right\} \psi \;,. \end{aligned} \tag{6.68}$$

Under local Lorentz transformations

$$D_\mu \psi \to (D_\mu \psi)' = S\, D_\mu \psi \;. \tag{6.69}$$

The Dirac equation in curved spacetime is

$$\mathcal{D}\psi = \Big\{ \gamma^a e_a{}^\mu D_\mu - m \Big\} \psi = 0 \ . \tag{6.70}$$

The corresponding action in n dimensions is given by

$$I = \int d^n x \sqrt{-g}\, \bar{\psi} \mathcal{D} \psi \ , \tag{6.71}$$

which is invariant under general coordinate transformations, local Lorentz transformations and gauge transformations.

6.6 Fermion fields in the RS space

Let us consider a Dirac field in the RS space. The topology of the RS space is the same as in the flat $M^4 \times (S^1/Z_2)$ spacetime, and zero modes of the Dirac field become chiral as seen in Section 5.4. Wave functions of KK modes are expressed in terms of Bessel functions. We denote Dirac matrices in five dimensions as Γ^a ($a = 0, 1, 2, 3, 5$, $\{\Gamma^a, \Gamma^b\} = 2\eta^{ab}$). The orbifold boundary condition satisfied by the Dirac field $\Psi(x, y)$ is given by

$$\Psi(x, y_j - y) = P_j \Gamma^5 \, \Psi(x, y_j + y) \ , \tag{6.72}$$

the same as (5.19). The action in the y coordinate is given by

$$I = \int d^4 x \int_0^L dy \sqrt{-g}\, \overline{\Psi} \Big\{ \Gamma^a e_a{}^M D_M - c\,\sigma'(y) \Big\} \Psi \ . \tag{6.73}$$

To avoid the complexity arising from the singular behavior of the integrand at the both edges in the y integral, one might write the y integral as $\frac{1}{2}\int_{-L+\epsilon}^{L+\epsilon} dy$. As in the case of $M^4 \times (S^1/Z_2)$, the boundary condition (6.72) implies that $\overline{\Psi}\Psi(x, y_j - y) = -\overline{\Psi}\Psi(x, y_j + y)$ so that a mass term $-m\overline{\Psi}\Psi$ is forbidden. However, the metric (6.11) and (6.20) in the RS space contains the function $\sigma(y)$ therefore it is natural for $\sigma'(y) = k\,\epsilon_{2L}(y)$ to appear in the action. Since $\sigma'(y_j - y) = -\sigma'(y_j + y)$, the last term in the action (6.73) has even parity and is allowed. This term is called the bulk mass term. The coefficient c is dimensionless, and is called the bulk mass parameter. The parameter c characterizes the wave function of the zero mode of Ψ. In the gauge-Higgs unification scenario discussed below, the value of c is related to the mass of a quark or lepton. The notable fact is that the value of c is $c = O(1)$ for all leptons and quarks. Typical values are $c_t = 0.3 \sim 0.4$ for the top quark and $c_e = 0.8 \sim 2.0$ for the electron. The huge mass hierarchy in quarks and leptons is explained by the bulk mass parameters of $O(1)$.

By making use of e_a and ω_{ab} in (6.13) and (6.14), it is easy to see that

$$\Gamma^a e_a{}^M D_M = e^\sigma (\gamma\partial)_{d=4} + \Gamma^5(\partial_y - 2\sigma') \ . \tag{6.74}$$

Here $(\gamma\partial)_{d=4}$ is $\gamma^\mu\partial_\mu$ in the four-dimensional Minkowski space. It is convenient to introduce $\check{\Psi} = e^{-2\sigma}\Psi$. As $(\partial_y - 2\sigma')\Psi = e^{2\sigma}\partial_y\check{\Psi}$, the action becomes

$$\begin{aligned} I &= \int d^4x \int_0^L dy\, \overline{\check{\Psi}} \Big\{ e^\sigma (\gamma\partial)_{d=4} + \Gamma^5\partial_y - c\sigma' \Big\} \check{\Psi} \\ &= \int d^4x \int_1^{z_L} \frac{dz}{k}\, \overline{\check{\Psi}} \Big\{ (\gamma\partial)_{d=4} + k\Big(\Gamma^5\partial_z - \frac{c}{z}\Big) \Big\}\, \check{\Psi} \ . \end{aligned} \tag{6.75}$$

With γ^a and $\Gamma^5 = \gamma^5$ in (A.7) and (A.9), the action takes the form

$$\begin{aligned} I &= \int d^4x \int_1^{z_L} \frac{dz}{k}\, \overline{\check{\Psi}} \begin{pmatrix} -kD_-(c) & \sigma^\mu\partial_\mu \\ \bar{\sigma}^\mu\partial_\mu & -kD_+(c) \end{pmatrix} \check{\Psi} \ , \\ \overline{\check{\Psi}} &= (-i\check{\Psi}_L^\dagger, i\check{\Psi}_R^\dagger), \quad \check{\Psi} = \begin{pmatrix} \check{\Psi}_R \\ \check{\Psi}_L \end{pmatrix}, \\ D_\pm(c) &= \pm\frac{d}{dz} + \frac{c}{z} \ . \end{aligned} \tag{6.76}$$

The equations of motion become

$$\begin{aligned} \bar{\sigma}^\mu\partial_\mu\check{\Psi}_R &= kD_+(c)\check{\Psi}_L \ , \\ \sigma^\mu\partial_\mu\check{\Psi}_L &= kD_-(c)\check{\Psi}_R \ , \end{aligned} \tag{6.77}$$

from which it follows that

$$\begin{aligned} \Box_4\check{\Psi}_R &= k^2\, D_+(c)D_-(c)\check{\Psi}_R \ , \\ \Box_4\check{\Psi}_L &= k^2\, D_-(c)D_+(c)\check{\Psi}_L \ . \end{aligned} \tag{6.78}$$

$\Box_4 = \eta^{\mu\nu}_{d=4}\partial_\mu\partial_\nu$ and the eigenvalue of the right-hand side in Eq. (6.78) corresponds to (mass)^2.

The condition (6.72) implies that Ψ_R and Ψ_L have opposite parity at $y = 0$ and L. If parity is odd $(-)$, $\check{\Psi}_{R/L} = 0$ there (Dirichlet condition, D). In the case when parity is even $(+)$, one needs to go back to the equations of motion to find the Neumann condition. If $\Psi_R(x, y_j - y) = +\Psi_R(x, y_j + y)$, $\Psi_L(x, y_j - y) = -\Psi_L(x, y_j + y)$ and $\check{\Psi}_L(x, y_j) = 0$, then the second equation in Eq. (6.77) implies that $D_-(c)\check{\Psi}_R = 0$.

To summarize, we find

$$\text{Neumann } (N) \text{ condition} \quad \begin{cases} \Psi_R : & D_-(c)\check{\Psi}_R = 0 \ , \\ \Psi_L : & D_+(c)\check{\Psi}_L = 0 \ . \end{cases} \tag{6.79}$$

In deriving eigenmodes of Eq. (6.77), we note that right-handed and left-handed two-component spinors in the four-dimensional part satisfy

$$\bar{\sigma}^{\mu}\partial_{\mu}u_R(x) = k\lambda\, u_L(x) \ , \quad \sigma^{\mu}\partial_{\mu}u_L(x) = k\lambda\, u_R(x) \ . \tag{6.80}$$

The mass of this mode is given by $m = k\lambda$. Let us write the eigenmode of the five-dimensional field as

$$\begin{aligned} \check{\Psi}_R(x,z) &= f_R(z)u_R(x) \ , \\ \check{\Psi}_L(x,z) &= f_L(z)u_L(x) \ . \end{aligned} \tag{6.81}$$

Then Eq. (6.77) becomes

$$\begin{aligned} D_+(c)\, f_L(z) &= \lambda f_R(z) \ , \\ D_-(c)\, f_R(z) &= \lambda f_L(z) \ , \end{aligned} \tag{6.82}$$

and the second-order differential equation (6.78) becomes

$$\begin{aligned} D_+(c)D_-(c)\, f_R(z) &= \lambda^2 f_R(z) \ , \\ D_-(c)D_+(c)\, f_L(z) &= \lambda^2 f_L(z) \ . \end{aligned} \tag{6.83}$$

For $\lambda \neq 0$, by making a change of variables $w = \lambda z$, Eq. (6.83) is reduced to the Bessel equation. The mode functions and the spectrum are expressed in terms of the basis functions $C_{R/L}(z;\lambda,c)$ and $S_{R/L}(z;\lambda,c)$ defined in Appendix C. For $\lambda \neq 0$ it is also confirmed, with the use of the equations of motion and the boundary conditions, that the relation $\int_1^{z_L} dz\, f_R(z)^2 = \int_1^{z_L} dz\, f_L(z)^2$ holds. The two independent solutions are $C_R(z;\lambda,c)$ and $S_R(z;\lambda,c)$ for $f_R(z)$, and $C_L(z;\lambda,c)$ and $S_L(z;\lambda,c)$ for $f_L(z)$. We also note that

$$\begin{aligned} D_{\pm}(-c) &= -D_{\mp}(c) \ , \\ C_{R/L}(z;\lambda,-c) &= +C_{L/R}(z;\lambda,c) \ , \\ S_{R/L}(z;\lambda,-c) &= -S_{L/R}(z;\lambda,c) \ . \end{aligned} \tag{6.84}$$

(i) Boundary condition $\Psi_R : (N,N)$, $\Psi_L : (D,D)$

A zero mode ($\lambda_0 = 0$) exists for Ψ_R. $D_-(c)f_R(z) = 0$ in the region $1 \leq z \leq z_L$ so that

$$\begin{aligned} &\lambda_0 = 0 \ , \\ &f_{0R}(z) = \begin{cases} \sqrt{\dfrac{2c+1}{z_L^{2c+1}-1}}\, z^c & \text{for } c \neq -\frac{1}{2}, \\ \dfrac{1}{\sqrt{\ln z_L}}\, z^{-1/2} & \text{for } c = -\frac{1}{2}, \end{cases} \\ &f_{0L}(z) = 0 \ . \end{aligned} \tag{6.85}$$

It follows from the boundary conditions at $z = z_R$ that solutions with $\lambda \neq 0$ take the form of $f_R \propto C_R$ and $f_L \propto S_L$. In order for Ψ_L to satisfy the boundary condition at $z = 1$, $S_L(1; \lambda, c) = 0$ must hold. Hence

$$\begin{aligned}
&\lambda_n \neq 0: \quad S_L(1; \lambda_n, c) = 0 , \\
&f_{nR}(z) = \frac{1}{\sqrt{r_{nR}}} C_R(z; \lambda_n, c), \ r_{nR} = \int_1^{z_L} dz\, C_R(z; \lambda_n, c)^2, \\
&f_{nL}(z) = \frac{1}{\sqrt{r_{nL}}} S_L(z; \lambda_n, c), \ r_{nL} = \int_1^{z_L} dz\, S_L(z; \lambda_n, c)^2, \\
&\int_1^{z_L} dz\, f_{nR}(z) f_{\ell R}(z) = \int_1^{z_L} dz\, f_{nL}(z) f_{\ell L}(z) = \delta_{n\ell} . \qquad (6.86)
\end{aligned}$$

The mass of each mode is $m_n = k\lambda_n$. The KK expansion of $\Psi(x, z)$ is given by

$$\begin{aligned}
&\Psi_R(x, z) = \sqrt{k}\, z^2 \sum_{n=0}^{\infty} \psi_R^{(n)}(x) f_{nR}(z) , \\
&\Psi_L(x, z) = \sqrt{k}\, z^2 \sum_{n=1}^{\infty} \psi_L^{(n)}(x) f_{nL}(z) , \\
&\psi^{(0)}(x) = \begin{pmatrix} \psi_R^{(0)}(x) \\ 0 \end{pmatrix}, \quad \psi^{(n\geq 1)}(x) = \begin{pmatrix} \psi_R^{(n)}(x) \\ \psi_L^{(n)}(x) \end{pmatrix} . \qquad (6.87)
\end{aligned}$$

$\psi^{(n)}(x)$ is a four-dimensional fermion field. The zero mode appears only in the right-handed component.

(ii) Boundary condition $\Psi_R : (D, D)$, $\Psi_L : (N, N)$

A zero mode ($\lambda_0 = 0$) exists for Ψ_L. $D_+(c) f_L(z) = 0$ in the region $1 \leq z \leq z_L$ so that

$$\begin{aligned}
&\lambda_0 = 0 , \\
&f_{0L}(z) = \begin{cases} \sqrt{\dfrac{2c-1}{1 - z_L^{1-2c}}}\, z^{-c} & \text{for } c \neq \frac{1}{2}, \\ \dfrac{1}{\sqrt{\ln z_L}}\, z^{-1/2} & \text{for } c = \frac{1}{2}, \end{cases} \\
&f_{0R}(z) = 0 . \qquad (6.88)
\end{aligned}$$

It follows from the boundary conditions at $z = z_R$ that solutions with $\lambda \neq 0$ take the form of $f_R \propto S_R$ and $f_L \propto C_L$. In order for Ψ_R to satisfy the boundary condition at $z = 1$, $S_R(1; \lambda, c) = 0$ must hold. Hence

$$\lambda_n \neq 0: \quad S_R(1; \lambda_n, c) = 0 \ ,$$
$$f_{nR}(z) = \frac{1}{\sqrt{r_{nR}}} S_R(z; \lambda_n, c), \ r_{nR} = \int_1^{z_L} dz\, S_R(z; \lambda_n, c)^2,$$
$$f_{nL}(z) = \frac{1}{\sqrt{r_{nL}}} C_L(z; \lambda_n, c), \ r_{nL} = \int_1^{z_L} dz\, C_L(z; \lambda_n, c)^2. \qquad (6.89)$$

The KK expansion of $\Psi(x,z)$ becomes

$$\Psi_R(x,z) = \sqrt{k}\, z^2 \sum_{n=1}^{\infty} \psi_R^{(n)}(x) f_{nR}(z) \ ,$$
$$\Psi_L(x,z) = \sqrt{k}\, z^2 \sum_{n=0}^{\infty} \psi_L^{(n)}(x) f_{nL}(z) \ ,$$
$$\psi^{(0)}(x) = \begin{pmatrix} 0 \\ \psi_L^{(0)}(x) \end{pmatrix}, \quad \psi^{(n\geq 1)}(x) = \begin{pmatrix} \psi_R^{(n)}(x) \\ \psi_L^{(n)}(x) \end{pmatrix}. \qquad (6.90)$$

The zero mode appears only in the left-handed component.

(iii) Boundary condition $\Psi_R : (N, D), \ \Psi_L : (D, N)$

There is no zero mode. It follows from the boundary conditions at $z = z_R$ that solutions with $\lambda \neq 0$ take the form of $f_R \propto S_R$ and $f_L \propto C_L$. In order for Ψ_L to satisfy the boundary condition at $z = 1$, $C_L(1; \lambda, c) = 0$ must hold. Hence

$$\lambda_n \neq 0: \quad C_L(1; \lambda_n, c) = 0 \ ,$$
$$f_{nR}(z) = \frac{1}{\sqrt{r_{nR}}} S_R(z; \lambda_n, c), \ r_{nR} = \int_1^{z_L} dz\, S_R(z; \lambda_n, c)^2,$$
$$f_{nL}(z) = \frac{1}{\sqrt{r_{nL}}} C_L(z; \lambda_n, c), \ r_{nL} = \int_1^{z_L} dz\, C_L(z; \lambda_n, c)^2. \qquad (6.91)$$

Note that the spectrum $\{\lambda_n\}$ differs from that in the case (ii). The KK expansion is given by

$$\Psi(x,z) = \begin{pmatrix} \Psi_R(x,z) \\ \Psi_L(x,z) \end{pmatrix} = \sqrt{k}\, z^2 \sum_{n=1}^{\infty} \begin{pmatrix} \psi_R^{(n)}(x) f_{nR}(z) \\ \psi_L^{(n)}(x) f_{nL}(z) \end{pmatrix}. \qquad (6.92)$$

(iv) Boundary condition $\Psi_R : (D, N), \ \Psi_L : (N, D)$

There is no zero mode. It follows from the boundary conditions at $z = z_R$ that solutions with $\lambda \neq 0$ take the form of $f_R \propto C_R$ and $f_L \propto S_L$.

Table 6.3 The spectrum and wave functions of Dirac fields in the RS space. Zero modes ($\lambda_0 = 0$) appear only when the boundary condition (BC) is (N, N). The functions $C_{R/L}$ and $S_{R/L}$ are defined in Appendix C.

	BC		Spectrum	Wave functions ($n \geq 1$)	
	Ψ_R	Ψ_L	$\{m_n = k\lambda_n\}$	$f_{nR}(z)$	$f_{nL}(z)$
(i)	(N,N)	(D,D)	$S_L(1;\lambda_n,c)=0$	$C_R(z;\lambda_n,c)$	$S_L(z;\lambda_n,c)$
(ii)	(D,D)	(N,N)	$S_R(1;\lambda_n,c)=0$	$S_R(z;\lambda_n,c)$	$C_L(z;\lambda_n,c)$
(iii)	(N,D)	(D,N)	$C_L(1;\lambda_n,c)=0$	$S_R(z;\lambda_n,c)$	$C_L(z;\lambda_n,c)$
(iv)	(D,N)	(N,D)	$C_R(1;\lambda_n,c)=0$	$C_R(z;\lambda_n,c)$	$S_L(z;\lambda_n,c)$

In order for Ψ_R to satisfy the boundary condition at $z = 1$, $C_R(1; \lambda, c) = 0$ must hold. Hence

$$
\begin{aligned}
&\lambda_n \neq 0 : \quad C_R(1;\lambda_n,c) = 0 \ , \\
&f_{nR}(z) = \frac{1}{\sqrt{r_{nR}}} C_R(z;\lambda_n,c), \ r_{nR} = \int_1^{z_L} dz\, C_R(z;\lambda_n,c)^2, \\
&f_{nL}(z) = \frac{1}{\sqrt{r_{nL}}} S_L(z;\lambda_n,c), \ r_{nL} = \int_1^{z_L} dz\, S_L(z;\lambda_n,c)^2.
\end{aligned} \tag{6.93}
$$

The KK expansion takes the same form as (6.92).

In Table 6.3 the spectrum determining equations and wave functions $f_{nR}(z)$ and $f_{nL}(z)$ are tabulated. The zero modes appear in cases (i) and (ii). For $c > +\frac{1}{2}$ the right-handed zero mode is localized near the IR brane ($z = z_L$), and the left-handed zero mode is localized near the UV brane ($z = 1$). For $c < -\frac{1}{2}$ the right-handed zero mode is localized near the UV brane ($z = 1$), and the left-handed zero mode is localized near the IR brane ($z = z_L$). For $-\frac{1}{2} < c < \frac{1}{2}$ the zero modes extend in the fifth dimension.

The fermion mass spectrum in the RS space differs from that in the flat $M^4 \times (S^1/Z_2)$ spacetime. In $M^4 \times (S^1/Z_2)$ $m_n/m_{\rm KK} = n$ for the boundary conditions (i) and (ii) in Table 6.3, and $m_n/m_{\rm KK} = n - \frac{1}{2}$ for the boundary conditions (iii) and (iv). In the RS space, as depicted in Fig. 6.3, the spectrum is quite different for $z_L \gg 1$ and $|c| > \frac{1}{2}$. Notice that for $c > \frac{1}{2}$ [$c < -\frac{1}{2}$] there appears the $n = 1$ mode with very small mass with the boundary condition (iv) [(iii)]. The spectrum of the KK excited modes ($n \geq 1$) is mostly determined by the boundary condition at the IR brane for $z_L \geq 10^4$.

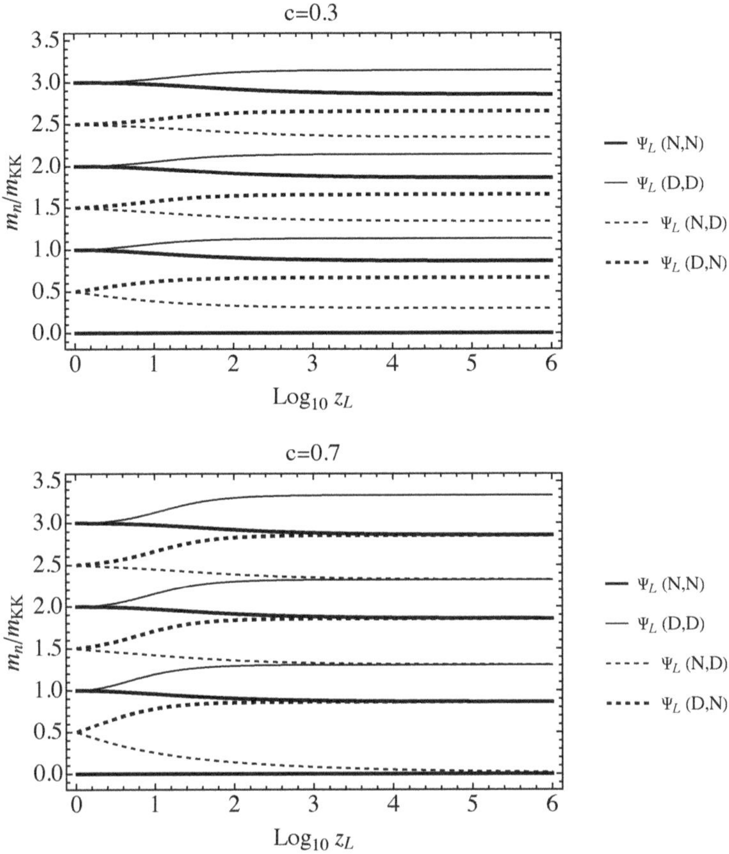

Fig. 6.3 The mass spectra of fermi fields Ψ with four types of boundary conditions in Table 6.3 are plotted as functions of z_L for $c = 0.3$ and $c = 0.7$. $m_n/m_{\rm KK} = \lambda_n(z_L - 1)/\pi$. The mass spectra for $c < 0$ are obtained by interchanging D and N in the figures. For instance, the spectrum in the case $\Psi_L(c < 0), (N, N)$ is the same as the spectrum in the case $\Psi_L(|c|), (D, D)$. Zero modes ($m_n = 0$) exist in the left-handed component of $\Psi_L, (N, N)$ and in the right-handed component of $\Psi_R, (N, N)$.

We note that the mass spectra of gauge fields, given in Table 6.3 and Fig. 6.2, are the same as the mass spectra of fermion fields with the bulk mass parameter $c = \frac{1}{2}$, as is confirmed by the relations in Eq. (C.21). The wave function of A_μ corresponds to that of Ψ_L with $c = \frac{1}{2}$.

Chapter 7

Gauge-Higgs unification in the RS space

The $SO(5) \times U(1)_X \times SU(3)_C$ gauge-Higgs unification model is introduced in the Randall-Sundrum warped space. The spectrum of quarks and leptons is reproduced and the electroweak symmetry is spontaneously broken by the Hosotani mechanism.

7.1 Goal of gauge-Higgs unification

The standard model (SM) is based on the $SU(3)_C \times SU(2)_L \times U(1)_Y (\equiv \mathcal{G}_{\rm SM})$ gauge symmetry, and its action $I = \int d^4x\, \mathcal{L}^{\rm SM}$ consists of four parts. The Lagrangian density of gauge fields, $\mathcal{L}^{\rm gauge}$, is uniquely determined once the gauge group is specified. $\mathcal{L}^{\rm fermion}$ is the kinetic term of quark and lepton fields. Once the representation in the group such as $SU(3)_C$ triplets and $SU(2)_L$ doublets are specified, derivatives with respect to spacetime coordinates x^μ in $\mathcal{L}^{\rm fermion}$ appear in the form of covariant derivatives D_μ from the requirement of the gauge invariance so that gauge interactions of quarks and leptons are uniquely fixed. The kinetic term of the Higgs field, $\mathcal{L}^{\rm Higgs}_{\rm kinetic}$, in $\mathcal{L}^{\rm Higgs}$ given by (2.36) is similarly determined by the gauge invariance. However, its potential term $\mathcal{L}^{\rm Higgs}_{\rm potential}$ is not fixed by the gauge invariance. It is customary to choose $m^2 < 0$ in (2.36) in order to spontaneously break the $SU(2)_L \times U(1)_Y$ gauge symmetry. $\mathcal{L}^{\rm Yukawa}$ in (2.37) represents the Yukawa interactions among quarks/leptons and the Higgs field. The Yukawa coupling constants appearing there are determined so as to reproduce the quark-lepton mass spectrum and the Kobayashi-Maskawa mixing angles. There remains the arbitrariness in $\mathcal{L}^{\rm Higgs}$ and $\mathcal{L}^{\rm Yukawa}$, which cannot be determined by the gauge symmetry alone.

In other words, there lacks a principle which governs the interactions of the Higgs field. Can there be a principle governing the Higgs interactions just as the gauge principle governs the gauge interactions? In the SM the Higgs field is an essential element for the unification of electromagnetic and weak interactions. The unification of interactions implies a larger and higher gauge symmetry, which is spontaneously broken by the Higgs mechanism in the SM. There must be an alternative mechanism of spontaneous gauge symmetry breaking, which enables us to have a principle governing the Higgs interactions.

Several proposals beyond the SM have been put forward to answer to this question. Among them is the gauge-Higgs unification (GHU) scenario. GHU models are defined as gauge theories in five (or higher) dimensions. The Higgs boson discovered in 2012 is identified with the zero mode of the fifth-dimensional component of the gauge fields. The correspondence between the SM and GHU models in five dimensions is depicted as

$$\begin{aligned} \int d^4x \left\{ \mathcal{L}^{\rm gauge} + \mathcal{L}^{\rm Higgs}_{\rm kinetic} \right\} &\Rightarrow \int d^5x \sqrt{-g}\, \mathcal{L}^{\rm gauge}_{\rm 5d} ~, \\ \int d^4x \left\{ \mathcal{L}^{\rm fermion} + \mathcal{L}^{\rm Yukawa} \right\} &\Rightarrow \int d^5x \sqrt{-g}\, \mathcal{L}^{\rm fermion}_{\rm 5d} ~, \\ - \int d^4x\, \mathcal{L}^{\rm Higgs}_{\rm potential} \qquad &\Rightarrow \int d^4x\, V^{\rm 1\,loop}_{\rm eff} ~. \end{aligned} \tag{7.1}$$

The Higgs field becomes a part of the gauge fields, being governed by the gauge principle. $\mathcal{L}^{\rm Higgs}_{\rm kinetic}$ is integrated into $\mathcal{L}^{\rm gauge}_{\rm 5d}$, and $\mathcal{L}^{\rm Yukawa}$ is integrated into $\mathcal{L}^{\rm fermion}_{\rm 5d}$. There is no Higgs potential at the tree level. The effective potential $V^{\rm 1\,loop}_{\rm eff}$ generated at the one loop level replaces the $\mathcal{L}^{\rm Higgs}_{\rm potential}$ term in the SM. $V^{\rm 1\,loop}_{\rm eff}$ is generated by matter fields and the electroweak gauge symmetry is dynamically broken. The Hosotani mechanism replaces the Higgs mechanism. Further $V^{\rm 1\,loop}_{\rm eff}$ is finite, and the generated Higgs boson mass becomes finite so that the gauge hierarchy problem is solved.

To be a realistic GHU model, several conditions must be satisfied.

(1) All fields and particles in the SM such as gauge fields, quarks, leptons and Higgs field, emerge as zero modes of 5D fields.
(2) Masses of all KK excited modes, other than the zero modes, must be sufficiently large.
(3) The gauge group $\mathcal{G}$ of the unified theory must contain $SU(2)_L \times U(1)_Y \times SU(3)_C$ and be larger than it. This is because the 4D Higgs field is an $SU(2)_L$ doublet, and the 4D Higgs field in GHU is a part of gauge fields in the adjoint representation of $\mathcal{G}$.
(4) Quarks and leptons must be chiral in four dimensions.

(5) The 4D Higgs field should have the custodial $SO(4)$ symmetry to the leading order as in the SM.
(6) $V_{\rm eff}^{\rm 1\,loop}$ must have the absolute minimum at a point where the $SU(2)_L \times U(1)_Y$ symmetry breaks down to $U(1)_{\rm EM}$. The electroweak symmetry must be spontaneously and dynamically broken.
(7) The mass spectrum of W boson, Z boson, quarks, leptons, and Higgs field, and coupling constants among those fields must be consistent with the observed values within experimental errors. Phenomenology at low energies must be nearly the same as that in the SM within experimental errors.

Conditions (1) to (4) seem obvious. Condition (5) is not an absolute necessity, but is desirable to have in order to satisfy the constraints coming from the precision measurements at low energies. Condition (6) is the essential part in the GHU scenario. It gives a severe constraint in the model building. In the $SU(3)$ GHU model considered in Section 5.10, for instance, the required gauge symmetry breaking does not occur. Condition (7) is also a very severe constraint, which requires a thorough examination of the model.

7.2 $SO(5) \times U(1)_X \times SU(3)_C$ GHU

$SO(5) \times U(1)_X \times SU(3)_C$ GHU in the Randall-Sundrum (RS) warped space is introduced as a promising candidate. [51–58] The electroweak interactions are contained in the $SO(5) \times U(1)_X$ part. $SU(3)_C$ represents the strong color interactions. Let $A_M(x, y)$, $B_M(x, y)$ and $C_M(x, y)$ be the gauge fields of $SO(5)$, $U(1)_X$ and $SU(3)_C$ in the coordinates (6.11) and (6.20) of the RS space, respectively.

Ten generators of the group $SO(5)$, $\{T^{jk} = -T^{kj}\,;\, j, k = 1 \sim 5\}$, satisfy the algebra

$$[T^{jk}, T^{\ell m}] = i(\delta^{j\ell} T^{km} - \delta^{jm} T^{k\ell} - \delta^{k\ell} T^{jm} + \delta^{km} T^{j\ell}) \ . \tag{7.2}$$

The vector representation $T^{jk}_{\rm vec}$ and spinor representation $T^{jk}_{\rm sp}$ of T^{jk} are given by (A.13) and (A.20) in Appendix A. They satisfy

$$\begin{aligned} {\rm Tr}\, T^{jk}_{\rm vec} T^{\ell m}_{\rm vec} &= 2(\delta^{j\ell}\delta^{km} - \delta^{jm}\delta^{k\ell}) \ , \\ {\rm Tr}\, T^{jk}_{\rm sp} T^{\ell m}_{\rm sp} &= \delta^{j\ell}\delta^{km} - \delta^{jm}\delta^{k\ell} \ . \end{aligned} \tag{7.3}$$

The gauge field A_M of $SO(5)$ is decomposed as

$$A_M = \frac{1}{\sqrt{2}} \sum_{1 \le j < k \le 5} A_M^{(jk)}\, T^{jk} \ , \tag{7.4}$$

and its field strengths are given by

$$\begin{aligned} F_{MN} &= A_{N;M} - A_{M;N} - ig_A[A_M, A_N] \\ &= \partial_M A_N - \partial_N A_N - ig_A[A_M, A_N] \\ &= \frac{1}{\sqrt{2}} \sum_{1\le j<k\le 5} F_{MN}^{(jk)} T^{jk} . \end{aligned} \tag{7.5}$$

Under gauge transformations $\Omega = \exp\left\{\frac{i}{2}\sum_{j\neq k}\omega^{jk}T^{jk}\right\}$ A_M and F_{MN} transform as

$$\begin{aligned} A'_M &= \Omega A_M \Omega^{-1} + \frac{i}{g_A}\Omega\,\partial_M \Omega^{-1}, \\ F'_{MN} &= \Omega F_{MN}\Omega^{-1}. \end{aligned} \tag{7.6}$$

The factor $1/\sqrt{2}$ has been inserted in the decomposition in (7.4) and (7.5) so that the 5D $SO(5)$ gauge coupling constant g_A and the 4D $SU(2)_L$ gauge coupling constant g_w are related by

$$g_w = \frac{g_A}{\sqrt{L}} \tag{7.7}$$

as seen below. Here L is the size of the fundamental region $(0 \le y \le L)$ of the RS space. In the vector representation

$$\frac{1}{4}\mathrm{Tr}\, F_{MN}F^{MN} = \sum_{j<k}\frac{1}{4}\mathrm{Tr}\, F_{MN}^{(jk)}F^{(jk)MN} . \tag{7.8}$$

In the spinor representation it is $\frac{1}{2}\mathrm{Tr}\, F_{MN}F^{MN}$.

The RS space with the identification (6.12) is an orbifold having the same space topology as $M^4 \times (S^1/Z_2)$ does. In the (x^μ, y) coordinates the fixed points are located at $y = y_j$ where $(y_0, y_1) = (0, L)$. The orbifold boundary conditions for the gauge fields are

$$\begin{aligned} &\begin{pmatrix} A_\mu \\ A_y \end{pmatrix}(x, y_j - y) = P_j \begin{pmatrix} A_\mu \\ -A_y \end{pmatrix}(x, y_j + y) P_j^{-1}, \\ &\qquad P_j \in SO(5), \quad P_j^2 = 1 , \\ &\begin{pmatrix} B_\mu \\ B_y \end{pmatrix}(x, y_j - y) = \begin{pmatrix} B_\mu \\ -B_y \end{pmatrix}(x, y_j + y) , \\ &\begin{pmatrix} C_\mu \\ C_y \end{pmatrix}(x, y_j - y) = \begin{pmatrix} C_\mu \\ -C_y \end{pmatrix}(x, y_j + y) . \end{aligned} \tag{7.9}$$

We take P_j given, in the vector representation, by

$$P_0^{\rm vec} = P_1^{\rm vec} = \begin{pmatrix} -1 & 0 & 0 & 0 & 0 \\ 0 & -1 & 0 & 0 & 0 \\ 0 & 0 & -1 & 0 & 0 \\ 0 & 0 & 0 & -1 & 0 \\ 0 & 0 & 0 & 0 & 1 \end{pmatrix} . \tag{7.10}$$

Note that

$$\begin{aligned} P_j T^{k\ell} P_j^{-1} &= +T^{k\ell} \quad \text{for } 1 \le k, \ell \le 4 \, , \\ P_j T^{k5} P_j^{-1} &= -T^{k5} \quad \text{for } 1 \le k \le 4 \, . \end{aligned} \tag{7.11}$$

With the help of (7.11), (A.20) and (A.25) it is seen that P_j in the spinor representation is given by

$$P_0^{\rm sp} = P_1^{\rm sp} = \begin{pmatrix} 1 & 0 & 0 & 0 \\ 0 & 1 & 0 & 0 \\ 0 & 0 & -1 & 0 \\ 0 & 0 & 0 & -1 \end{pmatrix} . \tag{7.12}$$

The boundary conditions (7.9) reduce the $SO(5)$ symmetry to $SO(4)$ symmetry. As seen in Chapter 6, zero modes in the RS space appear only in field components with parity + under the reflection at $y = 0, L$. The $SO(4)$ part of A_μ, $A_\mu^{(jk)}$ $(1 \le j, k \le 4)$, and the $SO(5)/SO(4)$ part of A_y, $A_y^{(j5)}$ $(1 \le j \le 4)$, have zero modes. As $SO(4) \simeq SU(2)_L \times SU(2)_R$, the $SO(5) \times U(1)_X$ gauge symmetry is broken to $SU(2)_L \times SU(2)_R \times U(1)_X$ gauge symmetry by the boundary conditions. To be a realistic theory, the $SU(2)_R \times U(1)_X$ symmetry must spontaneously break down to $U(1)_Y$. The zero modes of A_y form an $SO(4)$ vector, which is an $SU(2)_L$ doublet, corresponding to the Higgs field in the SM.

As seen in (A.15), the generators of $SU(2)_L$ and $SU(2)_R$ are given by

$$\begin{aligned} T_L^a &= \frac{1}{4}\epsilon^{abc} T^{bc} + \frac{1}{2} T^{a4} \, , \\ T_R^a &= \frac{1}{4}\epsilon^{abc} T^{bc} - \frac{1}{2} T^{a4} \, , \quad (a, b, c = 1 \sim 3). \end{aligned} \tag{7.13}$$

The remaining generators in $SO(5)/SO(4)$ are denoted as $\hat{T}^p = T^{p5}/\sqrt{2}$ $(p = 1 \sim 4)$. Canonically normalized gauge fields of $SU(2)_L$, $SU(2)_R$ and $SO(5)/SO(4)$ are

$$A_M^{a_L} = \frac{1}{\sqrt{2}} \Big(\frac{1}{2} \epsilon^{abc} A_M^{bc} + A_M^{a4} \Big) ,$$

$$A_M^{a_R} = \frac{1}{\sqrt{2}} \Big(\frac{1}{2} \epsilon^{abc} A_M^{bc} - A_M^{a4} \Big) ,$$

$$A_M^{\hat{p}} = A_M^{p5} . \tag{7.14}$$

Writing as $\{T^\alpha, \alpha = 1 \sim 10\} \equiv \{T_L^a, T_R^a (a = 1,2,3), \hat{T}^p (p = 1 \sim 4)\}$, we have the expression

$$A_M = \sum_{\alpha=1}^{10} A_M^\alpha T^\alpha . \tag{7.15}$$

As the field strengths F_{MN} are given by (7.5), g_A is the 5D $SU(2)_L$ gauge coupling constant, and the relation (7.7) is confirmed.

The action of the gauge fields takes the same form as in (6.45). Let us write as $A_M = A_M^{\rm c} + A_M^{\rm q}$ where $A_M^{\rm c}$ represents the background in the vacuum to be determined. In the conformal coordinates (x^μ, z)

$$\begin{aligned} I_{\rm gauge} = \int d^5x \sqrt{-g} \Big[&- {\rm Tr} \left\{ \frac{1}{4} F_{MN} F^{MN} + \frac{1}{2\xi} (f_{\rm gf}^{(A)})^2 + \mathcal{L}_{\rm gh}^{(A)} \right\} \\ &- \left\{ \frac{1}{4} F_{MN}^{(B)} F^{(B)MN} + \frac{1}{2\xi} (f_{\rm gf}^{(B)})^2 + \mathcal{L}_{\rm gh}^{(B)} \right\} \\ &- {\rm Tr} \left\{ \frac{1}{2} F_{MN}^{(C)} F^{(C)MN} + \frac{1}{\xi} (f_{\rm gf}^{(C)})^2 + \mathcal{L}_{\rm gh}^{(C)} \right\} \Big] , \end{aligned}$$

$$F_{MN}^{(B)} = \partial_M B_N - \partial_N B_M ,$$

$$F_{MN}^{(C)} = \partial_M C_N - \partial_N C_M - i g_C [C_M . C_N] ,$$

$$f_{\rm gf}^{(A)} = z^2 \Big\{ \eta^{\mu\nu} \mathcal{D}_\mu (A^{\rm c}) A_\nu^{\rm q} + \xi k^2 z \mathcal{D}_z (A^{\rm c}) \Big(\frac{1}{z} A_z^{\rm q} \Big) \Big\} ,$$

$$f_{\rm gf}^{(B)} = z^2 \Big\{ \eta^{\mu\nu} \partial_\mu B_\nu + \xi k^2 z \partial_z \Big(\frac{1}{z} B_z \Big) \Big\} ,$$

$$f_{\rm gf}^{(C)} = z^2 \Big\{ \eta^{\mu\nu} \partial_\mu C_\nu + \xi k^2 z \partial_z \Big(\frac{1}{z} C_z \Big) \Big\} ,$$

$$\mathcal{L}_{\rm gh}^{(A)} = \bar{c}_A \left\{ \eta^{\mu\nu} \mathcal{D}_\mu (A^{\rm c}) \mathcal{D}_\nu (A) + \xi k^2 z \, \mathcal{D}_z (A^{\rm c}) \frac{1}{z} \, \mathcal{D}_z (A) \right\} c_A ,$$

$$\mathcal{L}_{\rm gh}^{(B)} = \bar{c}_B \left\{ \eta^{\mu\nu} \partial_\mu \partial_\nu + \xi k^2 z \, \partial_z \frac{1}{z} \, \partial_z \right\} c_B ,$$

$$\mathcal{L}_{\rm gh}^{(C)} = \bar{c}_C \left\{ \eta^{\mu\nu} \partial_\mu \mathcal{D}_\nu(C) + \xi k^2 z \, \partial_z \frac{1}{z} \, \mathcal{D}_z(C) \right\} c_C \, ,$$

$$\begin{aligned}
&\mathcal{D}_M(A^{\rm c}) A_N^{\rm q} = \partial_M A_N^{\rm q} - i g_A [A_M^{\rm c}, A_N^{\rm q}] \, , \\
&\mathcal{D}_M(A) \, c_A = \partial_M c_A - i g_A [A_M, c_A] \, , \\
&\mathcal{D}_M(C) \, c_C = \partial_M c_C - i g_C [C_M, c_C] \, . \qquad (7.16)
\end{aligned}$$

7.3 Higgs boson and AB phase θ_H

With the orbifold boundary conditions (7.9) and (7.10), zero modes appear in the parity $(P_0, P_1) = (+,+)$ part of the $A_y^{(j5)}$ $(1 \le j \le 4)$ components of A_y. The wave function of the zero modes in the z coordinate is given by $h_0(z) \equiv u_H(z)$ in (6.58). As $A_y = kzA_z$, the wave function in the y coordinate in $0 \le y \le L$ is $\tilde{u}_H(y) = ke^{ky} u_H(e^{ky})$.

$$\begin{aligned}
u_H(z) &= \sqrt{\frac{2}{z_L^2 - 1}} \, z \, , \quad (1 \le z \le z_L) \\
\tilde{u}_H(y) &= \tilde{u}_H(-y) = \tilde{u}_H(y + 2L) \\
&= k \sqrt{\frac{2}{z_L^2 - 1}} \, e^{2ky} \quad \text{for } 0 \le y \le L \, . \qquad (7.17)
\end{aligned}$$

We expand $A_z^{(j5)}(x, z)$ $(1 \le j \le 4)$ in the KK modes as in (6.49). Writing the 4D zero modes as $A_z^{(j5)(0)}(x) = \phi_j(x)$, we find

$$A_z^{(j5)}(x, z) = \frac{1}{\sqrt{k}} \phi_j(x) u_H(z) + \cdots \, . \qquad (7.18)$$

Then

$$\Phi(x) = \frac{1}{\sqrt{2}} \begin{pmatrix} \phi_2 + i\phi_1 \\ \phi_4 - i\phi_3 \end{pmatrix} \qquad (7.19)$$

transforms as an $SU(2)_L$ doublet. (See (A.33) and (A.34).) $\Phi(x)$ in (7.19) plays the role of the Higgs field in the SM. From the evaluation of the effective potential it will be seen below that $\langle \Phi \rangle \ne 0$. Without loss of generality one can take $\langle \phi_1 \rangle, \langle \phi_2 \rangle, \langle \phi_3 \rangle = 0$ and $\langle \phi_4 \rangle \ne 0$. The 4D fluctuation mode of $\phi_4(x)$ is the field of the observed Higgs particle.

The RS warped space has the same spacetime topology as $M^4 \times (S^1/Z_2)$. There appears an AB phase along the fifth dimension as seen in Section 5.7.

As in (5.34), the eigenvalues of a line integral

$$\hat{W} = P\exp\left\{ig_A \int_{-L}^{L} dy\, A_y\right\} \cdot P_1 P_0 \tag{7.20}$$

are gauge invariant, and their phases become AB phases. As $A_y = (2k)^{-1/2}\phi_4(x)\tilde{u}_H(y)T^{45}$ and $P_1 P_0 = 1$, one finds

$$\begin{aligned}\hat{W} &= \exp\left\{i\hat{\theta}_H(x) \cdot 2T^{45}\right\} , \\ \hat{\theta}_H(x) &= \frac{g_A}{2}\sqrt{\frac{z_L^2 - 1}{k}}\, \phi_4(x) \,. \end{aligned}\tag{7.21}$$

Eigenvalues of $2T^{45}_{\rm sp}$ in the spinor representation are ± 1 so that $\hat{\theta}_H(x)$ is the AB phase in the current case.

The 2π periodicity of $\hat{\theta}_H(x)$ originates from the gauge invariance. It is seen in Section 5.7 that the orbifold boundary conditions P_j change to P_j' as in (5.32) under general gauge transformations. In particular under

$$\Omega(y;\alpha) = \exp\left\{ -i\frac{g_A \alpha}{\sqrt{2k}} \int_y^L dy\, \tilde{u}_H(y)\, T^{45}\right\} \tag{7.22}$$

P_0' and P_1' become

$$\begin{aligned} P_0' &= \Omega(-y;\alpha)P_0\Omega(y;\alpha)^{-1} \\ &= \Omega(-y;\alpha)\Omega(y;\alpha)P_0 \\ &= \Omega(0;2\alpha)P_0 \ , \\ P_1' &= \Omega(L-y;\alpha)P_1\Omega(L+y;\alpha)^{-1} \\ &= \Omega(L-y;\alpha)\Omega(L+y;\alpha)P_1 \\ &= P_1 \ . \end{aligned}\tag{7.23}$$

For

$$\frac{g_A \alpha}{2}\sqrt{\frac{z_L^2 - 1}{k}} = 2\pi n \quad (n: \text{an integer}), \tag{7.24}$$

$\Omega(0;2\alpha) = 1$ so that the boundary conditions remain unchanged. Under the transformation

$$\hat{\theta}_H(x) \to \hat{\theta}_H'(x) = \hat{\theta}_H(x) + 2\pi n \ . \tag{7.25}$$

In other words the periodicity of $\hat{\theta}_H$ is guaranteed by the gauge invariance.

The Higgs field $H(x)$ appears in $A_z^{(45)}(x,z)$, with $\theta_H = \langle \hat{\theta}_H(x)\rangle$, as

$$\begin{aligned} A_z^{(45)}(x,z) &= \frac{1}{\sqrt{k}}\left\{\theta_H f_H + H(x)\right\} u_H(z) + \cdots \ , \\ f_H &= \frac{2}{g_A}\sqrt{\frac{k}{z_L^2 - 1}} = \frac{2}{g_w}\sqrt{\frac{k}{L(z_L^2 - 1)}} \ . \end{aligned}\tag{7.26}$$

The 4D Higgs field is the fluctuation mode of the AB phase θ_H. The Higgs field is a part of the gauge fields. Its mass is generated by quantum effects, being finite as a result of the nature of the AB phase.

It will be shown below that the connection between the electric charge $eQ_{\rm EM}$ and the weak hypercharge Y, appearing in (2.33) in the SM, is given by

$$\begin{aligned} Q_{\rm EM} &= T_L^3 + T_R^3 + Q_X \ , \\ Y &= T_R^3 + Q_X \ . \end{aligned} \tag{7.27}$$

7.4 Quarks

In which representation of $SO(5)$ should quarks and leptons be introduced? When $SO(5) \times U(1)_X$ breaks down to $SU(2)_L \times U(1)_Y$, the left-handed components of quarks and leptons must be $SU(2)_L$ doublets, whereas the right-handed components must be $SU(2)_L$ singlets, at least to high accuracy. In the gauge-Higgs unification scenario, as will be shown below, the physical W boson contains a tiny portion of the $SU(2)_R$ component of the $SO(5)$ gauge fields when $\theta_H \neq 0$. The right-handed components of quarks and leptons should not couple to the W boson, or their W couplings must be negligibly small. The experimental fact is that the charged currents in weak interactions, namely the W couplings, are the $V - A$ (left-handed) type to very high accuracy.

A field multiplet in the spinor representation **4** of $SO(5)$ is decomposed into an $SU(2)_L$ doublet and an $SU(2)_R$ doublet; $\mathbf{4} = (\mathbf{2}, \mathbf{1}) \oplus (\mathbf{1}, \mathbf{2})$ where $(SU(2)_L, SU(2)_R)$ content is shown on the right-hand side. A field multiplet in the vector representation **5** of $SO(5)$ is decomposed into an $SU(2)_L \times SU(2)_R$ bispinor and a singlet; $\mathbf{5} = (\mathbf{2}, \mathbf{2}) \oplus (\mathbf{1}, \mathbf{1})$. (See Appendix A for properties of the bispinor representation.) The $(\mathbf{1}, \mathbf{2})$ or $(\mathbf{2}, \mathbf{2})$ component of the field couples to the $SU(2)_R$ part of W at $\theta_H \neq 0$. Quarks and leptons appear as zero modes of 5D fermion fields. Brane interactions at the UV brane (at $y = 0$) play an important role in avoiding the W coupling of the right-handed components of quarks and leptons.

In the early stage of the model building, quark and lepton multiplets are introduced in the vector representation of $SO(5)$ with additional brane fermions. This type of the $SO(5) \times U(1)_X$ GHU model is called the A model in the literature. Although the A model has been partially successful in reproducing the phenomenology at low energies, it is difficult to be embedded in the framework of grand unification.

An alternative model (the B model) has been proposed in which quark and lepton multiplets are introduced in the combination of the spinor and singlet representation of $SO(5)$. [56]. This model can be embedded in the $SO(11)$ gauge-Higgs grand unification scheme, [59] and is called the GUT inspired GHU model. It is a realistic candidate for describing electroweak and strong interactions, leading to many predictions. In the following we focus on the GUT inspired GHU model.

The matter field content of the GUT inspired GHU model is summarized in Table 7.1. In the bulk region ($0 \le y \le L$) quark and lepton multiplets and dark fermion multiplets are defined. In addition Majorana fermion fields and brane scalar field are introduced on the UV brane at $y = 0$ ($z = 1$).

Table 7.1 The matter field content in the GUT inspired GHU model. The $(SU(3)_C, SO(5))_{U(1)_X}$ content of each field is shown in the last column.

in the bulk	quark Ψ^q	$(\mathbf{3},\mathbf{4})_{\frac{1}{6}}$ $(\mathbf{3},\mathbf{1})^+_{-\frac{1}{3}}$ $(\mathbf{3},\mathbf{1})^-_{-\frac{1}{3}}$
$(1 \le z \le z_L)$	lepton Ψ^ℓ	$(\mathbf{1},\mathbf{4})_{-\frac{1}{2}}$
	dark fermion Ψ^{F_q}	$(\mathbf{3},\mathbf{4})_{\frac{1}{6}}$
	dark fermion Ψ^{F_ℓ}	$(\mathbf{1},\mathbf{4})_{-\frac{1}{2}}$
	dark fermion Ψ^V	$(\mathbf{1},\mathbf{5})^+_0$ $(\mathbf{1},\mathbf{5})^-_0$
on the UV brane	Majorana fermion $\hat{\chi}$	$(\mathbf{1},\mathbf{1})_0$
(at $z = 1$)	brane scalar $\hat{\Phi}$	$(\mathbf{1},\mathbf{4})_{\frac{1}{2}}$

The quark sector consists of three kinds of multiplets in $(\mathbf{3},\mathbf{4})_{\frac{1}{6}}$, $(\mathbf{3},\mathbf{1})^+_{-\frac{1}{3}}$ and $(\mathbf{3},\mathbf{1})^-_{-\frac{1}{3}}$ in the $(SU(3)_C, SO(5))_{U(1)_X}$ content. Each field satisfies the boundary conditions

$$\begin{aligned}
&\Psi^{q,\alpha}_{(\mathbf{3},\mathbf{4})}(x, y_j - y) = -P_j^{\rm sp}\gamma^5\Psi^{q,\alpha}_{(\mathbf{3},\mathbf{4})}(x, y_j + y) \ , \\
&\Psi^{q,\alpha}_{(\mathbf{3},\mathbf{1})^\pm}(x, y_j - y) = \mp\gamma^5\Psi^{q,\alpha}_{(\mathbf{3},\mathbf{1})^\pm}(x, y_j + y) \ , \ (\alpha = 1, 2, 3)
\end{aligned} \tag{7.28}$$

where $P_j^{\rm sp}$ is given in (7.12) and α denotes the generation. $\Psi^{q,1}_{(\mathbf{3},\mathbf{4})}(x, y)$, for instance, consists of four color-triplet fields (u, d, u', d'). The parity assignment and electric charge of the fields are tabulated in Table 7.2.

Table 7.2 Quark multiplets are introduced in three generations ($\alpha = 1, 2, 3$). In the column "Left" ("Right") the parity (P_0, P_1) of the left-handed (right-handed) components is shown.

Field	Name	T_L^3	T_R^3	Q_X	$Q_{\rm EM}$	Left	Right	zero mode
$\Psi^{q,\alpha}_{(\mathbf{3},\mathbf{4})}$	$u\ c\ t$	$+\frac{1}{2}$	0	$\frac{1}{6}$	$\frac{2}{3}$	$(+,+)$	$(-,-)$	$u_L\ c_L\ t_L$
	$d\ s\ b$	$-\frac{1}{2}$			$-\frac{1}{3}$			$d_L\ s_L\ b_L$
	$u'\ c'\ t'$	0	$+\frac{1}{2}$	$\frac{1}{6}$	$\frac{2}{3}$	$(-,-)$	$(+,+)$	$u'_R\ c'_R\ t'_R$
	$d'\ s'\ b'$		$-\frac{1}{2}$		$-\frac{1}{3}$			$d'_R\ s'_R\ b'_R$
$\Psi^{q,\alpha}_{(\mathbf{3},\mathbf{1})^+}$	$D_d^+\ D_s^+\ D_b^+$	0	0	$-\frac{1}{3}$	$-\frac{1}{3}$	$(+,+)$	$(-,-)$	$D_{dL}^+\ D_{sL}^+\ D_{bL}^+$
$\Psi^{q,\alpha}_{(\mathbf{3},\mathbf{1})^-}$	$D_d^-\ D_s^-\ D_b^-$	0	0	$-\frac{1}{3}$	$-\frac{1}{3}$	$(-,-)$	$(+,+)$	$D_{dR}^-\ D_{sR}^-\ D_{bR}^-$

The action for $\Psi^{q,\alpha}_{(\mathbf{3},\mathbf{4})}$ and $\Psi^{q,\alpha}_{(\mathbf{3},\mathbf{1})^\pm}$ in the bulk region of the RS space is obtained by adding gauge interactions to (6.73). The bulk mass parameters of $\Psi^{q,\alpha}_{(\mathbf{3},\mathbf{4})}$ are denoted as $(c_{q_1}, c_{q_2}, c_{q_3}) = (c_u, c_c, c_t)$. The bulk mass parameter of $\Psi^{q,\alpha}_{(\mathbf{3},\mathbf{1})^\pm}$ is denoted as $c_{D_\alpha^\pm}$. We take $c_{D_\alpha^+} = c_{D_\alpha^-} \equiv c_{D_\alpha}$ and write as $(c_{D_1}, c_{D_2}, c_{D_3}) = (c_{D_d}, c_{D_s}, c_{D_b})$. Then the action in the bulk region is

$$
\begin{aligned}
I_{\rm quark}^{\rm bulk} = \int d^5x \sqrt{-g} \bigg\{ \sum_{\alpha=1}^{3} \Big(&\overline{\Psi}^{q,\alpha}_{(\mathbf{3},\mathbf{4})} \mathcal{D}(c_{q_\alpha}) \Psi^{q,\alpha}_{(\mathbf{3},\mathbf{4})} \\
&+ \overline{\Psi}^{q,\alpha}_{(\mathbf{3},\mathbf{1})^+} \mathcal{D}(c_{D_\alpha}) \Psi^{q,\alpha}_{(\mathbf{3},\mathbf{1})^+} + \overline{\Psi}^{q,\alpha}_{(\mathbf{3},\mathbf{1})^-} \mathcal{D}(c_{D_\alpha}) \Psi^{q,\alpha}_{(\mathbf{3},\mathbf{1})^-} \Big) \\
&- \sum_{\alpha,\beta=1}^{3} \Big(m_D^{\alpha\beta} \overline{\Psi}^{q,\alpha}_{(\mathbf{3},\mathbf{1})^+} \Psi^{q,\beta}_{(\mathbf{3},\mathbf{1})^-} + m_D^{\alpha\beta *} \overline{\Psi}^{q,\beta}_{(\mathbf{3},\mathbf{1})^-} \Psi^{q,\alpha}_{(\mathbf{3},\mathbf{1})^+} \Big) \bigg\} ~, \\
\mathcal{D}(c_\alpha) = \Gamma^a {e_a}^M \Big(&\partial_M + \frac{1}{8} \omega_{bcM} [\Gamma^b, \Gamma^c] \\
&- i g_A Q_\alpha^{SO(5)} A_M - i g_B Q_{X\alpha} B_M - i g_C C_M \Big) - c_\alpha \sigma'(y) ~, \qquad (7.29)
\end{aligned}
$$

where $(Q_\alpha^{SO(5)}, Q_{X\alpha}) = (1, \frac{1}{6})$ for $\Psi^{q,\alpha}_{(\mathbf{3},\mathbf{4})}$ and $(0, -\frac{1}{3})$ for $\Psi^{q,\alpha}_{(\mathbf{3},\mathbf{1})^\pm}$. A_M is given in spinor representation. For $\check{\Psi}_\alpha = z^{-2} \Psi_\alpha$ the kinetic part of the action takes the form

$$I^{\text{kinetic}}_{\text{quark}} = \int d^4x \int_1^{z_L} \frac{dz}{k} \sum_\alpha \overline{\check{\Psi}}_\alpha \begin{pmatrix} -k\hat{D}_-(c_\alpha) & \sigma^\mu D_\mu \\ \bar{\sigma}^\mu D_\mu & -k\hat{D}_+(c_\alpha) \end{pmatrix} \check{\Psi}_\alpha ,$$

$$\hat{D}_\pm(c_\alpha) = \pm D_z + \frac{c_\alpha}{z} ,$$

$$D_M = \partial_M - ig_A Q^{SO(5)}_\alpha A_M - ig_B Q_{X\alpha} B_M - ig_C C_M . \tag{7.30}$$

The $m_D^{\alpha\beta}$ term in $I^{\text{bulk}}_{\text{quark}}$ in (7.29) gives Dirac-type masses to D_α^+ and D_α^- fields. In the absence of brane interactions discussed below, massless 4D fields appear, for instance in the first generation, in u_L, d_L, u'_R and d'_R for $\theta_H = 0$.

7.5 Leptons

The lepton sector consists of multiplets $\Psi^{\ell,\alpha}$ in $(\mathbf{1},\mathbf{4})_{-\frac{1}{2}}$ in the bulk region and Majorana brane fermions $\hat{\chi}^\alpha$ on the UV brane at $y = 0$. $\Psi^{\ell,\alpha}$ satisfies the boundary conditions

$$\Psi^{\ell,\alpha}_{(\mathbf{1},\mathbf{4})}(x, y_j - y) = -P_j^{\text{sp}} \gamma^5 \Psi^{\ell,\alpha}_{(\mathbf{1},\mathbf{4})}(x, y_j + y) , \ (\alpha = 1, 2, 3). \tag{7.31}$$

The brane fermion field $\hat{\chi}^\alpha(x)$ satisfies the Majorana condition $(\hat{\chi}^\alpha)^c = \hat{\chi}^\alpha$:

$$\hat{\chi}^\alpha = \begin{pmatrix} \xi_\alpha \\ \eta_\alpha \end{pmatrix}, \quad (\hat{\chi}^\alpha)^c = \begin{pmatrix} +(\eta_\alpha)^c \\ -(\xi_\alpha)^c \end{pmatrix} = e^{i\delta_C} \begin{pmatrix} +\sigma^2 \eta_\alpha^* \\ -\sigma^2 \xi_\alpha^* \end{pmatrix} . \tag{7.32}$$

The bulk mass parameters of $\Psi^{\ell,\alpha}_{(\mathbf{1},\mathbf{4})}$ are denoted as $(c_{\ell_1}, c_{\ell_2}, c_{\ell_3}) = (c_e, c_\mu, c_\tau)$. The action for $\Psi^{\ell,\alpha}_{(\mathbf{1},\mathbf{4})}$ in the bulk region is given by

$$I^{\text{bulk}}_{\text{lepton}} = \int d^5x \sqrt{-g} \sum_{\alpha=1}^{3} \overline{\Psi}^{\ell,\alpha}_{(\mathbf{1},\mathbf{4})} \mathcal{D}(c_{\ell_\alpha}) \Psi^{\ell,\alpha}_{(\mathbf{1},\mathbf{4})} ,$$

$$\mathcal{D}(c_{\ell_\alpha}) = \Gamma^a {e_a}^M \Big(\partial_M + \frac{1}{8} \omega_{bcM} [\Gamma^b, \Gamma^c] - ig_A A_M + \frac{i}{2} g_B B_M \Big) - c_{\ell_\alpha} \sigma'(y) . \tag{7.33}$$

In terms of $\check{\Psi}^{\ell,\alpha}_{(\mathbf{1},\mathbf{4})} = z^{-2} \Psi^{\ell,\alpha}_{(\mathbf{1},\mathbf{4})}$ the action takes the form of (7.30). The action for $\hat{\chi}^\alpha(x)$ is given by

$$I^{\chi}_{\text{brane}} = \int d^5x \sqrt{-g}\, \delta(y) \sum_{\alpha,\beta=1}^{3} \frac{1}{2} \Big\{ \delta_{\alpha\beta} \overline{\hat{\chi}}^\alpha \gamma^\mu \partial_\mu \hat{\chi}^\beta - M_{\alpha\beta} \overline{\hat{\chi}}^\alpha \hat{\chi}^\beta \Big\} . \tag{7.34}$$

Here the $M_{\alpha\beta} = M^*_{\beta\alpha}$ term represents Majorana masses.

The parity assignment and electric charge of the fields are tabulated in Table 7.3. As explained below, additional brane interactions mix the ν_α, ν'_α and η_α components to induce the inverse seesaw mechanism for neutrino masses.

Table 7.3 Lepton multiplets are introduced in three generations ($\alpha = 1, 2, 3$). In the column "Left" ("Right") the parity (P_0, P_1) of the left-handed (right-handed) components is shown.

Field	Name	T_L^3	T_R^3	Q_X	$Q_{\rm EM}$	Left	Right	zero mode
$\Psi^{\ell,\alpha}_{(\mathbf{1},\mathbf{4})}$	$\nu_e\ \nu_\mu\ \nu_\tau$	$+\frac{1}{2}$	0	$-\frac{1}{2}$	0	$(+,+)$	$(-,-)$	$\nu_{eL}\ \nu_{\mu L}\ \nu_{\tau L}$
	$e\ \ \mu\ \ \tau$	$-\frac{1}{2}$			-1			$e_L\ \ \mu_L\ \ \tau_L$
	$\nu'_e\ \nu'_\mu\ \nu'_\tau$	0	$+\frac{1}{2}$	$-\frac{1}{2}$	0	$(-,-)$	$(+,+)$	$\nu'_{eR}\ \nu'_{\mu R}\ \nu'_{\tau R}$
	$e'\ \ \mu'\ \ \tau'$		$-\frac{1}{2}$		-1			$e'_R\ \ \mu'_R\ \ \tau'_R$
$\hat{\chi}^{\ell,\alpha}$	$\eta_1\ \eta_2\ \eta_3$	0	0	0	0	——	——	——

7.6 Brane scalar and brane interactions

We have seen in Section 7.2 that $SO(5)\times U(1)_X$ gauge symmetry is reduced to $SO(4) \times U(1)_X \simeq SU(2)_L \times SU(2)_R \times U(1)_X$ by the orbifold boundary conditions (7.9). The brane scalar field $\hat{\Phi}(x)$ defined on the UV brane at $y = 0$ spontaneously breaks $SU(2)_R \times U(1)_X$ to $U(1)_Y$, resulting the SM gauge symmetry at this stage, namely with $\theta_H = 0$.

The $\hat{\Phi}(x)$ scalar field is introduced in the representation $(\mathbf{1},\mathbf{4})_{\frac{1}{2}}$. Its action is given by

$$I^{\Phi}_{\rm brane} = \int d^5x\sqrt{-g}\,\delta(y)\Big\{ - (D_\mu\hat{\Phi})^\dagger D^\mu\hat{\Phi} - \lambda_\Phi(\hat{\Phi}^\dagger\hat{\Phi} - |w|^2)^2\Big\},$$

$$D_\mu\hat{\Phi} = \Big(\partial_\mu - ig_A A_\mu - \frac{i}{2}g_B B_\mu\Big)\hat{\Phi}\ . \tag{7.35}$$

$\hat{\Phi}$ is decomposed as

$$\hat{\Phi} = \hat{\Phi}_{(\mathbf{1},\mathbf{4})} = \begin{pmatrix}\Phi_{[\mathbf{2},\mathbf{1}]}\\ \Phi_{[\mathbf{1},\mathbf{2}]}\end{pmatrix} \tag{7.36}$$

where $\Phi_{[\mathbf{2},\mathbf{1}]}$ and $\Phi_{[\mathbf{1},\mathbf{2}]}$ are in $(\mathbf{2},\mathbf{1})$ and $(\mathbf{1},\mathbf{2})$ representations of $SU(2)_L \times SU(2)_R$, respectively. The conjugate field of $\tilde{\Phi}$ in $(\mathbf{1},\mathbf{4})_{-\frac{1}{2}}$ is given by

$$\tilde{\Phi} = \tilde{\Phi}_{(\mathbf{1},\mathbf{4})} = \begin{pmatrix} i\sigma^2\Phi^*_{[\mathbf{2},\mathbf{1}]}\\ -i\sigma^2\Phi^*_{[\mathbf{1},\mathbf{2}]}\end{pmatrix}\ . \tag{7.37}$$

$\hat{\Phi}$ develops a nonvanishing VEV

$$\langle \hat{\Phi} \rangle = \begin{pmatrix} 0 \\ 0 \\ 0 \\ w \end{pmatrix} , \quad \langle \tilde{\Phi} \rangle = \begin{pmatrix} -w^* \\ 0 \\ 0 \\ 0 \end{pmatrix} . \tag{7.38}$$

Let us define $A_\mu^{3'_R}$ and B_μ^Y by

$$\begin{pmatrix} A_\mu^{3'_R} \\ B_\mu^Y \end{pmatrix} = \begin{pmatrix} c_\phi & -s_\phi \\ s_\phi & c_\phi \end{pmatrix} \begin{pmatrix} A_\mu^{3_R} \\ B_\mu \end{pmatrix} ,$$

$$c_\phi = \frac{g_A}{\sqrt{g_A^2 + g_B^2}} , \quad s_\phi = \frac{g_B}{\sqrt{g_A^2 + g_B^2}} . \tag{7.39}$$

Then the mass term generated by $w \neq 0$ is

$$\delta(y) \bigg[- \frac{|w|^2}{4} \Big\{ g_A^2 (A_\mu^{1_R} A^{1_R \mu} + A_\mu^{2_R} A^{2_R \mu}) + (g_A^2 + g_B^2) A_\mu^{3'_R} A^{3'_R \mu} \Big\} \bigg] . \tag{7.40}$$

The B_μ^Y fields remain massless, as the gauge fields of the $U(1)_Y$ gauge symmetry. The generator of the $U(1)_Y$ symmetry is confirmed to be $Y = T_R^3 + Q_Y$ as noted in (7.27). The gauge coupling in the $g_Y Y B_\mu^Y$ term appearing in the covariant derivative D_μ is

$$g_Y = \frac{g_A g_B}{\sqrt{g_A^2 + g_B^2}} = g_A s_\phi . \tag{7.41}$$

To see how the mass term (7.40) affects the spectrum of $A_\mu^{1_R}, A_\mu^{2_R}, A_\mu^{3'_R}$, we write the equations of motion in the region $|y| \leq L$;

$$\begin{gathered} \bigg\{ \eta^{\mu\nu} \bigg[\Box + \frac{\partial}{\partial y} e^{-2k|y|} \frac{\partial}{\partial y} - 2k\delta(y) \begin{pmatrix} \omega \\ \omega' \end{pmatrix} \bigg] \\ - \Big(1 - \frac{1}{\xi} \Big) \partial^\mu \partial^\nu \bigg\} \begin{pmatrix} A_\nu^{1_R, 2_R} \\ A_\nu^{3'_R} \end{pmatrix} = \cdots , \\ \omega = \frac{g_A^2 |w|^2}{4k} , \quad \omega' = \frac{(g_A^2 + g_B^2)|w|^2}{4k} \end{gathered} \tag{7.42}$$

The $\cdots$ term on the right-hand side of Eq. (7.42) represents terms of higher order in fields. The parity of $A_\mu^{1_R}, A_\mu^{2_R}, A_\mu^{3'_R}$ at $y = 0$ is $+$ so that $A_\mu^{1_R}, A_\mu^{2_R}, A_\mu^{3'_R}$ are even functions in y. By integrating the equation of $A_\mu^{1_R}$, for instance, over y from $-\epsilon$ to $+\epsilon$, one obtains $(\partial A_\mu^{1_R}/\partial y - k\omega A_\mu^{1_R})_{y=0} = 0$. In the z coordinate, at $z = 1^+$

$$\begin{aligned} D_{\rm eff}(\omega) : \quad & \Big(\frac{\partial}{\partial z} - \omega \Big) A_\mu^{1_R, 2_R} = 0 , \\ D_{\rm eff}(\omega') : \quad & \Big(\frac{\partial}{\partial z} - \omega' \Big) A_\mu^{3'_R} = 0 . \end{aligned} \tag{7.43}$$

Table 7.4 The boundary conditions for the gauge fields at $z=1$ and z_L are summarized. N and D stand for Neumann and Dirichlet conditions, respectively. $D_{\rm eff}$ stands for the effective Dirichlet condition specified in (7.43).

	Number of generators	A_μ	A_z
$SU(3)_C$	8	(N,N)	(D,D)
$SU(2)_L$	3	(N,N)	(D,D)
$U(1)_Y$	1	(N,N)	(D,D)
$(SU(2)_R\cup U(1)_X)/U(1)_Y$	3	$(D_{\rm eff},N)$	(D,D)
$SO(5)/(SU(2)_L\cup SU(2)_R)$	4	(D,D)	(N,N)

In other words, the $A_\mu^{1_R,2_R,3'_R}$ fields have a cusp at $y=0$. These fields originally satisfied the Neumann condition (6.53) at $z=1$. As a result of the interactions on the brane, the condition is changed to (7.43). For sufficiently large $|w|$, boundary conditions of $A_\mu^{1_R,2_R,3'_R}$ at $z=1$ are modified from the Neumann condition to the Dirichlet condition for low-lying modes in their KK towers. It will be seen below that this approximation is justified for $|w|=O(k)$. Boundary conditions of gauge fields are summarized in Table 7.4.

The brane scalar field $\hat\Phi$ also couples to the quark and lepton field multiplets. Recall that $\hat\Phi$ and $\tilde\Phi$ are in the $(\mathbf{1},\mathbf{4})_{\frac12}$ and $(\mathbf{1},\mathbf{4})_{-\frac12}$ representations, respectively. $SO(5)\times U(1)_X\times SU(3)_C$-invariant brane interactions on the UV brane are given by

$$
\begin{aligned}
I^{q,\ell}_{\rm brane} &= \int d^5x\sqrt{-\det G}\,\delta(y)\big(\mathcal{L}_1+\mathcal{L}_2+\mathcal{L}_3\big)\ ,\\
\mathcal{L}_1 &= -\Big\{\kappa_q^{\alpha\beta}\,\overline{\Psi}{}^{q,\alpha}_{(\mathbf{3},\mathbf{4})}\,\hat\Phi_{(\mathbf{1},\mathbf{4})}\cdot\Psi^{q,\beta}_{(\mathbf{3},\mathbf{1})^+}+{\rm H.c.}\Big\}\ ,\\
\mathcal{L}_2 &= -\Big\{\tilde\kappa_\ell^{\alpha\beta}\,\overline{\hat\chi}{}^{\beta}\,\tilde\Phi^\dagger_{(\mathbf{1},\mathbf{4})}\,\Psi^{\ell,\alpha}_{(\mathbf{1},\mathbf{4})}+{\rm H.c.}\Big\}\ ,\\
\mathcal{L}_3 &= -\Big\{\tilde\kappa'^{\alpha\beta}_{\ell,V}\,\overline{\Psi}{}^{\ell,\alpha}_{(\mathbf{1},\mathbf{4})}\,\Gamma^a\,\tilde\Phi_{(\mathbf{1},\mathbf{4})}\cdot\big(\Psi^{V,\beta}_{(\mathbf{1},\mathbf{5})^-}\big)_a+{\rm H.c.}\Big\}\ .
\end{aligned}
\tag{7.44}
$$

The $\mathcal{L}_3$ term connects the lepton multiplet to the dark fermion multiplet explained below. It plays an irrelevant role, and we set $\tilde\kappa'^{\alpha\beta}_{\ell,V}=0$. The $\mathcal{L}_3$ term is strictly forbidden if the global symmetry of the dark fermion number conservation is imposed. With $\langle\hat\Phi\rangle$ in (7.38), the interaction (7.44) generates mass terms given by

$$
\begin{aligned}
I_{\text{brane mass}}^{q,\ell} &= \int d^5x \sqrt{-\det G}\, \delta(y) \Big(\mathcal{L}_1^m + \mathcal{L}_2^m \Big) ~, \\
\mathcal{L}_1^m &= 2\mu_q^{\alpha\beta}\, \overline{\check{d}}{}_R'^{\alpha} \check{D}_L^{+\beta} + \text{H.c.} ~, \\
\mathcal{L}_2^m &= -\bigg\{ \frac{m_B^{\alpha\beta}}{\sqrt{k}}\, \overline{\chi}^{\beta} \check{\nu}_R'^{\alpha} + \text{H.c.} \bigg\} ~.
\end{aligned} \tag{7.45}
$$

Here $2\mu_q^{\alpha\beta} = \kappa_q^{\alpha\beta} w$, $m_B^{\alpha\beta} = \tilde{\kappa}_\ell^{\alpha\beta} w\sqrt{2k}$, and $(\check{d}_R'^1, \check{d}_R'^2, \check{d}_R'^3) = (\check{d}_R', \check{s}_R', \check{b}_R')$ etc. It will be seen below that $\mathcal{L}_1^m$ is necessary to reproduce the masses of down-type quarks and that $\mathcal{L}_2^m$ is necessary to explain tiny masses of neutrinos.

7.7 Dark fermions

As listed in Table 7.1, dark fermion fields $\Psi_{(\mathbf{3},\mathbf{4})}^{F_q,\alpha}$, $\Psi_{(\mathbf{1},\mathbf{4})}^{F_\ell,\alpha}$ and $\Psi_{(\mathbf{1},\mathbf{5})\pm}^{V,\alpha}$ are introduced. These fields satisfy the orbifold conditions

$$
\begin{aligned}
\Psi_{(\mathbf{3},\mathbf{4})}^{F_q,\alpha}(x, y_j - y) &= (-1)^j P_j^{\text{sp}} \gamma^5 \Psi_{(\mathbf{3},\mathbf{4})}^{F_q,\alpha}(x, y_j + y) ~, \\
\Psi_{(\mathbf{1},\mathbf{4})}^{F_\ell,\alpha}(x, y_j - y) &= (-1)^j P_j^{\text{sp}} \gamma^5 \Psi_{(\mathbf{1},\mathbf{4})}^{F_\ell,\alpha}(x, y_j + y) ~, \\
\Psi_{(\mathbf{1},\mathbf{5})\pm}^{V,\beta}(x, y_j - y) &= \mp P_j^{\text{vec}} \gamma^5 \Psi_{(\mathbf{1},\mathbf{5})\pm}^{V,\beta}(x, y_j + y) ~, \\
\alpha = 1, \cdots, n_F, &\quad \beta = 1, \cdots, n_V.
\end{aligned} \tag{7.46}
$$

The action for the dark fermion fields is given by

$$
\begin{aligned}
I_{\text{darkF}}^{\text{bulk}} &= \int d^5x \sqrt{-g} \\
&\times \bigg\{ \sum_{\alpha=1}^{n_F} \Big(\overline{\Psi}{}_{(\mathbf{3},\mathbf{4})}^{F_q,\alpha} \mathcal{D}(c_{F_q,\alpha}) \Psi_{(\mathbf{3},\mathbf{4})}^{F_q,\alpha} + \overline{\Psi}{}_{(\mathbf{1},\mathbf{4})}^{F_\ell,\alpha} \mathcal{D}(c_{F_\ell,\alpha}) \Psi_{(\mathbf{1},\mathbf{4})}^{F_\ell,\alpha} \Big) \\
&+ \sum_{\beta=1}^{n_V} \Big(\overline{\Psi}{}_{(\mathbf{1},\mathbf{5})+}^{V,\beta} \mathcal{D}(c_{V_\beta}) \Psi_{(\mathbf{1},\mathbf{5})+}^{V,\beta} + \overline{\Psi}{}_{(\mathbf{1},\mathbf{5})-}^{V,\beta} \mathcal{D}(c_{V_\beta}) \Psi_{(\mathbf{1},\mathbf{5})-}^{V,\beta} \Big) \\
&- \sum_{\alpha,\beta=1}^{n_V} \Big(m_V^{\alpha\beta} \overline{\Psi}{}_{(\mathbf{1},\mathbf{5})+}^{V,\alpha} \Psi_{(\mathbf{1},\mathbf{5})-}^{V,\beta} + m_V^{\beta\alpha *} \overline{\Psi}{}_{(\mathbf{1},\mathbf{5})-}^{V,\beta} \Psi_{(\mathbf{1},\mathbf{5})+}^{V,\alpha} \Big) \bigg\} ~.
\end{aligned} \tag{7.47}
$$

In the minimal case one may take $n_F = n_V = 1$. Due to the boundary conditions (7.46) both left-handed and right-handed components of $\Psi_{(\mathbf{3},\mathbf{4})}^{F_q,\alpha}$ and $\Psi_{(\mathbf{1},\mathbf{4})}^{F_\ell,\alpha}$ have the parity $(+,-)$ or $(-,+)$ so that no massless zero mode appears. The mass (m_V) terms of $O(k)$ in (7.47), connecting $\Psi^{V,\alpha}_{(\mathbf{1},\mathbf{5})^+}$ and $\Psi^{V,\alpha}_{(\mathbf{1},\mathbf{5})^-}$, give masses to the lowest modes of these fields. Dark fermion fields are necessary for having desired electroweak symmetry breaking at the one loop level.

7.8 Twisted gauge in the RS space

When the AB phase θ_H defined in Section 7.3 is not zero, the $SU(2)_L \times U(1)_Y$ symmetry breaks down to $U(1)_{\rm EM}$. The value of θ_H is determined by the location of the global minimum of the effective potential $V_{\rm eff}(\theta_H)$. At the one loop level $V_{\rm eff}(\theta_H)$ is determined by the mass spectrum of the gauge and fermion fields at $\theta_H \neq 0$. Unlike the case of the GHU model in $M^4 \times (S^1/Z_2)$ discussed in Section 5.10, it becomes very difficult to solve the equations of motion with nontrivial background fields $A_y \neq 0$ in the RS space. At this point the method of the twisted gauge explained in Section 5.11 becomes useful and indispensable.

As shown in (7.26)

$$A_z^{(45)}(x,z) = \frac{1}{\sqrt{k}}\,\theta_H f_H\, u_H(z) + \cdots \,,$$
$$A_y^{(45)}(x,y) = \frac{1}{\sqrt{k}}\,\theta_H f_H\, \tilde{u}_H(y) + \cdots \,, \tag{7.48}$$

for $\theta_H \neq 0$. Under a gauge transformation $\tilde{A}_M = \Omega A_M \Omega^{-1} - (i/g_A)\Omega\partial_M\Omega^{-1}$ where Ω is given by $\Omega(y;\alpha)$ in (7.22) with $\alpha = \theta_H f_H$, the AB phase is transformed to $\tilde{\theta}_H = 0$, that is, the classical background field vanishes. This gauge is called **the twisted gauge.** [47, 48] We denote quantities in the twisted gauge by adding tilde "~" on top. In the twisted gauge every field obey the free equation with $\tilde{\theta}_H = 0$ in the zeroth approximation in perturbation theory.

However, as shown in (7.23), the orbifold boundary conditions change to new θ_H-dependent conditions $\tilde{P}_j$. Since $\Omega(0; 2\theta_H f_H) = \exp\{-2i\theta_H T^{45}\}$, one finds, with (7.10) and (7.12), the new $\tilde{P}_j$ are given by

$$\tilde{P}_0^{\rm vec} = \begin{pmatrix} -1 & 0 & 0 & 0 & 0 \\ 0 & -1 & 0 & 0 & 0 \\ 0 & 0 & -1 & 0 & 0 \\ 0 & 0 & 0 & -\cos 2\theta_H & -\sin 2\theta_H \\ 0 & 0 & 0 & -\sin 2\theta_H & \cos 2\theta_H \end{pmatrix},$$

$$\tilde{P}_0^{\rm sp} = \begin{pmatrix} \cos\theta_H & 0 & i\sin\theta_H & 0 \\ 0 & \cos\theta_H & 0 & i\sin\theta_H \\ -i\sin\theta_H & 0 & -\cos\theta_H & 0 \\ 0 & -i\sin\theta_H & 0 & -\cos\theta_H \end{pmatrix},$$

$$\tilde{P}_1^{\rm vec} = P_1^{\rm vec}, \quad \tilde{P}_1^{\rm sp} = P_1^{\rm sp}. \tag{7.49}$$

In the matrix representation introduced in Appendix A.2, $T_{\rm sp}^{45} = \frac{1}{2}\sigma^0 \otimes \sigma^1$ and we have $\tilde{P}_0^{\rm sp} = \sigma^0 \otimes (\cos\theta_H \sigma^3 - \sin\theta_H \sigma^2)$.

The gauge transformation connecting the original gauge to the twisted gauge, $\Omega(y;\theta_H f_H)$ or $\Omega(z;\theta_H f_H) \equiv \widetilde{\Omega}(z)$, is given by

$$\widetilde{\Omega}(z) = e^{-i\theta(z)T^{45}}, \quad \theta(z) = \theta_H \frac{z_L^2 - z^2}{z_L^2 - 1}. \tag{7.50}$$

Note that

$$\frac{g_A}{\sqrt{2k}} \theta_H f_H u_H(z) = -\theta'(z). \tag{7.51}$$

In the vector and spinor representations

$$\widetilde{\Omega}(z)^{\rm vec} = \begin{pmatrix} 1 & 0 & 0 & 0 & 0 \\ 0 & 1 & 0 & 0 & 0 \\ 0 & 0 & 1 & 0 & 0 \\ 0 & 0 & 0 & \cos\theta(z) & -\sin\theta(z) \\ 0 & 0 & 0 & \sin\theta(z) & \cos\theta(z) \end{pmatrix},$$

$$\widetilde{\Omega}(z)^{\rm sp} = \begin{pmatrix} \cos\frac{1}{2}\theta(z) & 0 & -i\sin\frac{1}{2}\theta(z) & 0 \\ 0 & \cos\frac{1}{2}\theta(z) & 0 & -i\sin\frac{1}{2}\theta(z) \\ -i\sin\frac{1}{2}\theta(z) & 0 & \cos\frac{1}{2}\theta(z) & 0 \\ 0 & -i\sin\frac{1}{2}\theta(z) & 0 & \cos\frac{1}{2}\theta(z) \end{pmatrix}. \tag{7.52}$$

For $a = 1, 2, 3$

$$\begin{aligned} \widetilde{\Omega}(z) T^{a4} \widetilde{\Omega}(z)^{-1} &= \cos\theta(z) T^{a4} + \sin\theta(z) T^{a5}, \\ \widetilde{\Omega}(z) T^{a5} \widetilde{\Omega}(z)^{-1} &= -\sin\theta(z) T^{a4} + \cos\theta(z) T^{a5}. \end{aligned} \tag{7.53}$$

It follows that for the $SO(5)$ gauge fields

$$\begin{aligned} A_M^{a4} &= \cos\theta(z)\tilde{A}_M^{a4} + \sin\theta(z)\tilde{A}_M^{a5}, \\ A_M^{a5} &= -\sin\theta(z)\tilde{A}_M^{a4} + \cos\theta(z)\tilde{A}_M^{a5}, \\ A_z^{45} &= \tilde{A}_z^{45} - \frac{\sqrt{2}}{g_A}\,\theta'(z), \end{aligned} \tag{7.54}$$

and other components remain unchanged.

7.9 Mass spectrum of gauge fields

The mass spectra and wave functions of various fields in the RS space with $\theta_H \neq 0$ are easily found in the twisted gauge. In the twisted gauge specified in the previous section the orbifold condition on the IR brane, $\tilde{P}_1 = P_1$, remains unchanged from the original gauge, and the fields satisfy free equations in the RS space. The $SU(3)$ gauge fields C_M do not couple to θ_H so their spectrum and wave functions are the same as in Section 6.4.

7.9.1 *Gauge fields: A_μ and B_μ*

The $SO(5)$ gauge fields are decomposed into $SU(2)_L$ components $A_M^{a_L}$, $SU(2)_R$ components $A_M^{a_R}$ and $SO(5)/SO(4)$ components $A_M^{\hat{p}}$ as described in Section 7.2. In the bulk region $(1 < z < z_L)$, $\tilde{A}_\mu$ satisfies

$$\left\{\eta^{\mu\nu}(\Box + k^2\mathcal{P}_4) - \left(1 - \frac{1}{\xi}\right)\partial^\mu\partial^\nu\right\}\tilde{A}_\nu = 0 \, , \tag{7.55}$$

as shown in (6.46). The 4D mode with a mass $m = k\lambda$ is hence described by $\tilde{A}_\mu(x,z) = \tilde{A}_\mu(x)f(z)$, where $f(z)$ is given by (6.47). The mode function $f(z)$ satisfies, at $z = z_L$, the Neumann (N) condition for $\tilde{P}_1 = +1$ and the Dirichlet (D) condition for $\tilde{P}_1 = -1$. As summarized in Table 6.2, the mode function is given by

$$\tilde{A}_\mu, B_\mu \quad N : C(z,\lambda), \quad D : S(z;\lambda), \tag{7.56}$$

where $C(z,\lambda)$ and $S(z;\lambda)$ are defined in Appendix C.

(i) $(\tilde{A}_\mu^{a_L}, \tilde{A}_\mu^{a_R}, \tilde{A}_\mu^{\hat{a}})$ $(a = 1, 2)$

The boundary condition for $A_\mu^{a_R}$ at the UV brane becomes (7.43) as a result of the brane interaction with $\hat{\Phi}$. In the original gauge at $z = 1$ or $z = 1^+$

$$\partial_z A_\mu^{a_L} = 0 \ ,$$
$$(\partial_z - \omega) A_\mu^{a_R} = 0 \ , \quad \omega = \frac{g_A^2 |w|^2}{4k} \ ,$$
$$A_\mu^{\hat{a}} = 0 \ , \tag{7.57}$$

By making use of (7.14) and (7.54), the conditions for the $a = 1$ components in the twisted gauge become

$$\begin{aligned} &\partial_z \tilde{A}_\mu^{23} + \cos\theta_H \partial_z \tilde{A}_\mu^{14} + \sin\theta_H \partial_z \tilde{A}_\mu^{15} = 0 \ , \\ &(\partial_z - \omega)\tilde{A}_\mu^{23} - \cos\theta_H (\partial_z - \omega)\tilde{A}_\mu^{14} - \sin\theta_H (\partial_z - \omega)\tilde{A}_\mu^{15} = 0 \ , \\ &- \sin\theta_H \tilde{A}_\mu^{14} + \cos\theta_H \tilde{A}_\mu^{15} = 0 \ . \end{aligned} \tag{7.58}$$

In transforming the first and second equations in (7.57) there arise terms proportional to $\theta'(1)$, which, as a consequence of the Dirichlet condition for $A_\mu^{\hat{1}}$ (the third equation), vanishes.

At $z = z_L$ $\tilde{A}_\mu^{23}$ and $\tilde{A}_\mu^{14}$ have $\tilde{P}_1 = +1$, whereas $\tilde{A}_\mu^{15}$ has $\tilde{P}_1 = -1$. Hence each mode is expressed as

$$\begin{pmatrix} \tilde{A}_\mu^{23} \\ \tilde{A}_\mu^{14} \\ \tilde{A}_\mu^{15} \end{pmatrix} = \begin{pmatrix} \alpha_{23} C(z,\lambda) \\ \alpha_{14} C(z,\lambda) \\ \alpha_{15}\, S(z,\lambda) \end{pmatrix} A_\mu(x) \ . \tag{7.59}$$

Inserting (7.59) into (7.58), one finds

$$K \begin{pmatrix} \alpha_{23} \\ \alpha_{14} \\ \alpha_{15} \end{pmatrix} = 0 \ ,$$

$$K = \begin{pmatrix} C' & \cos\theta_H C' & \sin\theta_H S' \\ C' - \omega C & -\cos\theta_H (C' - \omega C) & -\sin\theta_H (S' - \omega S) \\ 0 & -\sin\theta_H C & \cos\theta_H S \end{pmatrix} . \tag{7.60}$$

Here $C = C(1;\lambda)$, $C' = C'(1;\lambda)$, $S = S(1;\lambda)$ and $S' = S'(1;\lambda)$. Similar formulas are obtained for $a = 2$. The spectrum $\{\lambda_n\}$ is determined, from $\det K = 0$, by

$$2C'(SC' + \lambda \sin^2\theta_H) - \omega C(2SC' + \lambda \sin^2\theta_H) = 0 \ . \tag{7.61}$$

For the low-lying KK excited modes $C(1;\lambda_n)/C'(1;\lambda_n) = O(z_L^2) \gg 1$ so that the spectrum is determined effectively by the second term in (7.61). We find that

$$\begin{aligned} &W \ : 2S(1;\lambda_{W^{(n)}})C'(1;\lambda_{W^{(n)}}) + \lambda_{W^{(n)}} \sin^2\theta_H = 0 \ (n \geq 0) \ , \\ &W_R : C(1;\lambda_{W_R^{(n)}}) = 0 \ (n \geq 1) \ . \end{aligned} \tag{7.62}$$

The lowest mode $W^{(0)}$ of the KK tower of W, or the W tower, is the W boson. Its mass is, with the use of (C.25), approximately given by

$$m_W = k\lambda_{W^{(0)}} \sim \sqrt{\frac{k}{L}}\, z_L^{-1} \sin\theta_H = \frac{\sin\theta_H}{\pi\sqrt{kL}}\, m_{\rm KK} \ . \tag{7.63}$$

There is no light mode in the W_R tower.

The approximation of replacing (7.61) by (7.62) is justified, provided

$$\omega\, \frac{C}{C'} \sim \frac{g_A^2 w^2}{4k}\, \frac{1}{\lambda^2 \ln z_L} = \frac{g_w^2}{4}\Big(\frac{w}{m_W}\Big)^2 \gg 1 \ . \tag{7.64}$$

The condition is satisfied for $w \gg m_{\rm KK} \gg m_W$. In the absence of the brane interaction $\omega = 0$ in (7.61), in which case the spectrum of the W tower is given by $SC' + \lambda \sin^2\theta_H = 0$ and that of the W_R tower is given by $C' = 0$.

(ii) $(\tilde{A}_\mu^{3_L}, \tilde{A}_\mu^{3_R}, \tilde{A}_\mu^{\hat{3}}, B_\mu)$

With (7.43) taken into account, the boundary conditions at $z = 1$ or $z = 1^+$ are given by

$$\begin{aligned}
&\partial_z A_\mu^{3_L} = 0 \ , \\
&(\partial_z - \omega') A_\mu^{3'_R} = 0 \ , \quad \omega' = \frac{(g_A^2 + g_B^2) w^2}{4k} \ , \\
&A_\mu^{\hat{3}} = 0 \ , \\
&\partial_z B_\mu^Y = 0 \ .
\end{aligned} \tag{7.65}$$

In the twisted gauge they become

$$\begin{aligned}
&\partial_z \tilde{A}_\mu^{12} + \cos\theta_H \partial_z \tilde{A}_\mu^{34} + \sin\theta_H \partial_z \tilde{A}_\mu^{35} = 0 \ , \\
&(\partial_z - \omega')\Big\{\tilde{A}_\mu^{12} - \cos\theta_H \tilde{A}_\mu^{34} - \sin\theta_H \tilde{A}_\mu^{35} - \sqrt{2}\frac{s_\phi}{c_\phi} B_\mu\Big\} = 0 \ , \\
&-\sin\theta_H \tilde{A}_\mu^{34} + \cos\theta_H \tilde{A}_\mu^{35} = 0 \ , \\
&\partial_z \tilde{A}_\mu^{12} + \frac{c_\phi}{\sqrt{2}\, s_\phi} \partial_z B_\mu = 0 \ .
\end{aligned} \tag{7.66}$$

At $z = z_L$ $\tilde{A}_\mu^{12}$, $\tilde{A}_\mu^{34}$ and B_μ have $\tilde{P}_1 = +1$, whereas $\tilde{A}_\mu^{35}$ has $\tilde{P}_1 = -1$. Therefore each mode is expressed as

$$\begin{pmatrix} \tilde{A}_\mu^{12} \\ \tilde{A}_\mu^{34} \\ \tilde{A}_\mu^{35} \\ B_\mu \end{pmatrix} = \begin{pmatrix} \alpha_{12} C(z,\lambda) \\ \alpha_{34} C(z,\lambda) \\ \alpha_{35}\, S(z,\lambda) \\ \alpha_B C(z,\lambda) \end{pmatrix} A_\mu(x) \ . \tag{7.67}$$

Inserting (7.67) into (7.66), one finds at $z = 1^+$

$$K \begin{pmatrix} \alpha_{12} \\ \alpha_{34} \\ \alpha_{35} \\ \alpha_B \end{pmatrix} = 0 \ .$$

$$K = \begin{pmatrix} C' & \cos\theta_H C' & \sin\theta_H S' & 0 \\ \hat{C}' & -\cos\theta_H \hat{C}' & -\sin\theta_H \hat{S}' & -\dfrac{\sqrt{2}s_\phi}{c_\phi}\hat{C}' \\ 0 & -\sin\theta_H C & \cos\theta_H S & 0 \\ C' & 0 & 0 & \dfrac{c_\phi}{\sqrt{2}s_\phi}C' \end{pmatrix} ,$$

$$\hat{C}' = C' - \omega' C \ , \quad \hat{S}' = S' - \omega' S \ . \tag{7.68}$$

The spectrum $\{\lambda_n\}$ is determined, from $\det K = 0$, by

$$2C'^2(SC' + \lambda\sin^2\theta_H) - \omega' CC'\{2SC' + (1 + s_\phi^2)\lambda\sin^2\theta_H\} = 0 \ . \tag{7.69}$$

As in the case of (i) above, the spectrum of the low-lying modes is determined effectively by the second term in (7.69).

$$\begin{aligned} &\gamma \ : C'(1; \lambda_{\gamma^{(n)}}) = 0 \ (n \geq 0) \ , \\ &Z \ : 2S(1; \lambda_{Z^{(n)}})C'(1; \lambda_{Z^{(n)}}) + (1 + s_\phi^2)\lambda_{Z^{(n)}} \sin^2\theta_H = 0 \ (n \geq 0) \ , \\ &Z_R : C(1; \lambda_{Z_R^{(n)}}) = 0 \ (n \geq 1) \ . \end{aligned} \tag{7.70}$$

The lowest mode ($\gamma^{(0)}$) of the γ tower is the photon (γ), which is massless $m_\gamma = k\lambda_{\gamma^{(0)}} = 0$. The lowest mode ($Z^{(0)}$) of the Z tower is the Z boson. Its mass is, with the use of (C.25), approximately given by

$$m_Z \sim \sqrt{1 + s_\phi^2}\sqrt{\frac{k}{L}}\, z_L^{-1} \sin\theta_H = \sqrt{1 + s_\phi^2}\, \frac{\sin\theta_H}{\pi\sqrt{kL}}\, m_{\rm KK} \ . \tag{7.71}$$

There is no light mode in the Z_R tower. In the absence of the brane interaction $\omega' = 0$ in (7.69), in which case the spectrum of the γ and Z_R towers is given by $C' = 0$, and the spectrum of the Z tower is given by $SC' + \lambda\sin^2\theta_H = 0$. The spectrum of the Z tower is depicted in Fig. 7.1. The spectrum for $n \geq 20$ is nearly the same as the one determined by $SC'(1, \lambda_n) = 0$.

We define the bare weak mixing angle θ_W^0 by

$$\begin{aligned} \sin\theta_W^0 &= \frac{g_Y}{\sqrt{g_A^2 + g_Y^2}} = \frac{g_B}{\sqrt{g_A^2 + 2g_B^2}} = \frac{s_\phi}{\sqrt{1 + s_\phi^2}} \ , \\ \cos\theta_W^0 &= \frac{1}{\sqrt{1 + s_\phi^2}} \ , \end{aligned} \tag{7.72}$$

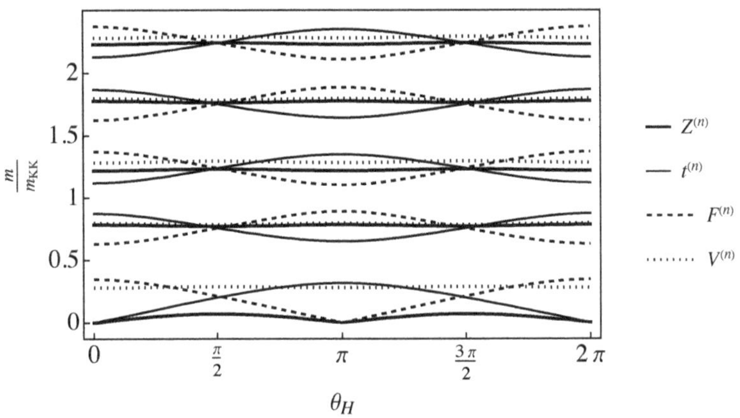

Fig. 7.1 The mass spectra of the Z and top quark (t) towers ($0 \leq n \leq 4$) and the dark fermion F and V towers ($1 \leq n \leq 5$) are shown as functions of θ_H for $m_{\rm KK} = 13$ TeV. $\sin^2\theta_W^0 = 0.2304$, $z_L = 3.832 \times 10^{11}$, and $c_t = -0.2745$. The parameters have been chosen such that $m_Z = 91.1876$ GeV, $m_t = 172.9$ GeV and $m_H = 125.1$ GeV at $\theta_H = 0.1$. The spectrum of the Z tower is determined by (7.70) and is periodic in θ_H with a period π. The spectrum of the top (t) tower is determined by (7.89) and is periodic in θ_H with a period 2π. The input parameters for the dark fermions are $n_F = 1$, $n_V = 3$, $c_V = 0$. The additional parameters are determined to be $c_F = 0.2359$ and $\tilde{m}_V = 0.0812$ such that the effective potential is minimized at $\theta_H = 0.1$ and $m_H = 125.1$ GeV.

where g_Y and s_ϕ are given in (7.41). Comparing (7.63) and (7.71), one finds that $m_W \sim m_Z \cos\theta_W^0$.

(iii) $\tilde{A}_\mu^{\hat{4}}$

$\tilde{A}_\mu^{\hat{4}} = A_\mu^{\hat{4}}$ and its parity is $(\tilde{P}_0, \tilde{P}_1) = (-,-)$ so that the mode function is given by $S(z;\lambda)$ and the spectrum is determined by

$$A^{\hat{4}} \ : S(1; \lambda_{A^{\hat{4}(n)}}) = 0 \ (n \geq 1) \ . \tag{7.73}$$

There is no light mode.

7.9.2 Gauge fields: A_z and B_z

A_z and B_z do not couple to the brane scalar field $\hat{\Phi}$. As seen in (6.46), $\tilde{A}_z$ satisfies in the bulk region ($1 < z < z_L$)

$$(\Box + \xi k^2 \mathcal{P}_z)\tilde{A}_z = 0 \ . \tag{7.74}$$

With $h(z)$ in (6.48) each mode can be expressed as $\tilde{A}_z(x,z) = \tilde{A}_z(x)h(z)$. The mode function $h(z)$ satisfies, at $z = z_L$, the Neumann (N) condition for $\tilde{P}_1 = +1$ and the Dirichlet (D) condition for $\tilde{P}_1 = -1$. As summarized in Table 6.2, $h(z)$ is given by

$$\tilde{A}_z, B_z \quad N : S'(z, \lambda), \quad D : C'(z; \lambda) \ . \tag{7.75}$$

Note that the Neumann (N) condition is given by (6.53).

(i) $(\tilde{A}_z^{12}, \tilde{A}_z^{23}, \tilde{A}_z^{31}, B_z)$

For these components $\tilde{A}_z^{jk} = A_z^{jk}$ and $\tilde{P}_0, \tilde{P}_1 = -1$ so that mode functions are given by $C'(z; \lambda_n)$. The spectrum is determined by

$$C'(1; \lambda_n) = 0 \ . \tag{7.76}$$

There is no zero mode. The mass is given by $m_n^2 = \xi(k\lambda_n)^2$, depending on the gauge parameter ξ. All modes are unphysical.

(ii) $(\tilde{A}_z^{a4}, \tilde{A}_z^{a5})$ $(a = 1, 2, 3)$

In the original gauge, A_z^{a4} and A_z^{a5} have parity $P_0 = -$ and $+$ so that

$$\begin{aligned} &A_z^{a4} = 0 \ , \\ &\frac{\partial}{\partial z}\Big(\frac{1}{z} A_z^{a5}\Big) = 0 \end{aligned} \tag{7.77}$$

at $z = 1$. In the twisted gauge

$$\begin{aligned} &\cos\theta_H \tilde{A}_z^{a4} + \sin\theta_H \tilde{A}_z^{a5} = 0 \ , \\ &-\sin\theta_H \frac{\partial}{\partial z}\frac{1}{z}\tilde{A}_z^{a4} + \cos\theta_H \frac{\partial}{\partial z}\frac{1}{z}\tilde{A}_z^{a5} = 0 \ . \end{aligned} \tag{7.78}$$

As $\tilde{A}_z^{a4}$ and $\tilde{A}_z^{a5}$ have parity $\tilde{P}_1 = -$ and $+$ at $z = z_L$, each mode can be expressed as

$$\begin{pmatrix} \tilde{A}_z^{a4} \\ \tilde{A}_z^{a5} \end{pmatrix} = \begin{pmatrix} \alpha_{a4} C'(z, \lambda) \\ \alpha_{a5} S'(z, \lambda) \end{pmatrix} A_z(x) \ . \tag{7.79}$$

Inserting (7.79) into (7.78) and using the relations in (C.15), one finds

$$K \begin{pmatrix} \alpha_{a4} \\ \alpha_{a5} \end{pmatrix} = 0 \ , \quad K = \begin{pmatrix} \cos\theta_H C' & \sin\theta_H S' \\ -\sin\theta_H C & \cos\theta_H S \end{pmatrix} . \tag{7.80}$$

The spectrum $\{\lambda_n\}$ is determined by $\det K = 0$, or

$$SC' + \lambda_n \sin^2\theta_H = 0 \ . \tag{7.81}$$

There is no zero mode. The mass is given by $m_n^2 = \xi(k\lambda_n)^2$, depending on the gauge parameter ξ. All modes are unphysical.

(iii) $\tilde{A}_z^{45}$

Parity of $\tilde{A}_z^{45}$ is + at both $z = 1$ and z_L. There exists one zero mode which is the 4D Higgs boson discussed in Section 7.3. Its wave function is given by (7.17). Wave functions of the KK excited modes are given by $S'(z, \lambda_n)$ and the spectrum $\{\lambda_n, n \geq 1\}$ is determined by

$$\text{Higgs}: \quad S(1; \lambda_n) = 0 . \tag{7.82}$$

The modes $n \geq 1$ are unphysical. In the viewpoint of 4D gauge theory all modes of A_z except for the zero mode are absorbed by the corresponding A_μ modes, leading to making them massive.

7.10 Mass spectrum of fermion fields

The mass spectrum of fermion fields for $\theta_H \neq 0$ is determined in the twisted gauge in a similar manner. The brane interactions affect the spectrum of down-type quark multiplets and neutrino multiplets.

7.10.1 *Quarks*

The components of $SO(5)$ spinor fermions $\Psi^{q,\alpha}_{(\mathbf{3},\mathbf{4})}$ and $\Psi^{\ell,\alpha}_{(\mathbf{1},\mathbf{4})}$ $(\alpha = 1, 2, 3)$ in the original and twisted gauges are related to each other by

$$\chi = \begin{pmatrix} \cos\frac{1}{2}\theta(z) & i\sin\frac{1}{2}\theta(z) \\ i\sin\frac{1}{2}\theta(z) & \cos\frac{1}{2}\theta(z) \end{pmatrix} \tilde{\chi} , \tag{7.83}$$

where χ is given, for the first generation $(\alpha = 1)$, by

$$\chi = \begin{pmatrix} u \\ u' \end{pmatrix}, \begin{pmatrix} d \\ d' \end{pmatrix}, \begin{pmatrix} \nu_e \\ \nu'_e \end{pmatrix}, \begin{pmatrix} e \\ e' \end{pmatrix}. \tag{7.84}$$

$T^{45} = \frac{1}{2}\sigma^1$ for these χ's.

A. Up-type quark

The equations of motion for the up-type quarks, $(\check{u}, \check{u}') = z^{-2}(u, u')$, in the twisted gauge are given, from Eq. (7.30), by

$$\begin{aligned} & - kD_-(c_u) \begin{pmatrix} \tilde{\check{u}}_R \\ \tilde{\check{u}}'_R \end{pmatrix} + \sigma^\mu D_\mu \begin{pmatrix} \tilde{\check{u}}_L \\ \tilde{\check{u}}'_L \end{pmatrix} = 0 , \\ & - kD_+(c_u) \begin{pmatrix} \tilde{\check{u}}_L \\ \tilde{\check{u}}'_L \end{pmatrix} + \bar{\sigma}^\mu D_\mu \begin{pmatrix} \tilde{\check{u}}_R \\ \tilde{\check{u}}'_R \end{pmatrix} = 0 . \end{aligned} \tag{7.85}$$

Parity of $(\tilde{u}, \tilde{u}')$ at $z = z_L$ is the same as that of (u, u') since $\theta(z_L) = 0$. In particular, $(\tilde{\check{u}}_R, \tilde{\check{u}}'_L)$ satisfies the Dirichlet (D) condition $(\tilde{\check{u}}_R, \tilde{\check{u}}'_L)_{z_L} = 0$. It follows from Eq. (7.85) that

$$\begin{bmatrix} \tilde{\check{u}}_R \\ D_-(c_u)\, \tilde{\check{u}}'_R \end{bmatrix} = 0 \,, \quad \begin{bmatrix} D_+(c_u)\, \tilde{\check{u}}_L \\ \tilde{\check{u}}'_L \end{bmatrix} = 0 \quad \text{at } z = z_L \,. \tag{7.86}$$

Then the mode functions satisfying (7.85) and (7.86) are expressed as

$$\begin{aligned} \begin{pmatrix} \tilde{\check{u}}_R \\ \tilde{\check{u}}'_R \end{pmatrix} &= \begin{pmatrix} \alpha_u S_R(z;\lambda,c_u) \\ \alpha_{u'} C_R(z;\lambda,c_u) \end{pmatrix} f_R(x) \,, \\ \begin{pmatrix} \tilde{\check{u}}_L \\ \tilde{\check{u}}'_L \end{pmatrix} &= \begin{pmatrix} \alpha_u C_L(z;\lambda,c_u) \\ \alpha_{u'} S_L(z;\lambda,c_u) \end{pmatrix} f_L(x) \,, \end{aligned} \tag{7.87}$$

where $\bar{\sigma}\partial f_R(x) = k\lambda f_L(x)$ and $\sigma\partial f_L(x) = k\lambda f_R(x)$. Both right- and left-handed modes have the same coefficients α_u and $\alpha_{u'}$ as a consequence of Eq. (7.85).

The boundary conditions at $z = 1$ are $D_+\check{u}_L = 0$, $\check{u}_R = 0$, $\check{u}'_L = 0$, and $D_-\check{u}'_R = 0$. Making use of (7.83) and applying (7.87) for $\check{u}_R = 0$ and $D_-\check{u}'_R = 0$, one finds

$$K_u \begin{pmatrix} \alpha_u \\ \alpha_{u'} \end{pmatrix} = \begin{pmatrix} \cos\frac{1}{2}\theta_H S^u_R & i\sin\frac{1}{2}\theta_H C^u_R \\ i\sin\frac{1}{2}\theta_H C^u_L & \cos\frac{1}{2}\theta_H S^u_L \end{pmatrix} \begin{pmatrix} \alpha_u \\ \alpha_{u'} \end{pmatrix} = 0 \,, \tag{7.88}$$

where $S^u_{L/R} = S_{L/R}(1,\lambda,c_u)$ etc. Here the relations $D_-C_R = \lambda S_L$ etc. in (C.17) have been made use of. The spectrum of the up (u) quark tower is determined by $\det K_u = 0$.

$$u \;:\; S_L S_R(1,\lambda_{u^{(n)}},c_u) + \sin^2\frac{\theta_H}{2} = 0 \quad (n \geq 0) \,. \tag{7.89}$$

The mass $m_Q = k\lambda_Q$ of the lowest mode of the up-type quark $Q = (u, c, t)$ is approximately given by

$$m_Q \sim \begin{cases} \pi^{-1}\sqrt{1-4c_Q^2}\, \sin\frac{1}{2}\theta_H\, m_{\rm KK} & \text{for } |c_Q| < \frac{1}{2} \,, \\ \pi^{-1}\sqrt{4c_Q^2 - 1}\, z_L^{-|c_Q|+0.5} \sin\frac{1}{2}\theta_H\, m_{\rm KK} & \text{for } |c_Q| > \frac{1}{2} \,. \end{cases} \tag{7.90}$$

Note that $S_L(z;\lambda,-c) = -S_R(z;\lambda,c)$ and $C_L(z;\lambda,-c) = C_R(z;\lambda,c)$. With given m_Q, there are two solutions to (7.89); $c_Q > 0$ and $c_Q < 0$. For $m_{\rm KK} = 13\,\text{TeV}$, $\theta_H = 0.1$ and $z_L = 3.83 \times 10^{11}$, one finds that $(|c_u|, |c_c|, |c_t|) = (0.859, 0.719, 0.275)$. The spectrum of the top (t) quark tower is depicted in Fig. 7.1. Notice that the mass m_Q is suppressed exponentially $(\propto z_L^{-|c_Q|+0.5})$ for $|c_Q| > \frac{1}{2}$, which naturally explains the large mass hierarchy in the quark sector.

B. Down-type quark

In the down-type quark sector, the fields $\Psi^{q,\alpha}_{(\mathbf{3},\mathbf{4})}$ and $\Psi^{q,\alpha}_{(\mathbf{3},\mathbf{1})^\pm}$ intertwine with each other through the brane interaction $\mathcal{L}_1$ in (7.44), namely $\mathcal{L}_1^m$ in (7.45). In the first generation d, d', D_d^+ and D_d^- fields intertwine. The equations of motion in the original gauge are given by

$$
\begin{aligned}
&\begin{matrix}(a)\\(b)\end{matrix} \quad -i\delta \begin{pmatrix} d_L^\dagger \\ d_L'^\dagger \end{pmatrix} : -k\hat{D}_-(c_u)\begin{pmatrix} \check{d}_R \\ \check{d}'_R \end{pmatrix} + \sigma^\mu\partial_\mu \begin{pmatrix} \check{d}_L \\ \check{d}'_L \end{pmatrix} = 0 \,, \\
&\begin{matrix}(c)\\(d)\end{matrix} \quad i\delta \begin{pmatrix} d_R^\dagger \\ d_R'^\dagger \end{pmatrix} : \overline{\sigma}^\mu\partial_\mu \begin{pmatrix} \check{d}_R \\ \check{d}'_R \end{pmatrix} - k\hat{D}_+(c_u)\begin{pmatrix} \check{d}_L \\ \check{d}'_L \end{pmatrix} = 2\mu_d\,\delta(y)\begin{pmatrix} 0 \\ \check{D}_{dL}^+ \end{pmatrix}, \\
&(e) \quad -i\delta D_{dL}^{+\dagger} : -k\hat{D}_-(c_{D_d})\check{D}_{dR}^+ + \sigma^\mu\partial_\mu\check{D}_{dL}^+ - \frac{m_{D_d}}{z}\check{D}_{dR}^- = 2\mu_d^*\,\delta(y)\check{d}'_R \,, \\
&(f) \quad i\delta D_{dR}^{+\dagger} : \overline{\sigma}^\mu\partial_\mu\check{D}_{dR}^+ - k\hat{D}_+(c_{D_d})\check{D}_{dL}^+ - \frac{m_{D_d}}{z}\check{D}_{dL}^- = 0 \,, \\
&(g) \quad -i\delta D_{dL}^{-\dagger} : -k\hat{D}_-(c_{D_d})\check{D}_{dR}^- + \sigma^\mu\partial_\mu\check{D}_{dL}^- - \frac{m_{D_d}^*}{z}\check{D}_{dR}^+ = 0 \,, \\
&(h) \quad i\delta D_{dR}^{-\dagger} : \overline{\sigma}^\mu\partial_\mu\check{D}_{dR}^- - k\hat{D}_+(c_{D_d})\check{D}_{dL}^- - \frac{m_{D_d}^*}{z}\check{D}_{dL}^+ = 0 \,.
\end{aligned} \tag{7.91}
$$

Here

$$
\hat{D}_\pm(c) = \pm\Big(\frac{\partial}{\partial z} - i\theta'(z)T^{45}\Big) + \frac{c}{z} \,, \tag{7.92}
$$

and we have assumed that the mass term and brane interaction are diagonal in generations, namely $m_D^{\alpha\beta} = \delta^{\alpha\beta}m_{D_\alpha}$ and $\mu_q^{\alpha\beta} = \delta^{\alpha\beta}\mu_\alpha$ $(\alpha = d, s, b)$.

Under the parity transformation around $y = 0$, $\Psi_+ = d_L, d'_R, D_{dL}^+, D_{dR}^-$ are parity even whereas $\Psi_- = d_R, d'_L, D_{dR}^+, D_{dL}^-$ are parity odd. Note that $\Psi_-(y)\big|_{-\epsilon}^{+\epsilon} = 2\Psi_-(+\epsilon)$ and

$$
D_\pm(c) = \frac{e^{-\sigma(y)}}{k}\Big\{\pm\frac{\partial}{\partial y} + c\sigma'(y)\Big\} \tag{7.93}
$$

in the y coordinate. We integrate the equations for parity odd fields, $(a), (d), (e), (h)$ in (7.91), from $y = -\epsilon$ to $+\epsilon$ to find

$$
\begin{aligned}
&(a) \ \Rightarrow \check{d}_R(\epsilon) = 0 \,, \\
&(d) \ \Rightarrow \ -2\check{d}'_L(\epsilon) - 2\mu_d\,\check{D}_{dL}^+(0) = 0 \,, \\
&(e) \ \Rightarrow 2\check{D}_{dR}^+(\epsilon) - 2\mu_d^*\,\check{d}'_R(0) = 0 \,, \\
&(h) \ \Rightarrow \check{D}_{dL}^-(\epsilon) = 0 \,.
\end{aligned} \tag{7.94}
$$

For parity-even fields, we evaluate the equations at $y = +\epsilon$ by using the relations (7.94):

$$\begin{aligned}
&(c) \ \Rightarrow \hat{D}_+(c_u)\check{d}_L = 0 \ , \\
&(b) \ \Rightarrow \mu_d \left[\hat{D}_-(c_{D_d})\check{D}^+_{dR} + \tilde{m}_{D_d}\check{D}^-_{dR}\right] + \hat{D}_-(c_u)\check{d}'_R = 0 \ , \\
&(f) \ \Rightarrow \mu_d^* \hat{D}_+(c_u)\check{d}'_L - \hat{D}_+(c_{D_d})\check{D}^+_{dL} = 0 \ , \\
&(g) \ \Rightarrow \hat{D}_-(c_{D_d})\check{D}^-_{dR} + \tilde{m}^*_{D_d}\mu_d^*\check{d}'_R = 0 \ ,
\end{aligned} \tag{7.95}$$

where the equations of motion (e) and (d) at $y = +\epsilon$ have been made use of. Relations (7.94) and (7.95) specify the boundary conditions at $z = 1^+$.

The BCs at $z = z_L$ are given by

$$\begin{cases} d_R = 0, \\ D_+(c_u)d_L = 0, \\ D_-(c_u)d'_R = 0, \\ d'_L = 0, \end{cases} \qquad \begin{cases} D^+_{dR} = 0, \\ D_+(c_{D_d})D^+_{dL} = 0, \\ D_-(c_{D_d})D^-_{dR} = 0, \\ D^-_{dL} = 0. \end{cases} \tag{7.96}$$

In the twisted gauge the fields obey free equations and the boundary conditions at $z = z_L$ are the same as in the original gauge. The mode functions for d and d' take the same form as those of u and u'. $D^{\pm}_d$ fields obey the equations with mass terms, being expressed in terms of $\mathcal{C}_{L/Rj}(z;\lambda,c,\tilde{m})$ and $\mathcal{S}_{L/Rj}(z;\lambda,c,\tilde{m})$ defined in (C.28) and (C.29). One finds that

$$\begin{pmatrix} \tilde{\check{d}}_R \\ \tilde{\check{d}}'_R \\ \tilde{\check{D}}^+_{dR} \\ \tilde{\check{D}}^-_{dR} \end{pmatrix} = \begin{pmatrix} \alpha_d S_R(z;\lambda,c_u) \\ \beta_d C_R(z;\lambda,c_u) \\ a_d \mathcal{S}_{R2}(z;\lambda,c_{D_d},\tilde{m}_{D_d}) + b_d \mathcal{S}_{R1}(z;\lambda,c_{D_d},\tilde{m}_{D_d}) \\ a_d \mathcal{C}_{R1}(z;\lambda,c_{D_d}),\tilde{m}_{D_d}) + b_d \mathcal{C}_{R2}(z;\lambda,c_{D_d},\tilde{m}_{D_d}) \end{pmatrix},$$

$$\begin{pmatrix} \tilde{\check{d}}_L \\ \tilde{\check{d}}'_L \\ \tilde{\check{D}}^+_L \\ \tilde{\check{D}}^-_L \end{pmatrix} = \begin{pmatrix} \alpha_d C_L(z;\lambda,c_u) \\ \beta_d S_L(z;\lambda,c_u) \\ a_d \mathcal{C}_{L2}(z;\lambda,c_{D_d},\tilde{m}_{D_d}) + b_d \mathcal{C}_{L1}(z;\lambda,c_{D_d},\tilde{m}_{D_d}) \\ a_d \mathcal{S}_{L1}(z;\lambda,c_{D_d},\tilde{m}_{D_d}) + b_d \mathcal{S}_{L2}(z;\lambda,c_{D_d},\tilde{m}_{D_d}) \end{pmatrix}, \tag{7.97}$$

where $\tilde{m}_{D_d} = m_{D_d}/k$ and α_d, β_d, a_d, b_d are parameters to be determined.

Boundary conditions at $z = 1^+$ for the left-handed fields $\check{d}_L, \check{d}'_L, \check{D}^+_L, \check{D}^-_L$ are found by inserting (7.97) into (c), (d), (f) and (h) of Eqs. (7.94) and (7.95). One finds

$$K_d \begin{pmatrix} \alpha_d \\ \beta_d \\ a_d \\ b_d \end{pmatrix} = 0 \, ,$$

$$K_d = \begin{pmatrix} \cos \frac{1}{2}\theta_H S_R^u & i \sin \frac{1}{2}\theta_H C_R^u & 0 & 0 \\ i \sin \frac{1}{2}\theta_H C_L^u & \cos \frac{1}{2}\theta_H S_L^u & \mu_d \mathcal{C}_{L2}^{D_d} & \mu_d \mathcal{C}_{L1}^{D_d} \\ i\mu_d^* \sin \frac{1}{2}\theta_H S_R^u & \mu_d^* \cos \frac{1}{2}\theta_H C_R^u & -\mathcal{S}_{R2}^{D_d} & -\mathcal{S}_{R1}^{D_d} \\ 0 & 0 & \mathcal{S}_{L1}^{D_d} & \mathcal{S}_{L2}^{D_d} \end{pmatrix} , \tag{7.98}$$

where $S_{L/R}^u = S_{L/R}(1;\lambda, c_u)$ and $\mathcal{S}_{L/Rj}^{D_d} = \mathcal{S}_{L/Rj}(1;\lambda, c_{D_d}, \tilde{m}_{D_d})$ etc. The spectrum of the down (d, D_d) quark tower is determined by $\det K_d = 0$:

$$\begin{aligned} d, D_d : \ \det K_d = & \Big(S_L^u S_R^u + \sin^2 \frac{\theta_H}{2} \Big) \big(\mathcal{S}_{L1}^{D_d} \mathcal{S}_{R1}^{D_d} - \mathcal{S}_{L2}^{D_d} \mathcal{S}_{R2}^{D_d} \big) \\ & + |\mu_d|^2 C_R^u S_R^u \big(\mathcal{S}_{L1}^{D_d} \mathcal{C}_{L1}^{D_d} - \mathcal{S}_{L2}^{D_d} \mathcal{C}_{L2}^{D_d} \big) = 0 \, . \end{aligned} \tag{7.99}$$

The spectrum contains both the d tower, $\{d^{(n)}, n \geq 0\}$, and the D_d tower, $\{D_d^{(n)}, n \geq 1\}$.

The brane interaction term with μ_α is necessary to lift the degeneracy between the up-type and down-type quark masses. Its coefficient contains the factor $C_R^Q = C_R(1;\lambda, c_Q)$. For the first and second generations $|c_Q| = |c_u|, |c_c| > \frac{1}{2}$. For $\lambda z_L \ll 1$, $C_R(1;\lambda, c) \sim z_L^{-c} \ll 1$ for $c > \frac{1}{2}$ and $C_R(1;\lambda, c) \gg 1$ for $c < -\frac{1}{2}$. The detailed study shows that with $c > \frac{1}{2}$ Eq. (7.99) necessarily yields the first KK mode with a mass much less than $m_{\rm KK}$, which contradicts experimental observation. One needs to take $c_u, c_c < 0$. For the third generation $|c_t| < \frac{1}{2}$, and this problem does not occur.

Typical values of the parameters reproducing the quark mass spectrum (except for m_d) are tabulated in Table 7.5. It is shown that generally the spectrum determined by (7.99) leads to $m_d < m_u$, $m_s < m_c$ and $m_b < m_t$. It is not clear whether or not radiative corrections yield the spectrum $m_d > m_u$. Additional interactions or more general masses $m_V^{\alpha\beta}$ for dark fermions may be necessary.

Table 7.5 Typical parameters which reproduce the spectrum of three generations of quarks ($\alpha = 1, 2, 3$) for $\theta_H = 0.1$ and $m_{\rm KK} = 13$ TeV. The values $m_u = 20$ MeV, $m_d = 2.90$ MeV, $m_s = 55$ MeV, $m_c = 0.619$ GeV, $m_b = 2.89$ GeV, and $m_t = 172.9$ GeV have been used. In the first generation the values $m_u > m_d$ have been set for the reason explained in the text. μ_α and $\tilde{m}_{D_\alpha}$ are input parameters. The predicted masses of the first KK modes of u_α, d_α and D_α are also shown.

Quarks	c_α	μ_α	c_{D_α}	$\tilde{m}_{D_\alpha}$	$m_{u_\alpha^{(1)}}$ (TeV)	$m_{d_\alpha^{(1)}}$ (TeV)	$m_{D_\alpha^{(1)}}$ (TeV)
(u, d)	-0.8591	0.1	0.6244	1.0	12.17	12.17	16.65
(c, s)	-0.7191	0.1	0.6563	1.0	11.32	11.32	16.87
(t, b)	-0.2745	1.0	0.8725	1.0	11.35	8.47	18.25

7.10.2 *Leptons*

The mass spectrum in the lepton sector is determined in a similar manner.

A. Charged lepton

In the absence of the brane interaction $\mathcal{L}_3$ in (7.44), the equations of motion for charged leptons take the same form as those for the up-type quarks. The parity (P_0, P_1) is also the same. The mode functions for the first generation (e, e') in the twisted gauge are expressed as

$$\begin{pmatrix} \check{\tilde{e}}_R \\ \check{\tilde{e}}'_R \end{pmatrix} = \begin{pmatrix} \alpha_e S_R(z; \lambda, c_e) \\ \alpha_{e'} C_R(z; \lambda, c_e) \end{pmatrix} f_R(x) \ ,$$

$$\begin{pmatrix} \check{\tilde{e}}_L \\ \check{\tilde{e}}'_L \end{pmatrix} = \begin{pmatrix} \alpha_e C_L(z; \lambda, c_e) \\ \alpha_{e'} S_L(z; \lambda, c_e) \end{pmatrix} f_L(x) \ , \tag{7.100}$$

as in (7.87). To satisfy the boundary conditions at $z = 1$

$$K_e \begin{pmatrix} \alpha_e \\ \alpha_{e'} \end{pmatrix} = \begin{pmatrix} \cos\frac{1}{2}\theta_H S_R^e & i\sin\frac{1}{2}\theta_H C_R^e \\ i\sin\frac{1}{2}\theta_H C_L^e & \cos\frac{1}{2}\theta_H S_L^e \end{pmatrix} \begin{pmatrix} \alpha_e \\ \alpha_{e'} \end{pmatrix} = 0 \ , \tag{7.101}$$

must hold, and therefore the spectrum of the electron (e) tower is determined by $\det K_e = 0$.

$$e \ : \ S_L S_R(1, \lambda_{e^{(n)}}, c_e) + \sin^2\frac{\theta_H}{2} = 0 \quad (n \geq 0) \ . \tag{7.102}$$

For $m_{\rm KK} = 13$ TeV, $\theta_H = 0.1$ and $z_L = 3.83 \times 10^{11}$, one finds that $(|c_e|, |c_\mu|, |c_\tau|) = (1.0068, 0.7930, 0.6754)$.

B. Neutrino

In the neutrino sector the (ν, ν') components of $\Psi^{\ell,\alpha}_{(\mathbf{1},\mathbf{4})}(x,y)$ and the brane fermion $\hat{\chi}^\alpha(x)$ intertwine with each other through the brane interaction $\mathcal{L}_2$ in (7.44). Let us consider the case in which the brane masses $m_B^{\alpha\beta}$ in (7.45) and the Majorana masses $M_{\alpha\beta}$ in (7.34) are diagonal in generations. We set $m_B^{\alpha\beta} = m_{B\alpha}\delta^{\alpha\beta}$ and $M_{\alpha\beta} = -M_\alpha \delta_{\alpha\beta}$. Recall that $\hat{\chi}^\alpha = (\xi_\alpha, \eta_\alpha)$ satisfies the Majorana condition $(\hat{\chi}^\alpha)^c = \hat{\chi}^\alpha$ and therefore $\xi_\alpha = (\eta_\alpha)^c = e^{i\delta_C}\sigma^2\eta_\alpha^*$.

Equations of motion in the first generation are given by

$$
\begin{aligned}
&\begin{matrix}(a)\\(b)\end{matrix} \quad -k\hat{D}_-(c_e)\begin{pmatrix}\check{\nu}_{eR}\\ \check{\nu}'_{eR}\end{pmatrix} + \sigma^\mu\partial_\mu\begin{pmatrix}\check{\nu}_{eL}\\ \check{\nu}'_{eL}\end{pmatrix} = 0 \,, \\
&\begin{matrix}(c)\\(d)\end{matrix} \quad \overline{\sigma}^\mu\partial_\mu\begin{pmatrix}\check{\nu}_{eR}\\ \check{\nu}'_{eR}\end{pmatrix} - k\hat{D}_+(c_e)\begin{pmatrix}\check{\nu}_{eL}\\ \check{\nu}'_{eL}\end{pmatrix} = \frac{2m_{Be}}{\sqrt{k}}\,\delta(y)\begin{pmatrix}0\\ \eta_e\end{pmatrix}, \\
&(e) \quad \Big\{\sigma^\mu\partial_\mu\eta_e - \frac{m_{Be}}{\sqrt{k}}\nu'_{eR} + M_e\eta_e^c\Big\}\,\delta(y) = 0 \,. \qquad (7.103)
\end{aligned}
$$

ν_{eR} and ν'_{eL} have odd parity at $y = 0$, whereas ν_{eL} and ν'_{eR} have even parity. We integrate the equations (a), (d) in the vicinity of $y = 0$ and evaluate the equations (b), (c) at $y = +\epsilon$ to find boundary conditions at $y = +\epsilon$ as

$$
\begin{aligned}
&(a) \;\Rightarrow\; \check{\nu}_{eR}(x,\epsilon) = 0 \,, \\
&(d) \;\Rightarrow\; -\check{\nu}'_{eL}(x,\epsilon) = +\frac{m_{Be}}{\sqrt{k}}\,\eta_e(x) \,, \\
&(b) \;\Rightarrow\; -\hat{D}_-(c_e)\check{\nu}'_{eR} - \frac{m_{Be}^2}{k^2}\check{\nu}'_{eR} + \frac{m_{Be}M_e}{k^{3/2}}\eta_e^c = 0 \,, \\
&(c) \;\Rightarrow\; \hat{D}_+(c_e)\check{\nu}_{eL} = 0 \,. \qquad (7.104)
\end{aligned}
$$

Boundary conditions at $z = z_L$ are given by $D_+(c_e)\,\check{\nu}_{eL} = \check{\nu}_{eR} = 0$ and $\check{\nu}'_{eL} = D_-(c_e)\,\check{\nu}'_{eR} = 0$.

Mode functions of these fields in the twisted gauge can be written as

$$\begin{pmatrix} \tilde{\check{\nu}}_{eR} \\ \tilde{\check{\nu}}'_{eR} \\ \eta_e^c \end{pmatrix} = \begin{pmatrix} \alpha_\nu S_R(z;\lambda,c_e) \\ i\alpha_{\nu'} C_R(z;\lambda,c_e) \\ \mp i\alpha_\eta^*/\sqrt{k} \end{pmatrix} f_{\pm R}(x) ,$$

$$\begin{pmatrix} \tilde{\check{\nu}}_{eL} \\ \tilde{\check{\nu}}'_{eL} \\ \eta_e \end{pmatrix} = \begin{pmatrix} \alpha_\nu C_L(z;\lambda,c_e) \\ i\alpha_{\nu'} S_L(z;\lambda,c_e) \\ i\alpha_\eta/\sqrt{k} \end{pmatrix} f_{\pm L}(x) ,$$

$$\begin{aligned} &\bar{\sigma}^\mu \partial_\mu f_{\pm R}(x) = k\lambda f_{\pm L}(x) , \\ &\sigma^\mu \partial_\mu f_{\pm L}(x) = k\lambda f_{\pm R}(x) , \\ &f_{\pm L}(x)^c = e^{i\delta_C} \sigma^2 f_{\pm L}(x)^* = \pm f_{\pm R}(x) . \end{aligned} \tag{7.105}$$

Note that there are two types of solutions with $f_{\pm L}(x)^c = \pm f_{\pm R}(x)$. One can take $\alpha_\nu, \alpha_{\nu'}, \alpha_\eta$ to be real. In this case $\sigma^\mu \partial_\mu \eta = \mp k\lambda \eta^c$ is satisfied so that the equation (e) in Eq. (7.103) implies that

$$\frac{m_{Be}}{\sqrt{k}} \check{\nu}'_{eR}\Big|_{y=0} - (M_e \mp k\lambda)\eta_e^c = 0 . \tag{7.106}$$

With this identity the third relation in Eq. (7.104) can be rewritten as

$$\hat{D}_-(c_e)\check{\nu}'_{eR} \mp \frac{m_{Be}\lambda}{\sqrt{k}} \eta_e^c = 0 . \tag{7.107}$$

Substituting (7.105) into (7.104), one finds

$$K_{\nu_e}^{\pm} \begin{pmatrix} \alpha_\nu \\ \alpha_{\nu'} \\ \alpha_\eta \end{pmatrix} = 0 ,$$

$$K_{\nu_e}^{\pm} = \begin{pmatrix} \cos\frac{\theta_H}{2} S_R^e & -\sin\frac{\theta_H}{2} C_R^e & 0 \\ \sin\frac{\theta_H}{2} C_L^e & \cos\frac{\theta_H}{2} S_L^e & \frac{m_{Be}}{k} \\ -m_{Be}\sin\frac{\theta_H}{2} S_R^e & -m_{Be}\cos\frac{\theta_H}{2} C_R^e & k\lambda \mp M_e \end{pmatrix} , \tag{7.108}$$

where $S_{L/R}^e = S_{L/R}(1;\lambda,c_e)$ etc. The spectrum in the neutrino sector is determined by $\det K_{\nu_e}^{\pm} = 0$;

$$\det K_{\nu_e}^{\pm} = (k\lambda \mp M_e)\Big\{ S_L^e S_R^e + \sin^2\frac{\theta_H}{2} \Big\} + \frac{m_{Be}^2}{k} S_R^e C_R^e = 0 . \tag{7.109}$$

One of the solutions with $f_{+R/L}(x)$ or $f_{-R/L}(x)$ allows a small mass eigenvalue $m_{\nu_e} = k\lambda_{\nu_e} > 0$. For $M_e > 0$, the neutrino mode is obtained with

$f_{+R/L}(x)$. Noting $\lambda z_L \ll 1$ and $k\lambda \ll M_e$, one finds that the neutrino mass is approximately given by

$$m_{\nu_e} \simeq \begin{cases} \dfrac{m_e^2 M_e z_L^{2c_e+1}}{(2c_e+1)m_{Be}^2} & \text{for } c_e > \frac{1}{2} \,, \\ \dfrac{m_e^2 M_e}{(2|c_e|-1)m_{Be}^2} & \text{for } c_e < -\frac{1}{2} \,. \end{cases} \tag{7.110}$$

The gauge-Higgs seesaw mechanism for tiny neutrino masses is at work. [56, 60] In particular for $c_\alpha < -\frac{1}{2}$ very light neutrino masses $m_{\nu_\alpha} \sim m_\alpha^2 M_\alpha / m_{B\alpha}^2$ ($\alpha = e, \mu, \tau$) are generated by moderate values of M_α and $m_{B\alpha}$. For $c_\alpha > \frac{1}{2}$, however, extremely large $m_{B\alpha}$ well above the Planck mass is necessary to have a tiny neutrino mass.

The gauge-Higgs seesaw mechanism is in the category of the inverse seesaw mechanism in grand unified theories. [61–65] Relevant left-handed 4D fields are ν_{0L}, $\nu_{0R}'^c$ and η. They have mass terms characterized by

$$\frac{i}{2}(\nu_{0L}^{c\dagger}, \nu_{0R}'^{\dagger}, \eta^{c\dagger}) \begin{pmatrix} 0 & m_\ell & 0 \\ m_\ell & 0 & m_B \\ 0 & m_B & M \end{pmatrix} \begin{pmatrix} \nu_{0L} \\ \nu_{0R}'^c \\ \eta \end{pmatrix} + \text{H.c.} \,, \tag{7.111}$$

where m_ℓ is a mass of the corresponding charged lepton and M is a Majorana mass of η. (7.111) leads to a light neutrino mass $m_\ell^2 M / m_B^2$.

For $c_e < 0$ and $M_e > 0$, the spectrum determined by $\det K_{\nu_e}^+ = 0$ gives charge-conjugation (C) even series $\{\lambda_{\nu_e^{+(n)}}; n \geq 0\}$, and the spectrum determined by $\det K_{\nu_e}^- = 0$ gives C odd series $\{\lambda_{\nu_e^{-(n)}}; n \geq 1\}$. The $\nu_e^{+(0)}$ mode corresponds to the observed neutrino. For $n \geq 1$ the $\nu_e^{+(n)}$ and $\nu_e^{-(n)}$ modes are almost degenerate and heavy; $m_{\nu_e^{\pm(n)}} \sim m_{e^{(n)}}$ ($n \geq 1$). Typical values of the parameters in the lepton sector are tabulated in Table 7.6.

For $c_e > 0$ the first KK excited modes become very light. The mass of $\nu_{e^{\pm(1)}}$ turns out to be several MeV, which contradicts with observation. In both of the quark and lepton sectors the bulk mass parameters of $\Psi_{(\mathbf{3},\mathbf{4})}^{q,\alpha}$ and $\Psi_{(\mathbf{1},\mathbf{4})}^{\ell,\alpha}$ need to be negative.

7.10.3 *Dark fermions*

Dark fermion fields are necessary to have proper dynamical electroweak symmetry breaking. These fields are expected from the viewpoint of grand unification (see Section 9.8).

Table 7.6 Typical parameters which reproduce the spectrum of three generations of leptons ($\alpha = 1, 2, 3$) for $\theta_H = 0.1$ and $m_{\rm KK} = 13$ TeV. The values $m_e = 0.4866$ MeV, $m_\mu = 102.7$ MeV and $m_\tau = 1.746$ GeV have been used. For neutrino masses we have temporarily set $m_{\nu_\alpha} = 1$ meV.

Leptons	c_α	M_α (GeV)	$m_{B\alpha}$ (GeV)	$m_{\nu_\alpha^{\pm(1)}}$ (TeV)	$m_{e_\alpha^{(1)}}$ (TeV)
(ν_e, e)	-1.0068	1×10^6	4.8×10^5	13.04	13.04
(ν_μ, μ)	-0.7930	1×10^6	1.3×10^8	11.77	11.77
(ν_τ, τ)	-0.6754	1×10^6	2.9×10^9	11.05	11.05

A. $\Psi^{F_q}_{(\mathbf{3},\mathbf{4})}$ and $\Psi^{F_\ell}_{(\mathbf{1},\mathbf{4})}$

Dark fermion fields $\Psi^{F_q}_{(\mathbf{3},\mathbf{4})}$ and $\Psi^{F_\ell}_{(\mathbf{1},\mathbf{4})}$ satisfy the boundary conditions (7.46) and their action is given by (7.47). The charge $(Q_X, Q_{\rm EM})$ assignments of $\Psi^{F_q}_{(\mathbf{3},\mathbf{4})}$ and $\Psi^{F_\ell}_{(\mathbf{1},\mathbf{4})}$ are the same as those of $\Psi^{q}_{(\mathbf{3},\mathbf{4})}$ and $\Psi^{\ell}_{(\mathbf{1},\mathbf{4})}$, respectively. Note that $(Q^{SO(5)}_\alpha, Q_{X\alpha}) = (1, \frac{1}{6})$ for $\Psi^{F_q,\alpha}_{(\mathbf{3},\mathbf{4})}$ and $(1, -\frac{1}{2})$ for $\Psi^{F_\ell,\alpha}_{(\mathbf{1},\mathbf{4})}$.

Let us denote the four components of the $SO(5)$ spinor $\Psi^{F_q}_{(\mathbf{3},\mathbf{4})}$ as $(F_u, F_d, F_{u'}, F_{d'})$. The equations of motion for the up-type dark fermions $(\check F_u, \check F_{u'}) = z^{-2}(F_u, F_{u'})$ in the twisted gauge are the same as (7.85) where (u, u') is replaced by $(F_u, F_{u'})$. From (7.46) the boundary conditions of $(\check F_u, \check F_{u'})$ at $z = z_L$ are the same as those of $(\check u, \check u')$ so that

$$\begin{bmatrix} \check{\tilde F}_{uR} \\ D_-(c_{F_u})\, \check{\tilde F}_{u'R} \end{bmatrix} = 0 \,, \quad \begin{bmatrix} D_+(c_{F_u})\, \check{\tilde F}_{uL} \\ \check{\tilde F}_{u'L} \end{bmatrix} = 0 \quad \text{at } z = z_L \,, \tag{7.112}$$

and the mode functions satisfying the equations of motion and (7.112) are expressed as

$$\begin{aligned} \begin{pmatrix} \check{\tilde F}_{uR} \\ \check{\tilde F}_{u'R} \end{pmatrix} &= \begin{pmatrix} \alpha_{F_u} S_R(z;\lambda, c_{F_u}) \\ \alpha_{F_{u'}} C_R(z;\lambda, c_{F_u}) \end{pmatrix} f_R(x) \,, \\ \begin{pmatrix} \check{\tilde F}_{uL} \\ \check{\tilde F}_{u'L} \end{pmatrix} &= \begin{pmatrix} \alpha_{F_u} C_L(z;\lambda, c_{F_u}) \\ \alpha_{F_{u'}} S_L(z;\lambda, c_{F_u}) \end{pmatrix} f_L(x) \,, \end{aligned} \tag{7.113}$$

The boundary conditions at $z = 1$ become $\check F_{uL} = 0$, $D_- \check F_{uR} = 0$, $D_+ \check F_{u'L} = 0$, and $\check F_{u'R} = 0$. Making use of (7.83) with $\chi = (F_u, F_{u'})$ and applying (7.113) for $D_-\check{F}_{uR} = 0$ and $\check{F}_{u'R} = 0$, one finds

$$K_{F_u} \begin{pmatrix} \alpha_{F_u} \\ \alpha_{F_{u'}} \end{pmatrix} = \begin{pmatrix} i \sin \frac{1}{2}\theta_H S_R^{F_u} & \cos \frac{1}{2}\theta_H C_R^{F_u} \\ \cos \frac{1}{2}\theta_H C_L^{F_u} & i \sin \frac{1}{2}\theta_H S_L^{F_u} \end{pmatrix} \begin{pmatrix} \alpha_{F_u} \\ \alpha_{F_{u'}} \end{pmatrix} = 0 \,, \tag{7.114}$$

where $S_{L/R}^{F_u} = S_{L/R}(1, \lambda, c_{F_u})$ etc. The spectrum of the up-type dark fermion (F_u) tower is determined by $\det K_{F_u} = 0$.

$$F_u \; : \; S_L S_R(1, \lambda_{F_u^{(n)}}, c_{F_u}) + \cos^2 \frac{\theta_H}{2} = 0 \quad (n \geq 1) \,. \tag{7.115}$$

The down-type dark fermions $(F_d, F_{d'})$ satisfy the same equations of motion and boundary conditions as the up-type dark fermions $(F_u, F_{u'})$. The mode functions are given by

$$\begin{pmatrix} \check{\tilde{F}}_{dR} \\ \check{\tilde{F}}_{d'R} \end{pmatrix} = \begin{pmatrix} \alpha_{F_d} S_R(z; \lambda, c_{F_u}) \\ \alpha_{F_{d'}} C_R(z; \lambda, c_{F_u}) \end{pmatrix} f_R(x) \,,$$

$$\begin{pmatrix} \check{\tilde{F}}_{dL} \\ \check{\tilde{F}}_{d'L} \end{pmatrix} = \begin{pmatrix} \alpha_{F_d} C_L(z; \lambda, c_{F_u}) \\ \alpha_{F_{d'}} S_L(z; \lambda, c_{F_u}) \end{pmatrix} f_L(x) \,,$$

$$K_{F_d} \begin{pmatrix} \alpha_{F_d} \\ \alpha_{F_{d'}} \end{pmatrix} = 0 \,, \quad K_{F_d} = K_{F_u} \,, \tag{7.116}$$

and therefore the spectrum $\{\lambda_{F_d^{(n)}}\}$ is the same as $\{\lambda_{F_u^{(n)}}\}$;

$$F_d \; : \; S_L S_R(1, \lambda_{F_d^{(n)}}, c_{F_u}) + \cos^2 \frac{\theta_H}{2} = 0 \quad (n \geq 1) \,. \tag{7.117}$$

Notice that the lowest modes, $F_u^{(1)}$ and $F_d^{(1)}$, become massless at $\theta_H = \pi$.

$\Psi_{(\mathbf{1},\mathbf{4})}^{F_\ell}$ satisfies the same form of the equations of motion and the same boundary conditions as $\Psi_{(\mathbf{3},\mathbf{4})}^{F_q}$. Let us denote the bulk mass parameter of $\Psi_{(\mathbf{1},\mathbf{4})}^{F_\ell,1} = (F_{\nu_e}, F_e, F_{\nu_e'}, F_{e'})$ as c_{F_e}. The mode functions are given in the same form as (7.113) and (7.116) where $(F_u, F_d, F_{u'}, F_{d'})$ and c_{F_u} are replaced by $(F_{\nu_e}, F_e, F_{\nu_e'}, F_{e'})$ and c_{F_e}. The spectrum $\{\lambda_{F_{\nu_e}^{(n)}}\} = \{\lambda_{F_e^{(n)}}\}$ is determined by

$$F_{\nu_e}, F_e \; : \; S_L S_R(1, \lambda_{F_e^{(n)}}, c_{F_e}) + \cos^2 \frac{\theta_H}{2} = 0 \quad (n \geq 1) \,. \tag{7.118}$$

B. $\Psi_{(\mathbf{1},\mathbf{5})^\pm}^{V}$

The action and boundary conditions of dark fermion fields $\Psi_{(\mathbf{1},\mathbf{5})^\pm}^{V,\alpha}$ are given by (7.47) and (7.46), respectively. We consider the case in which the mass terms are diagonal; $m_V^{\alpha\beta} = m_V^\alpha \delta^{\alpha\beta}$ ($m_V^\alpha = m_V^{\alpha *}$). Let us write as $\Psi^{V,\alpha}_{(\mathbf{1},\mathbf{5})^\pm} = \Psi^\pm$. $\tilde{\Psi}^\pm$ in the twisted gauge is related to $\Psi^\pm$ by

$$\Psi^\pm = \begin{pmatrix} \Psi_1^\pm \\ \Psi_2^\pm \\ \Psi_3^\pm \\ \Psi_4^\pm \\ \Psi_5^\pm \end{pmatrix} = \tilde{\Omega}(z)^{-1} \tilde{\Psi}^\pm = \begin{pmatrix} \tilde{\Psi}_1^\pm \\ \tilde{\Psi}_2^\pm \\ \tilde{\Psi}_3^\pm \\ \cos\theta(z) \tilde{\Psi}_4^\pm + \sin\theta(z) \tilde{\Psi}_5^\pm \\ -\sin\theta(z) \tilde{\Psi}_4^\pm + \cos\theta(z) \tilde{\Psi}_5^\pm \end{pmatrix} . \tag{7.119}$$

Equations of motion for $\check{\tilde{\Psi}}^\pm = z^{-2} \tilde{\Psi}^\pm$ are

$$\begin{aligned}
& - kD_-(c_V) \check{\tilde{\Psi}}_R^+ + \sigma^\mu \partial_\mu \check{\tilde{\Psi}}_L^+ - \frac{m_V}{z} \check{\tilde{\Psi}}_R^- = 0 , \\
& \overline{\sigma}^\mu \partial_\mu \check{\tilde{\Psi}}_R^+ - kD_+(c_V) \check{\tilde{\Psi}}_L^+ - \frac{m_V}{z} \check{\tilde{\Psi}}_L^- = 0 , \\
& - kD_-(c_V) \check{\tilde{\Psi}}_R^- + \sigma^\mu \partial_\mu \check{\tilde{\Psi}}_L^- - \frac{m_V}{z} \check{\tilde{\Psi}}_R^+ = 0 , \\
& \overline{\sigma}^\mu \partial_\mu \check{\tilde{\Psi}}_R^- - kD_+(c_V) \check{\tilde{\Psi}}_L^- - \frac{m_V}{z} \check{\tilde{\Psi}}_L^+ = 0 .
\end{aligned} \tag{7.120}$$

Here $c_V = c_{V_\alpha}$ and $m_V = m_V^\alpha$. Boundary conditions at $z = z_L$ are

$$\begin{aligned}
& D_-(c_V) \check{\tilde{\Psi}}_{jR}^+ = \check{\tilde{\Psi}}_{jL}^+ = 0 , \\
& \check{\tilde{\Psi}}_{jR}^- = D_+(c_V) \check{\tilde{\Psi}}_{jL}^- = 0 , \quad (j = 1, \cdots, 4), \\
& \check{\tilde{\Psi}}_{5R}^+ = D_+(c_V) \check{\tilde{\Psi}}_{5L}^+ = 0 , \\
& D_-(c_V) \check{\tilde{\Psi}}_{5R}^- = \check{\tilde{\Psi}}_{5L}^- = 0 ,
\end{aligned} \tag{7.121}$$

so that mode functions can be written as

$$\begin{aligned}
& \begin{pmatrix} \check{\tilde{\Psi}}_{jR}^+ \\ \check{\tilde{\Psi}}_{jR}^- \end{pmatrix} = \begin{pmatrix} a_j \mathcal{C}_{R1}(z; \lambda, c_V, \tilde{m}_V) + b_j \mathcal{C}_{R2}(z; \lambda, c_V, \tilde{m}_V) \\ a_j \mathcal{S}_{R2}(z; \lambda, c_V, \tilde{m}_V) + b_j \mathcal{S}_{R1}(z; \lambda, c_V, \tilde{m}_V) \end{pmatrix} , \\
& \begin{pmatrix} \check{\tilde{\Psi}}_{jL}^+ \\ \check{\tilde{\Psi}}_{jL}^- \end{pmatrix} = \begin{pmatrix} a_j \mathcal{S}_{L1}(z; \lambda, c_V, \tilde{m}_V) + b_j \mathcal{S}_{L2}(z; \lambda, c_V, \tilde{m}_V) \\ a_j \mathcal{C}_{L2}(z; \lambda, c_V, \tilde{m}_V) + b_j \mathcal{C}_{L1}(z; \lambda, c_V, \tilde{m}_V) \end{pmatrix} , \\
& \qquad (j = 1, \cdots, 4) , \\
& \begin{pmatrix} \check{\tilde{\Psi}}_{5R}^+ \\ \check{\tilde{\Psi}}_{5R}^- \end{pmatrix} = \begin{pmatrix} a_5 \mathcal{S}_{R2}(z; \lambda, c_V, \tilde{m}_V) + b_5 \mathcal{S}_{R1}(z; \lambda, c_V, \tilde{m}_V) \\ a_5 \mathcal{C}_{R1}(z; \lambda, c_V, \tilde{m}_V) + b_5 \mathcal{C}_{R2}(z; \lambda, c_V, \tilde{m}_V) \end{pmatrix} , \\
& \begin{pmatrix} \check{\tilde{\Psi}}_{5L}^+ \\ \check{\tilde{\Psi}}_{5L}^- \end{pmatrix} = \begin{pmatrix} a_5 \mathcal{C}_{L2}(z; \lambda, c_V, \tilde{m}_V) + b_5 \mathcal{C}_{L1}(z; \lambda, c_V, \tilde{m}_V) \\ a_5 \mathcal{S}_{L1}(z; \lambda, c_V, \tilde{m}_V) + b_5 \mathcal{S}_{L2}(z; \lambda, c_V, \tilde{m}_V) \end{pmatrix} ,
\end{aligned} \tag{7.122}$$

where $\tilde{m}_V = m_V/k$ and $\mathcal{C}_{R/Lj}(z; \lambda, c, \tilde{m})$ etc. are defined in (C.28).

Orbifold conditions at $z = 1$ in the original gauge are

$$
\begin{aligned}
&P_0 = + : \ \check{\Psi}^+_{jR}, \check{\Psi}^-_{jL}, \check{\Psi}^+_{5L}, \check{\Psi}^-_{5R} \\
&P_0 = - : \ \check{\Psi}^+_{jL}, \check{\Psi}^-_{jR}, \check{\Psi}^+_{5R}, \check{\Psi}^-_{5L}, \quad (j = 1, \cdots, 4).
\end{aligned} \tag{7.123}
$$

As $\check{\Psi}^\pm_j = \tilde{\check{\Psi}}^\pm_j$ $(j = 1, 2, 3)$, Dirichlet conditions for $\check{\Psi}^+_{jL}$ and $\check{\Psi}^-_{jR}$ lead to

$$
\begin{pmatrix} \mathcal{S}_{L1} & \mathcal{S}_{L2} \\ \mathcal{S}_{R2} & \mathcal{S}_{R1} \end{pmatrix} \begin{pmatrix} a_j \\ b_j \end{pmatrix} = 0 \quad (j = 1, 2, 3) \tag{7.124}
$$

where $\mathcal{S}_{L1} = \mathcal{S}_{L1}(1; \lambda, c_V, \tilde{m}_V)$ etc. The spectrum is determined by

$$
\begin{aligned}
&\Psi^{V\pm}_j : \ \big[\mathcal{S}_{L1}\mathcal{S}_{R1} - \mathcal{S}_{L2}\mathcal{S}_{R2}\big](1; \lambda_{V_j^{(n)}}, c_V, \tilde{m}_V) = 0 \ , \\
&\qquad\qquad (n \geq 1, \ j = 1, 2, 3).
\end{aligned} \tag{7.125}
$$

It is seen that the Neumann conditions $D_+(c_V)\check{\Psi}^-_{jL} = D_-(c_V)\check{\Psi}^+_{jR} = 0$ at $z = 1$ are satisfied by using (C.29) and (7.124).

For $\check{\Psi}^\pm_4$ and $\check{\Psi}^\pm_5$ we use the relation (7.119). Inserting the mode functions in (7.122) into the Dirichlet conditions $\check{\Psi}^+_{4L}, \check{\Psi}^-_{4R}, \check{\Psi}^-_{5R}, \check{\Psi}^+_{5L} = 0$ at $z = 1$, one finds that

$$
\begin{aligned}
&K_V \begin{pmatrix} a_4 \\ b_4 \\ a_5 \\ b_5 \end{pmatrix} = 0 \ , \ K_V = \begin{pmatrix} \cos\theta_H A & \sin\theta_H B \\ -\sin\theta_H B & \cos\theta_H A \end{pmatrix}, \\
&A = \begin{pmatrix} \mathcal{S}_{L1} & \mathcal{S}_{L2} \\ \mathcal{S}_{R2} & \mathcal{S}_{R1} \end{pmatrix}, \quad B = \begin{pmatrix} \mathcal{C}_{L2} & \mathcal{C}_{L1} \\ \mathcal{C}_{R1} & \mathcal{C}_{R2} \end{pmatrix},
\end{aligned} \tag{7.126}
$$

where $\mathcal{S}_{L1} = \mathcal{S}_{L1}(1; \lambda, c_V, \tilde{m}_V)$ etc. It is checked that the Neumann conditions for $\check{\Psi}^+_{4R}, \check{\Psi}^-_{4L}, \check{\Psi}^-_{5L}, \check{\Psi}^+_{5R}$ at $z = 1$ are satisfied with the conditions (7.126).

The spectrum is determined by

$$
\det \hat{K}_V = 0 \ , \quad \hat{K}_V = \cos^2\theta_H B^{-1}A + \sin^2\theta_H A^{-1}B \ . \tag{7.127}
$$

It is straightforward to show that

$$
\begin{aligned}
&A^{-1}B = \frac{1}{\mathcal{B}_0 - 2}\begin{pmatrix} \alpha & \beta \\ \gamma & -\alpha \end{pmatrix}, \quad B^{-1}A = \frac{1}{\mathcal{B}_0 + 2}\begin{pmatrix} \alpha & \beta \\ \gamma & -\alpha \end{pmatrix}, \\
&\alpha = S^+_L S^-_R - S^-_L S^+_R - C^+_L C^-_R + C^-_L C^+_R \ , \\
&\beta = 2(S^+_R C^-_L + S^-_R C^+_L) \ , \\
&\gamma = 2(S^+_L C^-_R + S^-_L C^+_R) \ , \\
&\mathcal{B}_0(1; \lambda, c_V, \tilde{m}_V) = S^+_L S^-_R + S^-_L S^+_R + C^+_L C^-_R + C^-_L C^+_R \ , \\
&S^\pm_{L/R} = S_{L/R}(1; \lambda, c_V \pm \tilde{m}_V), \ C^\pm_{L/R} = C_{L/R}(1; \lambda, c_V \pm \tilde{m}_V).
\end{aligned} \tag{7.128}
$$

Here we have made use of the relations in (C.31). As $\alpha^2 + \beta\gamma = \mathcal{B}_0^2 - 4$, one finds that $\det \hat{K}_V = -(\mathcal{B}_0 - 2\cos 2\theta_H)^2/(\mathcal{B}_0^2 - 4)$. The spectrum of $(\Psi_4^{V\pm}, \Psi_5^{V\pm}) \equiv \Psi_N^V$ is determined by

$$\Psi_N^V : \ \mathcal{B}_0(1; \lambda_{V_N^{(n)}}, c_V, \tilde{m}_V) - 2\cos 2\theta_H = 0 \quad (n \geq 1). \tag{7.129}$$

For each $\lambda_{V_N^{(n)}}$, there are two linearly independent degenerate modes.

7.11 Effective potential $V_{\rm eff}(\theta_H)$ and electroweak symmetry breaking

In Sections 7.9 and 7.10 the equations which determine the mass spectrum of the fields have been found, with which one can evaluate the effective potential $V_{\rm eff}(\theta_H)$. Unlike in flat spacetime such as $M^4 \times S^1$ and $M^4 \times (S^1/Z_2)$ the mass spectrum $\{m_n\}$ in the RS warped space is determined by the location of the zeros of the equation which are expressed in terms of Bessel functions so that $\{m_n\}$ cannot be expressed in a simple form. Nevertheless $V_{\rm eff}(\theta_H)$ can be evaluated by the method explained in Appendix B.3. [47,66]

In examining the electroweak symmetry breaking, relevant is the θ_H-dependent part of $V_{\rm eff}(\theta_H)$. At the one loop level only the θ_H-dependent mass spectrum is important. In the GUT inspired $SO(5) \times U(1)_X \times SU(3)_C$ GHU the W tower (7.62), Z tower (7.70), (A_z^{a4}, A_z^{a5}) tower (7.81), quark towers (7.89) and (7.99), lepton towers (7.102) and (7.109), and dark fermion towers (7.115), (7.118) and (7.129) have θ_H-dependent mass spectra.

In general the KK mass spectrum $\{m_n(\theta_H) = k\lambda_n(\theta_H)\}$ is determined by the location of zeros of a certain function $\rho(\lambda; \theta_H)$;

$$\rho(\lambda_n; \theta_H) = 0 \ . \tag{7.130}$$

The contribution of this KK tower to $V_{\rm eff}(\theta_H)$ is given by

$$V_{\rm eff}(\theta_H) = \sum_n \pm \frac{1}{2} \int \frac{d^4 p_E}{(2\pi)^4} \ln \{p_E^2 + m_n(\theta_H)^2\} \quad \text{for} \begin{cases} \text{B} \\ \text{F} \end{cases} \tag{7.131}$$

where the sign is $+$ $(-)$ for bosons (fermions). By making use of the formula (B.23) in Appendix B, it is expressed as

$$V_{\rm eff}(\theta_H) = \pm \frac{(kz_L^{-1})^4}{16\pi^2} \int_0^\infty dq\, q^3 \ln \rho(iqz_L^{-1}; \theta_H) \ . \tag{7.132}$$

For instance, for the W and top quark towers, one finds from (7.62) and (7.89) that

$$\rho_W(\lambda;\theta_H) = 1 - \frac{\cos 2\theta_H}{1 + (4/\lambda) S(1;\lambda) C'(1;\lambda)} ,$$

$$\rho_t(\lambda;\theta_H) = 1 - \frac{\cos \theta_H}{1 + 2 S_L S_R(1;\lambda, c_t)} . \tag{7.133}$$

These functions are analytically continued to

$$\rho_W(iqz_L^{-1};\theta_H) = 1 - \frac{\cos 2\theta_H}{1 + 4R_0(q,\frac{1}{2})} ,$$

$$\rho_t(iqz_L^{-1};\theta_H) = 1 - \frac{\cos \theta_H}{1 + 2R_0(q,c_t)} ,$$

$$R_0(q,c) = \hat{S}_L \hat{S}_R(q,c) . \tag{7.134}$$

$\hat{S}_{L/R}(q,c)$ and $\hat{C}_{L/R}(q,c)$ are defined in (C.36).

The effective potential $V_{\text{eff}}(\theta_H)$ in the GUT inspired $SO(5) \times U(1)_X \times SU(3)_C$ GHU is given by [58]

$$\begin{aligned} V_{\text{eff}}(\theta_H) &\simeq 2(3-\xi^2) I[Q_W] + (3-\xi^2) I[Q_Z] + 3\xi^2 I[Q_S] \\ &\quad - 12 I[Q_t] - 12 I[Q_b] - 16 n_F I[Q_F] - 8 n_V I[Q_V] , \end{aligned} \tag{7.135}$$

where

$$I[Q] = \frac{(kz_L^{-1})^4}{16\pi^2} \int_0^\infty dq\, q^3 \ln \Big\{ 1 + Q(q;\theta_H) \Big\} ,$$

$$\{Q_W, Q_Z, Q_S\} = \left\{ \frac{-\cos 2\theta_H}{1 + 4R_0(q,\frac{1}{2})}, \frac{-\cos 2\theta_H}{1 + 4\cos^2\theta_W^0 R_0(q,\frac{1}{2})}, \frac{-\cos 2\theta_H}{1 + 2R_0(q,\frac{1}{2})} \right\},$$

$$Q_t = \frac{-\cos\theta_H}{1 + 2R_0(q,c_t)},$$

$$Q_b = \frac{-\cos\theta_H}{1 + 2R_0(q,c_t) + 2|\mu_b|^2 \hat{C}_R \hat{S}_R(q,c_t) \dfrac{\hat{\mathcal{B}}_1(q;c_{D_b},\tilde{m}_{D_b})}{\hat{\mathcal{B}}_0(q;c_{D_b},\tilde{m}_{D_b}) - 2}},$$

$$Q_F = \frac{\cos\theta_H}{1 + 2R_0(q,c_F)}, \quad Q_V = \frac{-2\cos 2\theta_H}{\hat{\mathcal{B}}_0(q,c_V,\tilde{m}_V)}. \tag{7.136}$$

Here $\hat{\mathcal{B}}_0(q;c,\tilde{m})$ and $\hat{\mathcal{B}}_1(q;c,\tilde{m})$ are defined in (C.39). For the sake of simplicity we set degenerate bulk mass parameters c_F for $\Psi^{F_q}_{\mathbf{(3,4)}}$ and $\Psi^{F_\ell}_{\mathbf{(1,4)}}$, and degenerate masses $m_V = k\tilde{m}_V$ and degenerate bulk mass parameters c_V for $\Psi^V_{(\mathbf{1},\mathbf{5})^\pm}$. The quark multiplets in the first and second generations and the lepton multiplets give negligible contributions to $V_{\rm eff}(\theta_H)$ as their bulk mass parameters (c) satisfy $|c| > \frac{1}{2}$. The dependence on the gauge parameter ξ appears, reflecting the ξ-dependence of the action (6.46). The factor 3 in front of $I[Q_W]$ and $I[Q_Z]$ for $\xi = 0$ is expected as W and Z are massive spin 1 particles. Note that the contributions from gauge fields and $\Psi^V_{(\mathbf{1},\mathbf{5})^\pm}$ are periodic in θ_H with a period π, whereas the contributions from quarks, leptons, $\Psi^{F_q}_{(\mathbf{3},\mathbf{4})}$ and $\Psi^{F_\ell}_{(\mathbf{1},\mathbf{4})}$ are periodic with a period 2π. The integration over q in $I[Q]$ above is mostly saturated in the range $0 \le q \le 10$.

The function $R_0(q;c)$ in (7.134) is positive definite for $q > 0$. In the absence of fermions $V_{\rm eff}(\theta_H)$ takes the minimum value at $\theta_H = 0$ and π so that the electroweak (EW) symmetry remains unbroken. The fermion contributions to $V_{\rm eff}(\theta_H)$ have an opposite sign. The contribution from the top quark multiplet, $-12I[Q_t]$, is periodic with a period 2π and becomes smallest at $\theta_H = \pi$. Its magnitude turns out larger than that of the gauge field contributions. Without dark fermions $V_{\rm eff}(\theta_H)$ becomes minimum at $\theta_H = \pi$ and the $SU(2)_R \times U(1)_{Y'}$ gauge symmetry remains unbroken. The contribution of dark fermions $\Psi^{F_q}_{(\mathbf{3},\mathbf{4})}$ and $\Psi^{F_\ell}_{(\mathbf{1},\mathbf{4})}$, $-16n_F I[Q_F]$, has a period 2π and becomes minimum at $\theta_H = 0$. The contribution of dark fermions $\Psi^V_{(\mathbf{1},\mathbf{5})^\pm}$, $-8n_V I[Q_V]$, has a period π and becomes minimum at $\theta_H = \pm\frac{1}{2}\pi$. With all contributions taken into account $V_{\rm eff}(\theta_H)$ in (7.135) can take a global minimum at a desired value of θ_H, say at $\theta_H \sim 0.1$, and the EW symmetry is dynamically broken to $U(1)_{\rm EM}$.

The 4D Higgs field $H(x)$ is nothing but the fluctuation mode of θ_H in (7.26). θ_H and $H(x)$ appear in the combination of

$$\hat{\theta}_H(x) = \theta_H + \frac{H(x)}{f_H} \tag{7.137}$$

where f_H is given in (7.26). Suppose that $V_{\rm eff}(\theta_H)$ has a global minimum at $\theta_H = \theta_H^{\rm min}$. The effective potential for the Higgs field $H(x)$ is given by $V_{\rm eff}[\theta_H^{\rm min} + f_H^{-1}H]$, and therefore the Higgs boson mass m_H is given by

$$m_H^2 = \frac{1}{f_H^2}\frac{d^2V_{\rm eff}(\theta_H)}{d\theta_H^2}\bigg|_{\theta_H^{\rm min}} . \tag{7.138}$$

Once the parameters of the theory are specified, $V_{\rm eff}(\theta_H)$ is evaluated and the value of θ_H is determined from the location of the global minimum of $V_{\rm eff}(\theta_H)$. In GHU the parameters of the theory must be determined such that the masses of the observed particles are reproduced. For instance the masses of the Z boson and top quark must satisfy (7.71) and (7.90), which depend on the values of the parameters other than θ_H. In other words the parameters of the theory must be set self-consistently.

We take the following procedure for fixing the parameters. To simplify the expression we write the effective potential as $V_{\rm eff}(\theta)$ and $\theta_H^{\rm min}$ as θ_H below. The masses m_Z, m_q (for quarks), m_ℓ (for leptons), the fine structure constant $\alpha_{\rm EM}$, and the Fermi constant G_F are input parameters. It will be shown in Section 9.1 that the bare weak mixing angle θ_W^0 in (7.72) is related to G_F. Temporarily we take g_w and $\sin\theta_W^0$, in place of $\alpha_{\rm EM}$ and G_F, as input quantities. In GHU θ_H and $m_{\rm KK}$ are two relevant additional parameters. We are going to see that with given $m_{\rm KK}$ the allowed range of θ_H is restricted and many of important physical quantities are mostly determined by the value of θ_H.

[Step 1] We pick the value of θ_H. In other words we are going to adjust the parameters of the theory such that $V_{\rm eff}(\theta)$ has a global minimum at $\theta = \theta_H$.
[Step 2] We pick the value of $m_{\rm KK}$. Then, to reproduce $m_Z = k\lambda_{Z^{(0)}} =$ 91.1876 GeV from the formula (7.70), or

$$2S(1;\lambda_{Z^{(0)}})C'(1;\lambda_{Z^{(0)}}) + \frac{\lambda_{Z^{(0)}}\sin^2\theta_H}{\cos^2\theta_W^0} = 0 \ , \tag{7.139}$$

the values of k and z_L ($m_{\rm KK} = \pi k(z_L-1)^{-1}$) are determined. Note that $S(z;\lambda)$ and $C'(z;\lambda)$ depend on z, λ and z_L. With these k and z_L the W boson mass $m_W^{\rm tree} = k\lambda_{W^{(0)}}$ is evaluated from (7.62). Note that $m_W^{\rm tree}$ is the W boson mass at the tree level. There are radiative corrections to m_W which will be explained in Section 9.1 in detail.
[Step 3] The bulk mass parameters of $\Psi^{q,\alpha}_{(\mathbf{3},\mathbf{4})}$ and $\Psi^{\ell,\alpha}_{(\mathbf{1},\mathbf{4})}$ are fixed from the masses of up-type quarks and charged leptons from (7.89) and (7.102). In particular, c_t is determined by m_t.
[Step 4] The bulk mass parameters c_{D_α} of $\Psi^{q,\alpha}_{(\mathbf{3},\mathbf{1})^\pm}$ with assigned values of the Dirac-type masses $\tilde{m}_{D_\alpha}$ and brane interaction coefficients μ_α (as in Table 7.5) are determined so as to reproduce the masses of down-type quarks from (7.99). Similarly the brane interaction masses $m_{B\alpha}$ with assigned values of the Majorana masses M^α (as in Table 7.3) are determined so as to reproduce desired neutrino masses from (7.109). As remarked above, these parameters are numerically irrelevant for $V_{\rm eff}(\theta)$.
[Step 5] At this stage there remain five parameters to be determined: (n_F, c_F) of $\Psi^{F_q}_{(\mathbf{3},\mathbf{4})}$ and $\Psi^{F_\ell}_{(\mathbf{1},\mathbf{4})}$ and $(n_V, c_V, \tilde{m}_V)$ of $\Psi^{V}_{(\mathbf{1},\mathbf{5})^\pm}$. There are two

conditions to be satisfied;

$$(a): \quad \frac{dV_{\text{eff}}}{d\theta}\bigg|_{\theta=\theta_H} = 0 ,$$

$$(b): \quad m_H^2 = \frac{1}{f_H^2}\frac{d^2V_{\text{eff}}}{d\theta^2}\bigg|_{\theta=\theta_H} , \tag{7.140}$$

where $m_H = 125.1$ GeV. We first fix, for instance, (n_F, n_V, c_V) and determine $(c_F, \tilde{m}_V)$ by (7.140). Typical values of the parameters are $(c_F, \tilde{m}_V) = (0.2359, 0.0812)$ for $\theta_H = 0.1$, $m_{\text{KK}} = 13$ TeV, $\sin^2\theta_W^0 = 0.2304$ and $(n_F, n_V, c_V) = (1, 3, 0)$. The resultant mass spectrum of the KK towers of the dark fermions is depicted in Fig. 7.1.

One may wonder whether the arbitrary choice of the parameters in the last step affects the predictive power of the theory. It will be seen later that many of the physical quantities do not depend on such details of the parameter choice, being determined solely by θ_H. There appears the θ_H universality which will be explained in the next chapter.

As mentioned above, with the given m_{KK} the allowed range of θ_H is limited to $\theta_H^{\min} \leq \theta_H \leq \theta_H^{\max}$. For $\theta_H < \theta_H^{\min}$ the top quark mass m_t cannot be reproduced. For $\theta_H \gtrsim \theta_H^{\max}$ the parameter k of the AdS curvature $\Lambda = -6k^2$ becomes larger than the Planck mass scale and the μ-e universality is spoiled. This issue will be discussed in Section 9.1 in connection with the W boson mass.

In Fig. 7.2 $V_{\text{eff}}(\theta)$ is plotted for $\theta_H = 0.1$, $m_{\text{KK}} = 13$ TeV, $\sin^2\theta_W^0 = 0.2304$, $(n_F, n_V) = (1, 3)$ and $(c_F, c_V, \tilde{m}_V) = (0.2359, 0, 0.0812)$. It is seen that the EW symmetry is spontaneously broken to $U(1)_{\text{EM}}$.

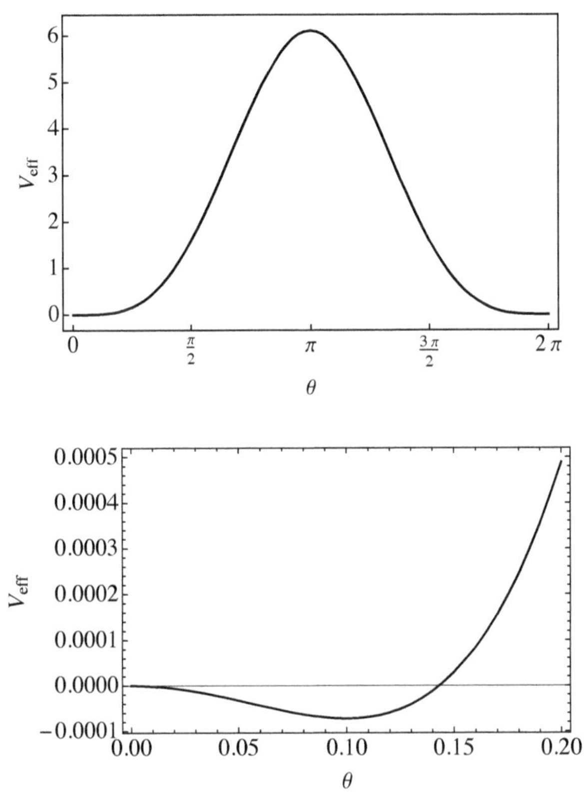

Fig. 7.2 The effective potential $V_{\rm eff}(\theta)$ (in the unit of $(kz_L^{-1})^4/16\pi^2$) is depicted for $\theta_H = 0.1$, $m_{\rm KK} = 13$ TeV, $\sin^2\theta_W^0 = 0.2304$, $(n_F, n_V) = (1, 3)$ and $(c_F, c_V, \tilde{m}_V) = (0.2359, 0, 0.0812)$. The global minimum of $V_{\rm eff}(\theta)$ is located at $\theta = \theta_H = 0.1$. The global maximum is located at $\theta = \pi$.

Chapter 8

New particles and interactions

Gauge-Higgs unification (GHU), particularly, the GUT-inspired $SO(5) \times U(1)_X \times SU(3)_C$ GHU in the RS space, leads to many predictions. Gauge couplings of quarks and leptons are nearly the same as in the SM. Higgs couplings of quarks and leptons become smaller than those in the SM, being suppressed by a factor $\cos^2 \frac{1}{2}\theta_H$ in comparison with the SM couplings. KK excited modes of the SM fields and particles emerge as new particles. Large parity violation results in the couplings of quarks and leptons to the first excited gauge bosons.

8.1 θ_H universality

GHU leads to many consequences. In particular the GUT-inspired $SO(5) \times U(1)_X \times SU(3)_C$ GHU in the RS space leads to nearly the same phenomenology at low energies as the SM. In order to confirm the GHU scenario it is necessary to clarify where deviations from the SM can be seen and how new particles can be observed in experiments.

We need to first evaluate the couplings among the fields and particles in question. To do so, it is necessary to have the wave functions of those fields and particles. In the present chapter we are going to show how those wave functions and couplings are determined.

Everything is determined once the parameters of the theory are fixed. However there remains arbitrariness in the choice of the parameters as mentioned in Section 7.11. One might wonder if this arbitrariness affects the predictions of GHU.

Let us focus on the $SO(5) \times U(1)_X$ part of the GUT-inspired GHU. In addition to the spacetime parameters k and $z_L = e^{kL}$ and the gauge couplings g_A and g_B, the model contains the bulk mass parameters of the quark-lepton multiplets which are related to the masses of quarks and leptons. Furthermore there are five parameters $(n_F, n_V, c_F, c_V, \tilde{m}_V)$ in the dark fermion sector. With these parameters specified the effective potential $V_{\rm eff}(\theta)$ for θ_H is evaluated, and the location of the global minimum of $V_{\rm eff}(\theta)$ determines the value of θ_H and the Higgs boson mass m_H is evaluated.

We have taken the following procedure to have the model consistent with phenomenology at low energies. We first choose the desired value θ_H. Input parameters (from observation) are the fine structure constant $\alpha_{\rm EM}$, the Fermi constant G_F determined from the muon decay, and the Z boson mass m_Z. In practice we specify $\sin^2\theta_W^0$ (or equivalently g_A/g_B), and later adjust its value such that the observed G_F is reproduced. There is one more parameter $m_{\rm KK}$ we may choose. It will turn out that the allowed range of $m_{\rm KK}$ is restricted once θ_H is given, or vice versa. With θ_H, $m_{\rm KK}$, m_Z, $\alpha_{\rm EM}$ and $\sin^2\theta_W^0$ given, k, $z_L = e^{kL}$, g_A and g_B are fixed. The parameters in the quark-lepton multiplet sector are fixed from the quark-lepton mass spectrum. Additional free parameters in the down-type quark sector and the neutrino sector do not significantly affect the low energy phenomenology. Finally we choose the values of three of the five parameters $(n_F, n_V, c_F, c_V, \tilde{m}_V)$ in the dark fermion sector, say, (n_F, n_V, c_V), and determine the values of the rest two $(c_F, \tilde{m}_V)$ such that $V_{\rm eff}(\theta)$ is minimized at the desired value θ_H and $m_H = 125.1$ GeV is reproduced.

Once $V_{\rm eff}(\theta)$ is given, one can evaluate the cubic and quartic self-couplings of the Higgs boson $H(x)$ in

$$\mathcal{L}_{\rm eff} = -\frac{1}{2}\partial_\mu H \partial^\mu H - \frac{1}{2}m_H^2 H^2 - \lambda_3 H^3 - \lambda_4 H^4 - \cdots ,$$

$$\lambda_3 = \frac{1}{3! f_H^3} \frac{d^3 V_{\rm eff}(\theta)}{d\theta^3}\bigg|_{\theta_H} , \quad \lambda_4 = \frac{1}{4! f_H^4} \frac{d^4 V_{\rm eff}(\theta)}{d\theta^4}\bigg|_{\theta_H} . \tag{8.1}$$

The couplings λ_3 and λ_4 determined in the procedure described above, in general, depend on θ_H, $m_{\rm KK}$, and (n_F, n_V, c_V). Note that in the SM $\lambda_3^{\rm SM} = m_H^2/2v = 31.8\,{\rm GeV}$ and $\lambda_4^{\rm SM} = m_H^2/8v^2 = 0.0323$ at the tree level.

In Table 8.1 we list the values of (λ_3, λ_4) for $\theta_H = 0.1$ with various choices of (n_F, n_V, c_V). Although the values of c_F, $\tilde{m}_V$, and therefore the masses $m_{F^{(1)}}$ and $m_{V^{(1)}}$ depend on the choice of (n_F, n_V, c_V), the values of λ_3 and λ_4 are universal to high accuracy. λ_3 and λ_4 are determined as functions of θ_H only. We observe that the $\boldsymbol{\theta_H}$ **universality** in the Higgs self-couplings. At $\theta_H = 0.1$, λ_3 and λ_4 turn out smaller than in the SM by 11% and 39% respectively.

Table 8.1 θ_H universality in λ_3 and λ_4 for $\theta_H = 0.1$ and $m_{\rm KK} = 13$ TeV. With given (n_F, n_V, c_V), c_F and $\tilde{m}_V$ are determined to satisfy the conditions in Eq. (7.140), and λ_3 and λ_4 are evaluated by (8.1). In the SM $\lambda_3^{\rm SM} = 31.8$ GeV and $\lambda_4^{\rm SM} = 0.0323$. The resultant $m_{F^{(1)}}$ and $m_{V^{(1)}}$ are also listed.

n_F	n_V	c_V	c_F	$\tilde{m}_V$	λ_3 (GeV)	λ_4	$m_{F^{(1)}}$ (TeV)	$m_{V^{(1)}}$ (TeV)
1	3	0	0.2359	0.0812	28.131	0.01972	4.504	3.641
1	3	0.2	0.2360	0.0788	28.130	0.01972	4.504	3.342
1	6	0	0.2352	0.0949	28.131	0.01973	4.512	3.749
2	3	0	0.3957	0.0800	28.129	0.01971	2.742	3.631
2	6	0	0.3955	0.0937	28.130	0.01972	2.745	3.741

The θ_H universality in λ_3 and λ_4 is understood in the following manner. The important observation is that λ_3 and λ_4 are determined by the local behavior of the effective potential $V_{\rm eff}(\theta)$ in the vicinity of the global minimum at $\theta = \theta_H$, and the universality reflects the local, but not global, behavior of $V_{\rm eff}(\theta)$.

The effective potential $V_{\rm eff}(\theta)$ in Eq. (7.135) is decomposed into three parts:

$$
\begin{aligned}
&V_{\rm eff}(\theta) - V_{\rm eff}(\theta_H) \\
&= \frac{(kz_L^{-1})^4}{(4\pi)^2} \Big\{ h_0(\theta) + n_F h_F(\theta; c_F, z_L) + n_V h_V(\theta; c_V, \tilde{m}_V, z_L) \Big\} \qquad (8.2)
\end{aligned}
$$

where $h_0(\theta)$ represents the contributions from gauge and top quark fields. With θ_H, $m_{\rm KK}$, $\sin^2\theta_W^0$ and ξ specified, k and z_L are determined from m_Z and c_t is subsequently determined by m_t so that $h_0(\theta)$ is fixed. All other parameters associated with quarks and leptons are irrelevant for $V_{\rm eff}(\theta)$. There remain five parameters $(n_F, c_F, n_V, c_V, \tilde{m}_V)$ to be specified in (8.2). They must be adjusted such that the two conditions in (7.140) are satisfied. The important feature in the RS space with $z_L \gg 1$ is that the θ dependence of $h_F(\theta; c_F, z_L)$ and $h_V(\theta; c_V, \tilde{m}_V, z_L)$ factorizes near $\theta = \theta_H$

$$
\begin{aligned}
h_F(\theta; c_F, z_L) &\simeq \alpha_F(c_F, z_L) \tilde{h}_F(\theta) \ , \\
h_V(\theta; c_V, \tilde{m}_V, z_L) &\simeq \alpha_V(c_V, \tilde{m}_V, z_L) \tilde{h}_V(\theta) \ , \qquad (8.3)
\end{aligned}
$$

to high accuracy. This can be confirmed numerically. The relation (8.3) implies, for instance, that the ratio $h_F(\theta; c_F^{(1)}, z_L)/h_F(\theta; c_F^{(2)}, z_L)$ is θ-independent near θ_H. For $\theta_H = 0.1$, $m_{\rm KK} = 13\,$TeV (therefore $z_L =$

3.83×10^{11}) and $(c_F^{(1)}, c_F^{(2)}) = (0.2, 0.4)$, the ratio varies from 2.18312 to 2.18317 in the range $0.09 \le \theta \le 0.11$. The variation is only $O(10^{-5})$. We stress that the factorization in Eq. (8.3) is only valid locally, namely near $\theta = \theta_H$, and that $\tilde{h}_F(\theta)$ and $\tilde{h}_V(\theta)$ depend on θ_H. The z_L dependence of $h_F(\theta; c_F, z_L)$ and $h_V(\theta; c_V, \tilde{m}_V, z_L)$ is also tiny for large $z_L \gg 1$.

Let us pick a set of values for (n_F, n_V, c_V) and determine $(c_F, \tilde{m}_V)$ by (7.140). Making use of (8.3), one finds

$$\left[\frac{dh_0}{d\theta} + n_F \alpha_F(c_F, z_L) \frac{d\tilde{h}_F}{d\theta} + n_V \alpha_V(c_V, \tilde{m}_V, z_L) \frac{d\tilde{h}_V}{d\theta} \right]_{\theta=\theta_H} = 0 \,,$$

$$\left[\frac{d^2h_0}{d\theta^2} + n_F \alpha_F(c_F, z_L) \frac{d^2\tilde{h}_F}{d\theta^2} + n_V \alpha_V(c_V, \tilde{m}_V, z_L) \frac{d^2\tilde{h}_V}{d\theta^2} \right]_{\theta=\theta_H} = \frac{(4\pi)^2 m_H^2 f_H^2}{(k z_L^{-1})^4} \,. \tag{8.4}$$

We perform this procedure for two sets: $(n_F, n_V, c_V) = (n_F^{(1)}, n_V^{(1)}, c_V^{(1)})$ and $(n_F^{(2)}, n_V^{(2)}, c_V^{(2)})$. Then (8.4) implies that

$$\begin{aligned} n_F^{(1)} \alpha_F(c_F^{(1)}, z_L) &= n_F^{(2)} \alpha_F(c_F^{(2)}, z_L) \equiv \beta_F \,, \\ n_V^{(1)} \alpha_V(c_V^{(1)}, \tilde{m}_V^{(1)}, z_L) &= n_V^{(2)} \alpha_V(c_V^{(2)}, \tilde{m}_V^{(2)}, z_L) \equiv \beta_V \,. \end{aligned} \tag{8.5}$$

Although values of $(c_F, \tilde{m}_V)$ depend on the choice of (n_F, n_V, c_V), $\beta_F = n_F \alpha_F(c_F, z_L)$ and $\beta_V = n_V \alpha_V(c_V, \tilde{m}_V, z_L)$ are universal, provided solutions exist. Consequently one obtains

$$\begin{aligned} V_{\rm eff}(\theta) - V_{\rm eff}(\theta_H) &\simeq \frac{(k z_L^{-1})^4}{(4\pi)^2} \, \bar{h}(\theta) \,, \\ \bar{h}(\theta) &= h_0(\theta) + \beta_F \tilde{h}_F(\theta) + \beta_V \tilde{h}_V(\theta) \,. \end{aligned} \tag{8.6}$$

It follows that

$$\begin{aligned} \lambda_3(\theta_H) &\sim \frac{g_w m_H^2 \sin\theta_H}{12 m_W} \frac{\bar{h}^{(3)}(\theta_H)}{\bar{h}^{(2)}(\theta_H)} \,, \\ \lambda_4(\theta_H) &\sim \frac{g_w^2 m_H^2 \sin^2\theta_H}{96 m_W^2} \frac{\bar{h}^{(4)}(\theta_H)}{\bar{h}^{(2)}(\theta_H)} \,, \end{aligned} \tag{8.7}$$

where Eq. (7.63) has been used. (8.7) explains the θ_H universality in λ_3 and λ_4. The θ_H universality is observed in other physical quantities such as the Higgs boson couplings to W, Z, quarks and leptons as explained below.

8.2 Effective interactions of the Higgs field

Gauge and Higgs couplings among the 4D fields appearing in the KK expansion are obtained by inserting the wave functions of the KK modes into the 5D action and integrating over the fifth-dimensional coordinate. Before going into the details, we would like to point out that some qualitative features of the Higgs couplings are deduced from the gauge invariance of the theory.

The Higgs field $H(x)$ is the fluctuation mode of θ_H, and appears in the combination of $\hat{\theta}_H(x) = \theta_H + (H/f_H)$, (7.137), in the effective interactions. As recognized in Section 7.3, the gauge invariance implies that the theory is periodic in $\hat{\theta}_H(x)$ with a period 2π. The effective interactions of the Higgs couplings to W, Z, quarks and leptons are written in the form

$$\mathcal{L}_{\rm eff} = -m_W(\hat{\theta}_H)^2 W_\mu^\dagger W^\mu - \frac{1}{2} m_Z(\hat{\theta}_H)^2 Z_\mu Z^\mu - m_\psi(\hat{\theta}_H)\bar{\psi}\psi \tag{8.8}$$

where the mass functions satisfy $m_W(\hat{\theta}_H + 2\pi) = m_W(\hat{\theta}_H)$ etc. As seen in Sections 7.9 and 7.10,

$$\begin{aligned} m_W &= m_W(\theta_H) \sim \frac{1}{2} g_w f_H \sin\theta_H \ , \\ m_Z &= m_Z(\theta_H) \sim \frac{1}{2\cos\theta_W^0} g_w f_H \sin\theta_H \ , \\ m_\psi &= m_\psi(\theta_H) \sim a_\psi \sin\tfrac{1}{2}\theta_H \ , \end{aligned} \tag{8.9}$$

in a good approximation. The vacuum expectation value of the Higgs field in the SM $v_{\rm SM}$ (~ 246 GeV) corresponds to

$$v_{\rm SM} = f_H \sin\theta_H \ . \tag{8.10}$$

The vacuum expectation value of the Higgs field in GHU is $\theta_H f_H$, but relevant for m_W in GHU is $f_H \sin\theta_H$.

With the mass functions in (8.9) the effective Lagrangian $\mathcal{L}_{\rm eff}$ in Eq. (8.8) leads to

$$\begin{aligned} \mathcal{L}_{\rm eff} = &- \Big\{ m_W^2 + g_{WWH} H + \cdots \Big\} W_\mu^\dagger W^\mu \\ &- \frac{1}{2}\Big\{ m_Z^2 + g_{ZZH} H + \cdots \Big\} Z_\mu Z^\mu \\ &- \Big\{ m_\psi + y_\psi H + \cdots \Big\} \bar{\psi}\psi \end{aligned} \tag{8.11}$$

where

$$g_{WWH} \sim g_w m_W \cos\theta_H = g_{WWH}^{\rm SM} \cos\theta_H ~,$$

$$g_{ZZH} \sim \frac{g_w}{\cos\theta_W^0} m_Z \cos\theta_H = g_{ZZH}^{\rm SM} \cos\theta_H ~,$$

$$y_\psi \sim \frac{m_\psi}{f_H \sin\theta_H} \cos^2 \tfrac{1}{2}\theta_H = y_\psi^{\rm SM} \cos^2 \tfrac{1}{2}\theta_H ~. \tag{8.12}$$

The Higgs couplings to W and Z are suppressed, compared with those in the SM, by a factor $\cos\theta_H$, and the Yukawa couplings of quarks and leptons are suppressed by a factor $\cos^2 \frac{1}{2}\theta_H$. This property does not depend on the details in the dark fermion sector, and represents the θ_H universality.

8.3 Wave functions

Wave functions of all KK modes of gauge bosons, quarks and leptons, and dark fermions are determined as eigenvectors with zero eigenvalues of the K matrices which appeared in determining the mass spectra in Sections 7.9 and 7.10.

8.3.1 *Gauge fields*

(i) W and W_R towers

Wave functions in the twisted gauge take the form in Eq. (7.59) where the coefficients α_{23}, α_{14} and α_{15} are determined as eigenvectors with the zero eigenvalue of K in Eq. (7.60). The spectrum is determined by (7.62) for $\omega|C/C'| \gg 1$. For the W tower $2SC'(1;\lambda_{W^{(n)}}) + \lambda_{W^{(n)}} \sin^2\theta_H = 0$ and

$$K_W = \begin{pmatrix} C' & c_H C' & s_H S' \\ -C & c_H C & s_H S \\ 0 & -s_H C & c_H S \end{pmatrix} ,~ \begin{pmatrix} \alpha_{23} \\ \alpha_{14} \\ \alpha_{15} \end{pmatrix} \propto \begin{pmatrix} 1 \\ c_H \\ s_H (C/S) \end{pmatrix} \tag{8.13}$$

where $c_H = \cos\theta_H$ and $s_H = \sin\theta_H$, and the relation $CS' - SC' = \lambda_{W^{(n)}}$ in Eq. (C.15) has been used. For the W_R tower $C(1;\lambda_{W_R^{(n)}}) = 0$, and

$$K_{W_R} = \begin{pmatrix} C' & c_H C' & s_H S' \\ 0 & 0 & s_H S \\ 0 & 0 & c_H S \end{pmatrix} ,~ \begin{pmatrix} \alpha_{23} \\ \alpha_{14} \\ \alpha_{15} \end{pmatrix} \propto \begin{pmatrix} -c_H \\ 1 \\ 0 \end{pmatrix} . \tag{8.14}$$

Similar formulas are obtained for $(\alpha_{31}, \alpha_{24}, \alpha_{25})$. In terms of $A_\mu^{a_L} = (\frac{1}{2}\epsilon^{abc} A_\mu^{bc} + A_\mu^{a4})/\sqrt{2}$, $A_\mu^{a_R} = (\frac{1}{2}\epsilon^{abc} A_\mu^{bc} - A_\mu^{a4})/\sqrt{2}$ and $A_\mu^{\hat{a}} = A_\mu^{a5}$ ($a =$

1, 2, 3) we have

$$\frac{1}{\sqrt{2k}}\begin{pmatrix}\tilde{A}^{1_L}_\mu + i\tilde{A}^{2_L}_\mu \\ \tilde{A}^{1_R}_\mu + i\tilde{A}^{2_R}_\mu \\ \tilde{A}^{\hat{1}}_\mu + i\tilde{A}^{\hat{2}}_\mu\end{pmatrix} = \begin{pmatrix}\mathring{W}^L_\mu \\ \mathring{W}^R_\mu \\ \hat{\mathring{W}}_\mu\end{pmatrix} + \begin{pmatrix}\mathring{W}^L_{R\mu} \\ \mathring{W}^R_{R\mu} \\ 0\end{pmatrix},$$

$$\begin{pmatrix}\mathring{W}^L_\mu \\ \mathring{W}^R_\mu \\ \hat{\mathring{W}}_\mu\end{pmatrix} = \sum_{n=0}^{\infty} W^{(n)}_\mu(x)\begin{pmatrix}h^L_{W^{(n)}}(z) \\ h^R_{W^{(n)}}(z) \\ \hat{h}_{W^{(n)}}(z)\end{pmatrix},$$

$$\begin{pmatrix}\mathring{W}^L_{R\mu} \\ \mathring{W}^R_{R\mu}\end{pmatrix} = \sum_{n=1}^{\infty} W^{(n)}_{R\mu}(x)\begin{pmatrix}h^L_{W_R^{(n)}}(z) \\ h^R_{W_R^{(n)}}(z)\end{pmatrix}, \tag{8.15}$$

where

$$\begin{pmatrix}h^L_{W^{(n)}}(z) \\ h^R_{W^{(n)}}(z) \\ \hat{h}_{W^{(n)}}(z)\end{pmatrix} = \frac{1}{\sqrt{2\,r_{W^{(n)}}}}\begin{pmatrix}(1+c_H)\,C(z;\lambda_{W^{(n)}}) \\ (1-c_H)\,C(z;\lambda_{W^{(n)}}) \\ \sqrt{2}\,s_H\check{S}(z;\lambda_{W^{(n)}})\end{pmatrix},$$

$$\begin{pmatrix}h^L_{W_R^{(n)}}(z) \\ h^R_{W_R^{(n)}}(z)\end{pmatrix} = \frac{1}{\sqrt{2\,r_{W_R^{(n)}}}}\begin{pmatrix}1-c_H \\ -1-c_H\end{pmatrix} C(z;\lambda_{W_R^{(n)}}),$$

$$r_{W^{(n)}} = \int_1^{z_L}\frac{dz}{z}\Big\{(1+c_H^2)C(z;\lambda_{W^{(n)}})^2 + s_H^2\check{S}(z;\lambda_{W^{(n)}})^2\Big\},$$

$$r_{W_R^{(n)}} = \int_1^{z_L}\frac{dz}{z}\,(1+c_H^2)C(z;\lambda_{W_R^{(n)}})^2\,, \tag{8.16}$$

and $\check{S}(z;\lambda)$ is defined in Eq. (C.22). At $\theta_H = 0$ the W tower is in pure $SU(2)_L$, whereas at $\theta_H \neq 0$ it contains $SU(2)_R$ and $SO(5)/SO(4)$ components. $W^{(0)}_\mu$ is massless for $\theta_H = 0$ and π. It is the $SU(2)_L$ gauge boson at $\theta_H = 0$, and is transformed to the $SU(2)_R$ gauge boson at $\theta_H = \pi$. As $\lambda_{W^{(0)}} z_L \ll 1$ for any θ_H, it follows from (C.23) that $C(z;\lambda_{W^{(0)}}) \sim z_L$ and $\check{S}(z;\lambda_{W^{(0)}}) \sim z_L(1 - z^2/z_L^2)$. The $SU(2)_L$ and $SU(2)_R$ components of the wave function of $W^{(0)}$ are nearly constant, which turns out important to reproduce nearly the same gauge couplings as those in the SM.

The normalization of the wave functions of each KK mode which has multiple components $\{h_n^\alpha(z)\}$ in the $SO(5)\times U(1)_X$ space is determined by

$$\int_1^{z_L}\frac{dz}{z}\sum_\alpha |h_n^\alpha(z)|^2 = 1\,. \tag{8.17}$$

One can recognize the orthonormality relations

$$\langle W^{(n)}|W^{(\ell)}\rangle = \int_1^{z_L} \frac{dz}{z} \sum_{\alpha=L,R,\wedge} h^{\alpha\,*}_{W^{(n)}} h^{\alpha}_{W^{(\ell)}} = \delta^{n\ell},$$

$$\langle W_R^{(n)}|W_R^{(\ell)}\rangle = \int_1^{z_L} \frac{dz}{z} \sum_{\alpha=L,R} h^{\alpha\,*}_{W_R^{(n)}} h^{\alpha}_{W_R^{(\ell)}} = \delta^{n\ell},$$

$$\langle W^{(n)}|W_R^{(\ell)}\rangle = \int_1^{z_L} \frac{dz}{z} \sum_{\alpha=L,R} h^{\alpha\,*}_{W^{(n)}} h^{\alpha}_{W_R^{(\ell)}} = 0\,, \tag{8.18}$$

where $h^{\wedge}_{W^{(n)}} = \hat{h}_{W^{(n)}}(z)$.

(ii) γ, Z and Z_R towers

Wave functions take the form in Eq. (7.67) where the coefficients α_{12}, α_{34}, α_{35} and α_B are determined as eigenvectors with the zero eigenvalue of K in Eq. (7.68). The spectrum is determined by (7.70) for $\omega'|C/C'| \gg 1$. For the γ tower $C''(1;\lambda_{\gamma^{(n)}}) = 0$, and

$$K_\gamma = \begin{pmatrix} 0 & 0 & s_H S' & 0 \\ -C & c_H C & s_H S & \dfrac{\sqrt{2}s_\phi}{c_\phi} C \\ 0 & -s_H C & c_H S & 0 \\ 0 & 0 & 0 & 0 \end{pmatrix}, \quad \begin{pmatrix} \alpha_{12} \\ \alpha_{34} \\ \alpha_{35} \\ \alpha_B \end{pmatrix} \propto \begin{pmatrix} \sqrt{2}\, s_\phi \\ 0 \\ 0 \\ c_\phi \end{pmatrix}. \tag{8.19}$$

For the Z tower $2SC'(1;\lambda_{Z^{(n)}}) + (1+s_\phi^2)\lambda_{Z^{(n)}} \sin^2\theta_H = 0$ and

$$K_Z = \begin{pmatrix} C' & c_H C' & s_H S' & 0 \\ -C & c_H C & s_H S & \dfrac{\sqrt{2}s_\phi}{c_\phi} C \\ 0 & -s_H C & c_H S & 0 \\ C' & 0 & 0 & \dfrac{c_\phi}{\sqrt{2}s_\phi} C' \end{pmatrix}, \quad \begin{pmatrix} \alpha_{12} \\ \alpha_{34} \\ \alpha_{35} \\ \alpha_B \end{pmatrix} \propto \begin{pmatrix} \dfrac{c_\phi^2}{1+s_\phi^2} \\ c_H \\ s_H (C/S) \\ -\dfrac{\sqrt{2}s_\phi c_\phi}{1+s_\phi^2} \end{pmatrix}. \tag{8.20}$$

For the Z_R tower $C(1;\lambda_{Z_R^{(n)}}) = 0$, and

$$K_{Z_R} = \begin{pmatrix} C' & c_H C' & s_H S' & 0 \\ 0 & 0 & s_H S & 0 \\ 0 & 0 & c_H S & 0 \\ C' & 0 & 0 & \dfrac{c_\phi}{\sqrt{2}s_\phi} C' \end{pmatrix}, \quad \begin{pmatrix} \alpha_{12} \\ \alpha_{34} \\ \alpha_{35} \\ \alpha_B \end{pmatrix} \propto \begin{pmatrix} -c_\phi c_H \\ c_\phi \\ 0 \\ \sqrt{2}s_\phi c_H \end{pmatrix}. \tag{8.21}$$

Hence we have the expansion

$$\frac{1}{\sqrt{k}}\begin{pmatrix}\tilde{A}^{3_L}_\mu\\ \tilde{A}^{3_R}_\mu\\ \tilde{A}^{\hat{3}}_\mu\\ B_\mu\end{pmatrix} = \begin{pmatrix}\mathring{\gamma}^L_\mu\\ \mathring{\gamma}^R_\mu\\ 0\\ \mathring{\gamma}^B_\mu\end{pmatrix} + \begin{pmatrix}\mathring{Z}^L_\mu\\ \mathring{Z}^R_\mu\\ \hat{\mathring{Z}}_\mu\\ \mathring{Z}^B_\mu\end{pmatrix} + \begin{pmatrix}\mathring{Z}^L_{R\mu}\\ \mathring{Z}^R_{R\mu}\\ 0\\ \mathring{Z}^B_{R\mu}\end{pmatrix},$$

$$\begin{pmatrix}\mathring{\gamma}^L_\mu\\ \mathring{\gamma}^R_\mu\\ \mathring{\gamma}^B_\mu\end{pmatrix} = \sum_{n=0}^\infty \gamma^{(n)}_\mu(x)\begin{pmatrix}h^L_{\gamma^{(n)}}(z)\\ h^R_{\gamma^{(n)}}(z)\\ h^B_{\gamma^{(n)}}(z)\end{pmatrix},$$

$$\begin{pmatrix}\mathring{Z}^L_\mu\\ \mathring{Z}^R_\mu\\ \hat{\mathring{Z}}_\mu\\ \mathring{Z}^B_\mu\end{pmatrix} = \sum_{n=0}^\infty Z^{(n)}_\mu(x)\begin{pmatrix}h^L_{Z^{(n)}}(z)\\ h^R_{Z^{(n)}}(z)\\ \hat{h}_{Z^{(n)}}(z)\\ h^B_{Z^{(n)}}(z)\end{pmatrix},$$

$$\begin{pmatrix}\mathring{Z}^L_{R\mu}\\ \mathring{Z}^R_{R\mu}\\ \mathring{Z}^B_{R\mu}\end{pmatrix} = \sum_{n=1}^\infty Z^{(n)}_{R\mu}(x)\begin{pmatrix}h^L_{Z_R^{(n)}}(z)\\ h^R_{Z_R^{(n)}}(z)\\ h^B_{Z_R^{(n)}}(z)\end{pmatrix}, \tag{8.22}$$

where

$$\begin{pmatrix}h^L_{\gamma^{(n)}}(z)\\ h^R_{\gamma^{(n)}}(z)\\ h^B_{\gamma^{(n)}}(z)\end{pmatrix} = \begin{pmatrix}\sin\theta^0_W\\ \sin\theta^0_W\\ \sqrt{1-2\sin^2\theta^0_W}\end{pmatrix} h_{\gamma^{(n)}}(z)\ ,$$

$$h_{\gamma^{(n)}}(z) = \begin{cases}\dfrac{1}{\sqrt{r_{\gamma^{(n)}}}}C(z,\lambda_{\gamma^{(n)}}) & \text{for } n\ge 1\ ,\\[2mm] \dfrac{1}{\sqrt{kL}} & \text{for } n=0\ ,\end{cases}$$

$$r_{\gamma^{(n)}} = \int_1^{z_L}\frac{dz}{z}\,C(z,\lambda_{\gamma^{(n)}})^2\ ,$$

$$\begin{pmatrix}h^L_{Z^{(n)}}(z)\\ h^R_{Z^{(n)}}(z)\\ \hat{h}_{Z^{(n)}}(z)\\ h^B_{Z^{(n)}}(z)\end{pmatrix} = \begin{pmatrix}h^{L,su2}_{Z^{(n)}}(z)\\ h^{R,su2}_{Z^{(n)}}(z)\\ \hat{h}^{su2}_{Z^{(n)}}(z)\\ 0\end{pmatrix} - \sin\theta^0_W\begin{pmatrix}\sin\theta^0_W\\ \sin\theta^0_W\\ 0\\ \sqrt{1-2\sin^2\theta^0_W}\end{pmatrix} h^{em}_{Z^{(n)}}(z),$$

$$
\begin{pmatrix} h_{Z^{(n)}}^{L,su2}(z) \\ h_{Z^{(n)}}^{R,su2}(z) \\ \hat{h}_{Z^{(n)}}^{su2}(z) \end{pmatrix} = \frac{1}{\sqrt{2\,r_{Z^{(n)}}}} \begin{pmatrix} (1+c_H)\,C(z,\lambda_{Z^{(n)}}) \\ (1-c_H)\,C(z,\lambda_{Z^{(n)}}) \\ \sqrt{2}s_H \check{S}(z,\lambda_{Z^{(n)}}) \end{pmatrix} ,
$$

$$
h_{Z^{(n)}}^{em}(z) = \sqrt{\frac{2}{r_{Z^{(n)}}}}\, C(z,\lambda_{Z^{(n)}}) \, ,
$$

$$
r_{Z^{(n)}} = \int_1^{z_L} \frac{dz}{z} \Big\{ (1 - 2\sin^2\theta_W^0 + c_H^2) C(z,\lambda_{Z^{(n)}})^2 + s_H^2 \check{S}(z,\lambda_{Z^{(n)}})^2 \Big\} ,
$$

$$
\begin{pmatrix} h_{Z_R^{(n)}}^{L}(z) \\ h_{Z_R^{(n)}}^{R}(z) \\ h_{Z_R^{(n)}}^{B}(z) \end{pmatrix} = \frac{1}{\sqrt{2\,r_{Z_R^{(n)}}}} \begin{pmatrix} (1-c_H) \\ -(1+c_H) \\ \dfrac{2\,c_H \sin\theta_W^0}{\sqrt{1-2\sin^2\theta_W^0}} \end{pmatrix} C(z;\lambda_{Z_R^{(n)}}),
$$

$$
r_{Z_R^{(n)}} = \int_1^{z_L} \frac{dz}{z} \left\{ 1 + \frac{c_H^2}{1 - 2\sin^2\theta_W^0} \right\} C(z,\lambda_{Z_R^{(n)}})^2 \, . \tag{8.23}
$$

Here $\sin\theta_W^0$ has been defined in Eq. (7.72). Indeed, by retaining the photon mode $A_\mu^\gamma = \gamma_\mu^{(0)}$ one finds that

$$
\begin{aligned}
A_\mu^\gamma &= \sin\theta_W^0 (\tilde{A}_\mu^{3L} + \tilde{A}_\mu^{3R}) + \sqrt{1-2\sin^2\theta_W^0}\, B_\mu \\
&= \sin\theta_W^0 \tilde{A}_\mu^{3L} + \cos\theta_W^0 B_\mu^Y \, , \\
B_\mu^Y &= \frac{1}{\cos\theta_W^0} \Big\{ \sin\theta_W^0 \tilde{A}_\mu^{3R} + \sqrt{1-2\sin^2\theta_W^0}\, B_\mu \Big\} \\
&= s_\phi \tilde{A}_\mu^{3R} + c_\phi B_\mu \, .
\end{aligned} \tag{8.24}
$$

It is recognized that $\sin\theta_W^0$ corresponds to the weak mixing angle $\sin\theta_W^{\rm SM}$ in the SM.

Notice that the wave functions of each KK mode $Z^{(n)}$ in the Z tower are decomposed into the $SU(2)$ and $U(1)_{\rm EM}$ parts. The $SU(2)$ part has the same structure as in the W tower, and the $U(1)_{\rm EM}$ part has the same structure as in the photon tower. This property becomes very useful in evaluating the Z couplings of various fields. We also note the orthonormality relations:

$$\begin{aligned}
\langle \gamma^{(n)} | \gamma^{(\ell)} \rangle &= \int_1^{z_L} \frac{dz}{z} \sum_{\alpha=L,R,B} h^{\alpha\,*}_{\gamma^{(n)}} h^{\alpha}_{\gamma^{(\ell)}} = \delta^{n\ell} , \\
\langle Z^{(n)} | Z^{(\ell)} \rangle &= \int_1^{z_L} \frac{dz}{z} \sum_{\alpha=L,R,\wedge,B} h^{\alpha\,*}_{Z^{(n)}} h^{\alpha}_{Z^{(\ell)}} = \delta^{n\ell} , \\
\langle Z_R^{(n)} | Z_R^{(\ell)} \rangle &= \int_1^{z_L} \frac{dz}{z} \sum_{\alpha=L,R,B} h^{\alpha\,*}_{Z_R^{(n)}} h^{\alpha}_{Z_R^{(\ell)}} = \delta^{n\ell} , \\
\langle \gamma^{(n)} | Z^{(\ell)} \rangle &= \langle \gamma^{(n)} | Z_R^{(\ell)} \rangle = \langle Z^{(n)} | Z_R^{(\ell)} \rangle = 0 \,.
\end{aligned} \tag{8.25}$$

(iii) $A^{\hat{4}}$ tower

The spectrum of the $A^{\hat{4}}$ tower is determined by Eq. (7.73), $S(1; \lambda_{A^{\hat{4}(n)}}) = 0$. The KK expansion is given by

$$\begin{aligned}
A_\mu^{\hat{4}} &= \tilde{A}_\mu^{\hat{4}} = \sqrt{k} \sum_{n=1}^{\infty} A_\mu^{\hat{4}(n)}(x) \hat{h}_{A^{\hat{4}(n)}}(z) \,, \\
\hat{h}_{A^{\hat{4}(n)}}(z) &= \frac{1}{\sqrt{r_{A^{\hat{4}(n)}}}} S(z; \lambda_{A^{\hat{4}(n)}}) \,, \\
r_{A^{\hat{4}(n)}} &= \int_1^{z_L} \frac{dz}{z} S(z; \lambda_{A^{\hat{4}(n)}})^2 \,.
\end{aligned} \tag{8.26}$$

(iv) Higgs tower

$\tilde{A}_z^{\hat{4}} = \tilde{A}_z^{45}$ contains the zero mode $H^{(0)}(x) \equiv H(x)$ corresponding to the neutral Higgs boson in the SM. The spectrum of the KK modes $\{H^{(n)}\}$ $(n \geq 1)$ is determined by Eq. (7.82), $S(1; \lambda_{H^{(n)}}) = 0$. The KK expansion is given by

$$\begin{aligned}
\tilde{A}_z^{\hat{4}} &= \frac{1}{\sqrt{k}} \sum_{n=0}^{\infty} H^{(n)}(x) u_{H^{(n)}}(z) \,, \\
u_{H^{(n)}}(z) &= \begin{cases} \sqrt{\dfrac{2}{z_L^2 - 1}}\, z & \text{for } n = 0, \\[2ex] \dfrac{1}{\sqrt{r_{H^{(n)}}}} S'(z; \lambda_{H^{(n)}}) & \text{for } n \geq 1, \end{cases} \\
r_{H^{(n)}} &= \int_1^{z_L} \frac{dz}{z} S'(z; \lambda_{H^{(n)}})^2 \,.
\end{aligned} \tag{8.27}$$

Only the zero mode $H(x)$ is physical. All other modes, including KK modes of other $\tilde{A}_z^{ab}$ components, are unphysical. We note that

$$\langle H^{(n)}|H^{(\ell)}\rangle = \int_1^{z_L} \frac{dz}{z} u^*_{H^{(n)}} u_{H^{(\ell)}} = \delta_{n\ell} \ . \tag{8.28}$$

The field $H(x)$ acquires a finite mass m_H by quantum effects as we have seen in the previous chapter.

8.3.2 *Quarks*

The wave functions of the fermion fields are determined in a similar manner. It is convenient to write the KK expansion for $\check{\Psi} = z^{-2}\Psi$ fields for fermions.

(i) Up-type quark tower

The spectrum of the up-type quark tower is determined by Eq. (7.89), $S_L S_R(1, \lambda_{u^{(n)}}, c_u) + \sin^2 \frac{1}{2}\theta_H = 0$. The K matrix for $(\check{u}, \check{u}')$ is

$$K_u = \begin{pmatrix} \bar{c}_H S_R^u & i\bar{s}_H C_R^u \\ i\bar{s}_H C_L^u & \bar{c}_H S_L^u \end{pmatrix}, \ \begin{pmatrix} \alpha_u \\ \alpha_{u'} \end{pmatrix} \propto \begin{pmatrix} \bar{c}_H \\ -i\bar{s}_H (C_L^u / S_L^u) \end{pmatrix}, \tag{8.29}$$

where $(\bar{c}_H, \bar{s}_H) = (\cos \frac{1}{2}\theta_H, \sin \frac{1}{2}\theta_H)$. The KK expansion becomes

$$\begin{pmatrix} \check{u} \\ \check{u}' \end{pmatrix} = \sqrt{k} \sum_{n=0}^{\infty} \left\{ u_L^{(n)}(x) \begin{pmatrix} f_L^{u^{(n)}}(z) \\ g_L^{u^{(n)}}(z) \end{pmatrix} + u_R^{(n)}(x) \begin{pmatrix} f_R^{u^{(n)}}(z) \\ g_R^{u^{(n)}}(z) \end{pmatrix} \right\},$$

$$\begin{pmatrix} f_L^{u^{(n)}}(z) \\ g_L^{u^{(n)}}(z) \end{pmatrix} = \frac{1}{\sqrt{r_{u^{(n)}L}}} \begin{pmatrix} \bar{c}_H C_L(z, \lambda_{u^{(n)}}, c_u) \\ -i\bar{s}_H \check{S}_L(z, \lambda_{u^{(n)}}, c_u) \end{pmatrix},$$

$$\begin{pmatrix} f_R^{u^{(n)}}(z) \\ g_R^{u^{(n)}}(z) \end{pmatrix} = \frac{1}{\sqrt{r_{u^{(n)}R}}} \begin{pmatrix} \bar{c}_H S_R(z, \lambda_{u^{(n)}}, c_u) \\ -i\bar{s}_H \check{C}_R(z, \lambda_{u^{(n)}}, c_u) \end{pmatrix}. \tag{8.30}$$

Here $\check{S}_{L/R}(z; \lambda, c)$ and $\check{C}_{L/R}(z; \lambda, c)$ are defined in Eq. (C.22), and the normalization factor for each mode is determined by the condition

$$\int_1^{z_L} dz \left\{ |f_n(z)|^2 + |g_n(z)|^2 \right\} = 1 \quad \text{for} \ \begin{pmatrix} f_n(z) \\ g_n(z) \end{pmatrix}. \tag{8.31}$$

It is easy to show that $r_{u^{(n)}L} = r_{u^{(n)}R} \equiv r_{u^{(n)}}$ for $\theta_H \neq 0$ as a result of the relations $S_L S_R(1, \lambda_{u^{(n)}}, c_u) + \bar{s}_H^2 = 0$ and (C.17). We also note that the wave functions can be written as

$$\begin{pmatrix} f_L^{u^{(n)}}(z) \\ g_L^{u^{(n)}}(z) \end{pmatrix} = \frac{1}{\sqrt{r'_{u^{(n)}}}} \begin{pmatrix} \bar{s}_H \check{C}_L(z, \lambda_{u^{(n)}}, c_u) \\ i\bar{c}_H S_L(z, \lambda_{u^{(n)}}, c_u) \end{pmatrix},$$

$$\begin{pmatrix} f_R^{u^{(n)}}(z) \\ g_R^{u^{(n)}}(z) \end{pmatrix} = \frac{1}{\sqrt{r'_{u^{(n)}}}} \begin{pmatrix} \bar{s}_H \check{S}_R(z, \lambda_{u^{(n)}}, c_u) \\ i\bar{c}_H C_R(z, \lambda_{u^{(n)}}, c_u) \end{pmatrix}. \tag{8.32}$$

(ii) Down-type quark tower

The spectrum of the down-type quark tower is determined by Eq. (7.99) with the K matrix in Eq. (7.98). The four fields d, d', D^+ and D^- intertwine with each other. It is straightforward to find eigenvectors (with zero eigenvalue) of

$$K_d = \begin{pmatrix} \bar{c}_H S_R^u & i\bar{s}_H C_R^u & 0 & 0 \\ i\bar{s}_H C_L^u & \bar{c}_H S_L^u & \mu_d \mathcal{C}_{L2}^{D_d} & \mu_d \mathcal{C}_{L1}^{D_d} \\ i\mu_d^* \bar{s}_H S_R^u & \mu_d^* \bar{c}_H C_R^u & -\mathcal{S}_{R2}^{D_d} & -\mathcal{S}_{R1}^{D_d} \\ 0 & 0 & \mathcal{S}_{L1}^{D_d} & \mathcal{S}_{L2}^{D_d} \end{pmatrix}. \tag{8.33}$$

The KK expansion is given by

$$\begin{pmatrix} \check{d} \\ \check{d}' \\ \check{D}^+ \\ \check{D}^- \end{pmatrix} = \sqrt{k} \sum_{n=0}^{\infty} \left\{ d_L^{(n)}(x) \begin{pmatrix} f_L^{d^{(n)}}(z) \\ g_L^{d^{(n)}}(z) \\ h_L^{d^{(n)}}(z) \\ k_L^{d^{(n)}}(z) \end{pmatrix} + d_R^{(n)}(x) \begin{pmatrix} f_R^{d^{(n)}}(z) \\ g_R^{d^{(n)}}(z) \\ h_R^{d^{(n)}}(z) \\ k_R^{d^{(n)}}(z) \end{pmatrix} \right\}$$

$$+ \sqrt{k} \sum_{n=1}^{\infty} \left\{ D_L^{(n)}(x) \begin{pmatrix} f_L^{D^{(n)}}(z) \\ g_L^{D^{(n)}}(z) \\ h_L^{D^{(n)}}(z) \\ k_L^{D^{(n)}}(z) \end{pmatrix} + D_R^{(n)}(x) \begin{pmatrix} f_R^{D^{(n)}}(z) \\ g_R^{D^{(n)}}(z) \\ h_R^{D^{(n)}}(z) \\ k_R^{D^{(n)}}(z) \end{pmatrix} \right\} \tag{8.34}$$

where the wave functions are normalized by

$$\int_1^{z_L} dz \left\{ |f|^2 + |g|^2 + |h|^2 + |k|^2 \right\} = 1 \tag{8.35}$$

in each mode. The KK excited states $d^{(n)}$ $(n \geq 1)$ are mostly contained in the d and d' components, whereas the KK excited states $D^{(n)}$ $(n \geq 1)$ are mostly contained in the D^+ and D^- components. Their masses for $\theta_H = 0.1$ and $m_{\rm KK} = 13$ TeV are $(m_{d^{(1)}}, m_{d^{(2)}}, m_{d^{(3)}}, \cdots) = (12.2,\ 17.8,\ 25.1, \cdots)$ TeV and $(m_{D^{(1)}}, m_{D^{(2)}}, m_{D^{(3)}}, \cdots) = (8.4,\ 16.7,\ 22.8, \cdots)$ TeV.

The wave functions for the $d^{(n)}$ modes are given by

$$\begin{pmatrix} f_L^{d^{(n)}}(z) \\ g_L^{d^{(n)}}(z) \\ h_L^{d^{(n)}}(z) \\ k_L^{d^{(n)}}(z) \end{pmatrix} = \frac{1}{\sqrt{r_{d^{(n)}}}} \begin{pmatrix} \alpha_{d^{(n)}} C_L(z) \\ \beta_{d^{(n)}} S_L(z) \\ a_{d^{(n)}} \mathcal{C}_{L2}(z) + b_{d^{(n)}} \mathcal{C}_{L1}(z) \\ a_{d^{(n)}} \mathcal{S}_{L1}(z) + b_{d^{(n)}} \mathcal{S}_{L2}(z) \end{pmatrix},$$

$$\begin{pmatrix} f_R^{d^{(n)}}(z) \\ g_R^{d^{(n)}}(z) \\ h_R^{d^{(n)}}(z) \\ k_R^{d^{(n)}}(z) \end{pmatrix} = \frac{1}{\sqrt{r_{d^{(n)}}}} \begin{pmatrix} \alpha_{d^{(n)}} S_R(z) \\ \beta_{d^{(n)}} C_R(z) \\ a_{d^{(n)}} \mathcal{S}_{R2}(z) + b_{d^{(n)}} \mathcal{S}_{R1}(z) \\ a_{d^{(n)}} \mathcal{C}_{R1}(z) + b_{d^{(n)}} \mathcal{C}_{R2}(z) \end{pmatrix},$$

$$\begin{aligned} C_{L/R}(z) &= C_{L/R}(z; \lambda_{d^{(n)}}, c_u), \\ S_{L/R}(z) &= S_{L/R}(z; \lambda_{d^{(n)}}, c_u), \\ \mathcal{C}_{L/Rj}(z) &= \mathcal{C}_{L/Rj}(z; \lambda_{d^{(n)}}, c_{D_d}, \tilde{m}_{D_d}), \\ \mathcal{S}_{L/Rj}(z) &= \mathcal{S}_{L/Rj}(z; \lambda_{d^{(n)}}, c_{D_d}, \tilde{m}_{D_d}), \end{aligned} \tag{8.36}$$

where

$$\begin{aligned} \beta_{d^{(n)}} &= i\, \frac{\bar{c}_H S_R(1)}{\bar{s}_H C_R(1)}\, \alpha_{d^{(n)}}\,, \\ a_{d^{(n)}} &= -i\, \frac{\mu_d S_R(1)\, \mathcal{S}_{L2}(1)}{\bar{s}_H [\mathcal{S}_{L1} \mathcal{S}_{R1}(1) - \mathcal{S}_{L2} \mathcal{S}_{R2}(1)]}\, \alpha_{d^{(n)}}\,, \\ b_{d^{(n)}} &= -\frac{\mathcal{S}_{L1}(1)}{\mathcal{S}_{L2}(1)}\, a_{d^{(n)}}\,. \end{aligned} \tag{8.37}$$

For the $D^{(n)}$ mode the formulas are obtained by replacing $d^{(n)}$ by $D^{(n)}$ in (8.36) and (8.37).

8.3.3 *Leptons*

(i) Charged lepton tower

The spectrum of the charged lepton tower is determined by Eq. (7.102), and the K matrix is given by (7.101) which has the same structure as for the up-type quark tower. The KK expansion for the electron tower ($\{e^{(n)}\}$) is given by

$$
\begin{aligned}
\begin{pmatrix} \check{e} \\ \check{e}' \end{pmatrix} &= \sqrt{k} \sum_{n=0}^{\infty} \left\{ e_L^{(n)}(x) \begin{pmatrix} f_L^{e^{(n)}}(z) \\ g_L^{e^{(n)}}(z) \end{pmatrix} + e_R^{(n)}(x) \begin{pmatrix} f_R^{e^{(n)}}(z) \\ g_R^{e^{(n)}}(z) \end{pmatrix} \right\}, \\
\begin{pmatrix} f_L^{e^{(n)}}(z) \\ g_L^{e^{(n)}}(z) \end{pmatrix} &= \frac{1}{\sqrt{r_{e^{(n)}}}} \begin{pmatrix} \bar{c}_H C_L(z, \lambda_{e^{(n)}}, c_e) \\ -i\bar{s}_H \check{S}_L(z, \lambda_{e^{(n)}}, c_e) \end{pmatrix}, \\
\begin{pmatrix} f_R^{e^{(n)}}(z) \\ g_R^{e^{(n)}}(z) \end{pmatrix} &= \frac{1}{\sqrt{r_{e^{(n)}}}} \begin{pmatrix} \bar{c}_H S_R(z, \lambda_{e^{(n)}}, c_e) \\ -i\bar{s}_H \check{C}_R(z, \lambda_{e^{(n)}}, c_e) \end{pmatrix}.
\end{aligned} \tag{8.38}
$$

(ii) Neutrino tower

The spectrum in the neutrino sector is determined by Eq. (7.109), $\det K_{\nu_e}^{\pm} = (k\lambda \mp M_e)(S_L^e S_R^e + \bar{s}_H^2) + (m_{B_e}^2/k) S_R^e C_R^e = 0$ where the K matrix is given by Eq. (7.108),

$$
K_{\nu_e}^{\pm} = \begin{pmatrix} \bar{c}_H S_R^e & -\bar{s}_H C_R^e & 0 \\ \bar{s}_H C_L^e & \bar{c}_H S_L^e & m_{B_e}/k \\ -m_{B_e} \bar{s}_H S_R^e & -m_{B_e} \bar{c}_H C_R^e & k\lambda \mp M_e \end{pmatrix} \tag{8.39}
$$

for the first generation. Noting the property $f_{\pm L}(x)^c = \pm f_{\pm R}(x)$ in Eq. (7.105), one finds the KK expansion

$$
\begin{aligned}
\begin{pmatrix} \check{\nu}_e \\ \check{\nu}_e' \end{pmatrix} &= \sqrt{k} \sum_{n=0}^{\infty} \left\{ \nu_{eL}^{+(n)}(x) \begin{pmatrix} f_L^{\nu_e^{+(n)}}(z) \\ g_L^{\nu_e^{+(n)}}(z) \end{pmatrix} + \nu_{eR}^{+(n)}(x) \begin{pmatrix} f_R^{\nu_e^{+(n)}}(z) \\ g_R^{\nu_e^{+(n)}}(z) \end{pmatrix} \right\} \\
&\quad + \sqrt{k} \sum_{n=1}^{\infty} \left\{ \nu_{eL}^{-(n)}(x) \begin{pmatrix} f_L^{\nu_e^{-(n)}}(z) \\ g_L^{\nu_e^{-(n)}}(z) \end{pmatrix} + \nu_{eR}^{-(n)}(x) \begin{pmatrix} f_R^{\nu_e^{-(n)}}(z) \\ g_R^{\nu_e^{-(n)}}(z) \end{pmatrix} \right\}, \\
\eta_e &= \sum_{n=0}^{\infty} \nu_{eL}^{+(n)}(x)\, h^{\nu_e^{+(n)}} + \sum_{n=1}^{\infty} \nu_{eL}^{-(n)}(x)\, h^{\nu_e^{-(n)}} , \\
\nu_{eR}^{\pm(n)} &= \pm(\nu_{eL}^{\pm(n)})^c = \pm e^{i\delta_C} \sigma^2 (\nu_{eL}^{\pm(n)})^* ,
\end{aligned} \tag{8.40}
$$

where

$$
\begin{aligned}
\begin{pmatrix} f_L^{\nu_e^{\pm(n)}}(z) \\ g_L^{\nu_e^{\pm(n)}}(z) \\ f_R^{\nu_e^{\pm(n)}}(z) \\ g_R^{\nu_e^{\pm(n)}}(z) \end{pmatrix} &= \frac{1}{\sqrt{r_{\nu_e^{\pm(n)}}}} \begin{pmatrix} \bar{s}_H C_L(z)/S_R(1) \\ i\bar{c}_H S_L(z)/C_R(1) \\ \bar{s}_H S_R(z)/S_R(1) \\ i\bar{c}_H C_L(z)/C_R(1) \end{pmatrix}, \\
C_{L/R}(z) &= C_{L/R}(z; \lambda_{\nu_e^{\pm(n)}}, c_e) , \\
S_{L/R}(z) &= S_{L/R}(z; \lambda_{\nu_e^{\pm(n)}}, c_e) , \\
h^{\nu_e^{\pm(n)}} &= \frac{1}{\sqrt{r_{\nu_e^{\pm(n)}}}} \frac{i m_{B_e}}{k\lambda_{\nu_e^{\pm(n)}} \mp M} ,
\end{aligned} \tag{8.41}
$$

and each mode function is normalized by

$$\int_1^{z_L} dz \left\{ |f_L|^2 + |g_L|^2 + |f_R|^2 + |g_R|^2 \right\} + |h|^2 = 1 \,. \tag{8.42}$$

We note that the KK excited modes $\nu_e^{+(n)}$ and $\nu_e^{-(n)}$ ($n \geq 1$) have nearly equal masses.

8.3.4 *Dark fermions*

(i) Dark fermion F tower

Dark fermion fields $\Psi^{F_q}_{(\mathbf{3},\mathbf{4})}$ and $\Psi^{F_\ell}_{(\mathbf{1},\mathbf{4})}$ are treated in a manner similar to that for the up-type quark and charged lepton towers. Let $\Psi^F = (F_u, F_d, F_{u'}, F_{d'})$ represent $\Psi^{F_q}_{(\mathbf{3},\mathbf{4})}$ and $\Psi^{F_\ell}_{(\mathbf{1},\mathbf{4})}$ collectively. The spectrum is determined by (7.115), $S_L S_R(1, \lambda_{F^{(n)}}, c_F) + \bar{c}_H^2 = 0$. The K matrix for $(F_u, F_{u'})$ and $(F_d, F_{d'})$ is given by (7.114)

$$\begin{aligned}
&K_F = \begin{pmatrix} i\bar{s}_H S_R^F & \bar{c}_H C_R^F \\ \bar{c}_H C_L^F & i\bar{s}_H S_L^F \end{pmatrix}, \\
&\begin{pmatrix} \alpha_{F_u} \\ \alpha_{F_{u'}} \end{pmatrix}, \begin{pmatrix} \alpha_{F_d} \\ \alpha_{F_{d'}} \end{pmatrix} \propto \begin{pmatrix} \bar{c}_H (C_R^F / S_R^F) \\ -i\bar{s}_H \end{pmatrix}.
\end{aligned} \tag{8.43}$$

The KK expansion for $(F_u, F_{u'})$ is given by

$$\begin{aligned}
&\begin{pmatrix} \tilde{\check{F}}_u \\ \tilde{\check{F}}_{u'} \end{pmatrix} = \sqrt{k} \sum_{n=1}^{\infty} \left\{ F_{uL}^{(n)}(x) \begin{pmatrix} f_L^{F_u^{(n)}}(z) \\ g_L^{F_u^{(n)}}(z) \end{pmatrix} + F_{uR}^{(n)}(x) \begin{pmatrix} f_R^{F_u^{(n)}}(z) \\ g_R^{F_u^{(n)}}(z) \end{pmatrix} \right\}, \\
&\begin{pmatrix} f_L^{F_u^{(n)}}(z) \\ g_L^{F_u^{(n)}}(z) \end{pmatrix} = \frac{1}{\sqrt{r_{F_u^{(n)}}}} \begin{pmatrix} \bar{c}_H \check{C}_L(z, \lambda_{F_u^{(n)}}, c_F) \\ -i\bar{s}_H S_L(z, \lambda_{F_u^{(n)}}, c_F) \end{pmatrix}, \\
&\begin{pmatrix} f_R^{F_u^{(n)}}(z) \\ g_R^{F_u^{(n)}}(z) \end{pmatrix} = \frac{1}{\sqrt{r_{F_u^{(n)}}}} \begin{pmatrix} \bar{c}_H \check{S}_R(z, \lambda_{F_u^{(n)}}, c_F) \\ -i\bar{s}_H C_R(z, \lambda_{F_u^{(n)}}, c_F) \end{pmatrix}.
\end{aligned} \tag{8.44}$$

The KK expansion for $(F_d, F_{d'})$ takes the same form as (8.44) where $F_u^{(n)}$ is replaced by $F_d^{(n)}$.

(ii) Dark fermion V tower

The dark fermion $V^\pm = \Psi^V_{(\mathbf{1},\mathbf{5})^\pm}$ fields are in the vector representation of $SO(5)$. The spectrum of the upper three components $V_j^\pm$ ($j = 1, 2, 3$) is determined by (7.125), $(\mathcal{S}_{L1}\mathcal{S}_{R1} - \mathcal{S}_{L2}\mathcal{S}_{R2})(1; \lambda_{V_j^{(n)}}, c_V, \tilde{m}_V) = 0$. The K matrix is given by (7.124) and

$$K_{V0} = \begin{pmatrix} \mathcal{S}_{L1} & \mathcal{S}_{L2} \\ \mathcal{S}_{R2} & \mathcal{S}_{R1} \end{pmatrix}, \quad \begin{pmatrix} a_j \\ b_j \end{pmatrix} \propto \begin{pmatrix} 1 \\ -\mathcal{S}_{L1}/\mathcal{S}_{L2} \end{pmatrix}. \tag{8.45}$$

The KK expansion for $V_j^\pm$ $(j=1,2,3)$ is given by

$$\begin{pmatrix} \tilde{V}_j^+ \\ \tilde{V}_j^- \end{pmatrix} = \sqrt{k} \sum_{n=1}^{\infty} \left\{ V_{jL}^{(n)}(x) \begin{pmatrix} f_L^{+(n)}(z) \\ f_L^{-(n)}(z) \end{pmatrix} + V_{jR}^{(n)}(x) \begin{pmatrix} f_R^{+(n)}(z) \\ f_R^{-(n)}(z) \end{pmatrix} \right\},$$

$$\begin{pmatrix} f_L^{+(n)}(z) \\ f_L^{-(n)}(z) \end{pmatrix} = \frac{1}{\sqrt{r_{V^{(n)}}}} \begin{pmatrix} \mathcal{S}_{L1}(z) - \check{\mathcal{S}}_{L2}(z) \\ \mathcal{C}_{L2}(z) - \check{\mathcal{C}}_{L1}(z) \end{pmatrix},$$

$$\begin{pmatrix} f_R^{+(n)}(z) \\ f_R^{-(n)}(z) \end{pmatrix} = \frac{1}{\sqrt{r_{V^{(n)}}}} \begin{pmatrix} \mathcal{C}_{R1}(z) - \check{\mathcal{C}}_{R2}(z) \\ \mathcal{S}_{R2}(z) - \check{\mathcal{S}}_{R1}(z) \end{pmatrix}, \tag{8.46}$$

where $(\mathcal{S},\mathcal{C})_{L/Rj}(z) = (\mathcal{S},\mathcal{C})_{L/Rj}(z;\lambda_{V_j^{(n)}}, c_V, \tilde{m}_V)$,

$$\begin{pmatrix} \check{\mathcal{S}}_{L2}(z) \\ \check{\mathcal{C}}_{L1}(z) \end{pmatrix} = \frac{\mathcal{S}_{L1}(1)}{\mathcal{S}_{L2}(1)} \begin{pmatrix} \mathcal{S}_{L2}(z) \\ \mathcal{C}_{L1}(z) \end{pmatrix}, \begin{pmatrix} \check{\mathcal{C}}_{R2}(z) \\ \check{\mathcal{S}}_{R1}(z) \end{pmatrix} = \frac{\mathcal{S}_{R2}(1)}{\mathcal{S}_{R1}(1)} \begin{pmatrix} \mathcal{C}_{R2}(z) \\ \mathcal{S}_{R1}(z) \end{pmatrix}, \tag{8.47}$$

and each mode is normalized by $\int_1^{z_L} dz\{(f_{L/R}^+)^2 + (f_{L/R}^-)^2\} = 1$. Making use of $\mathcal{S}_{L1}(1)/\mathcal{S}_{L2}(1) = \mathcal{S}_{R2}(1)/\mathcal{S}_{R1}(1)$ and the relations in Eq. (C.29), one can show that

$$\begin{aligned} &\int_1^{z_L} dz \Big\{ \big(\mathcal{S}_{L1}(z) - \check{\mathcal{S}}_{L2}(z)\big)^2 + \big(\mathcal{C}_{L2}(z) - \check{\mathcal{C}}_{L1}(z)\big)^2 \Big\} \\ &= \int_1^{z_L} dz \Big\{ \big(\mathcal{C}_{R1}(z) - \check{\mathcal{C}}_{R2}(z)\big)^2 + \big(\mathcal{S}_{R2}(z) - \check{\mathcal{S}}_{R1}(z)\big)^2 \Big\}. \end{aligned} \tag{8.48}$$

The spectrum of the $V_4^\pm$ and $V_5^\pm$ components is determined by Eq. (7.129), $\mathcal{B}_0(1;\lambda_{V_N^{(n)}}, c_V, \tilde{m}_V) = 2\cos 2\theta_H$. For each $\lambda_{V_N^{(n)}}$ there are doubly degenerate states. With the use of (7.122), (7.126) and (7.128), the KK expansion is found to be

$$\begin{pmatrix} \tilde{V}_4^+ \\ \tilde{V}_4^- \\ \tilde{V}_5^+ \\ \tilde{V}_5^- \end{pmatrix} = \sqrt{k} \sum_{n=1}^{\infty} \sum_{p=1}^{2} \left\{ V_L^{p(n)}(x) \begin{pmatrix} f_L^{+p(n)}(z) \\ f_L^{-p(n)}(z) \\ g_L^{+p(n)}(z) \\ g_L^{-p(n)}(z) \end{pmatrix} + V_R^{p(n)}(x) \begin{pmatrix} f_R^{+p(n)}(z) \\ f_R^{-p(n)}(z) \\ g_R^{+p(n)}(z) \\ g_R^{-p(n)}(z) \end{pmatrix} \right\},$$

$$\begin{pmatrix} f_L^{+p(n)}(z) \\ f_L^{-p(n)}(z) \\ g_L^{+p(n)}(z) \\ g_L^{-p(n)}(z) \end{pmatrix} = \frac{1}{\sqrt{r_{V^{p(n)}}}} \begin{pmatrix} a_4^{p(n)} \mathcal{S}_{L1}(z) + b_4^{p(n)} \mathcal{S}_{L2}(z) \\ a_4^{p(n)} \mathcal{C}_{L2}(z) + b_4^{p(n)} \mathcal{C}_{L1}(z) \\ a_5^{p(n)} \mathcal{C}_{L2}(z) + b_5^{p(n)} \mathcal{C}_{L1}(z) \\ a_5^{p(n)} \mathcal{S}_{L1}(z) + b_5^{p(n)} \mathcal{S}_{L2}(z) \end{pmatrix},$$

$$\begin{pmatrix} f_R^{+p(n)}(z) \\ f_R^{-p(n)}(z) \\ g_R^{+p(n)}(z) \\ g_R^{-p(n)}(z) \end{pmatrix} = \frac{1}{\sqrt{r_{V^{p(n)}}}} \begin{pmatrix} a_4^{p(n)} \mathcal{C}_{R1}(z) + b_4^{p(n)} \mathcal{C}_{R2}(z) \\ a_4^{p(n)} \mathcal{S}_{R2}(z) + b_4^{p(n)} \mathcal{S}_{R1}(z) \\ a_5^{p(n)} \mathcal{S}_{R2}(z) + b_5^{p(n)} \mathcal{S}_{R1}(z) \\ a_5^{p(n)} \mathcal{C}_{R1}(z) + b_5^{p(n)} \mathcal{C}_{R2}(z) \end{pmatrix}, \tag{8.49}$$

where $(\mathcal{S},\mathcal{C})_{L/Rj}(z) = (\mathcal{S},\mathcal{C})_{L/Rj}(z;\lambda_{V_N^{(n)}},c_V,\tilde{m}_V)$ and the coefficients $a_{4,5}$ and $b_{4,5}$ satisfy

$$\begin{pmatrix} a_4^{p(n)} \\ b_4^{p(n)} \end{pmatrix} = M \begin{pmatrix} a_5^{p(n)} \\ b_5^{p(n)} \end{pmatrix}, \quad \begin{pmatrix} a_5^{p(n)} \\ b_5^{p(n)} \end{pmatrix} = -M \begin{pmatrix} a_4^{p(n)} \\ b_4^{p(n)} \end{pmatrix},$$
$$M = \frac{1}{2\sin 2\theta_H}\begin{pmatrix} \alpha & \beta \\ \gamma & -\alpha \end{pmatrix}, \quad M^2 = -I \ . \tag{8.50}$$

Here α,β,γ are defined in Eq. (7.128) with $\lambda = \lambda_{V_N^{(n)}}$. The normalization factors are determined by

$$\int_1^{z_L} dz \Big\{ \big|f_{L/R}^{+p(n)}\big|^2 + \big|f_{L/R}^{-p(n)}\big|^2 + \big|g_{L/R}^{+p(n)}\big|^2 + \big|g_{L/R}^{-p(n)}\big|^2 \Big\} = 1 \ . \tag{8.51}$$

The two degenerate states for each $\lambda_{V_N^{(n)}}$ must be chosen to be orthogonal to each other.

8.3.5 *Chiral structure in the fermion zero modes*

In the SM the W and Z couplings to quarks and leptons are chiral. In GHU the couplings slightly depend on θ_H, as will be seen explicitly in the next section. In this subsection we examine the wave functions of the fermion zero modes with $m/m_{\rm KK} \ll 1$ for $|\theta_H| \lesssim 0.1$.

Note that all quarks and leptons are massless at $\theta_H = 0$. Even for $|\theta_H| \lesssim 0.1$, one sees that $\lambda^{(0)} z_L \ll 1$ for the $u^{(0)}$, $d^{(0)}$, $e^{(0)}$ and $\nu_e^{+(0)}$ modes. The wave functions of the $u^{(0)}$ and $d^{(0)}$ modes at $\theta_H = 0$ are given by

$$\begin{pmatrix} f_L^{u^{(0)}}(z) \\ g_L^{u^{(0)}}(z) \end{pmatrix} \propto \begin{pmatrix} z^{-c_u} \\ 0 \end{pmatrix}, \quad \begin{pmatrix} f_R^{u^{(0)}}(z) \\ g_R^{u^{(0)}}(z) \end{pmatrix} \propto \begin{pmatrix} 0 \\ z^{c_u} \end{pmatrix},$$
$$\begin{pmatrix} f_L^{d^{(0)}}(z) \\ g_L^{d^{(0)}}(z) \\ h_L^{d^{(0)}}(z) \\ k_L^{d^{(0)}}(z) \end{pmatrix} \propto \begin{pmatrix} z^{-c_u} \\ 0 \\ 0 \\ 0 \end{pmatrix}, \quad \begin{pmatrix} f_R^{d^{(0)}}(z) \\ g_R^{d^{(0)}}(z) \\ h_R^{d^{(0)}}(z) \\ k_R^{d^{(0)}}(z) \end{pmatrix} \propto \begin{pmatrix} 0 \\ \mathcal{C}_R(z) \\ a_d \mathcal{S}_{R2}(z) \\ a_d \mathcal{C}_{R1}(z) \end{pmatrix},$$
$$a_d = \mu_d \mathcal{C}_R(1)/\mathcal{S}_{R2}(1) \ ,$$
$$\begin{pmatrix} \mathcal{C}_R(z) \\ \mathcal{S}_{R2}(z) \\ \mathcal{C}_{R1}(z) \end{pmatrix} = \lim_{\lambda\to 0} \begin{pmatrix} \mathcal{C}_R(z;\lambda,c_u) \\ \mathcal{S}_{R2}(z;\lambda,c_{D_d},\tilde{m}_{D_d}) \\ \mathcal{C}_{R1}(z;\lambda,c_{D_d},\tilde{m}_{D_d}) \end{pmatrix} . \tag{8.52}$$

Note that $c_u < -\frac{1}{2}$ and $C_R(z) \propto z^{c_u}$. Similarly the wave functions of $e^{(0)}$ and $\nu_e^{+(0)}$ modes at $\theta_H = 0$ are given by

$$\begin{pmatrix} f_L^{e^{(0)}}(z) \\ g_L^{e^{(0)}}(z) \end{pmatrix} \propto \begin{pmatrix} z^{-c_e} \\ 0 \end{pmatrix}, \quad \begin{pmatrix} f_R^{e^{(0)}}(z) \\ g_R^{e^{(0)}}(z) \end{pmatrix} \propto \begin{pmatrix} 0 \\ z^{c_e} \end{pmatrix},$$

$$\begin{pmatrix} f_L^{\nu_e^{+(0)}}(z) \\ g_L^{\nu_e^{+(0)}}(z) \\ f_R^{\nu_e^{+(0)}}(z) \\ g_R^{\nu_e^{+(0)}}(z) \\ h^{\nu_e^{+(0)}} \end{pmatrix} \propto \begin{pmatrix} z^{-c_e} \\ 0 \\ 0 \\ 0 \\ 0 \end{pmatrix}. \tag{8.53}$$

The chiral structure is evident. The left-handed components $(u_L^{(0)}, d_L^{(0)})$ and $(\nu_{eL}^{+(0)}, e_L^{(0)})$ form $SU(2)_L$ doublets, whereas the right-handed components $u_R^{(0)}$, $d_R^{(0)}$ and $e_R^{(0)}$ are $SU(2)_L$ singlets.

At $\theta_H \neq 0$ the mass eigenstates contain contributions from other components, but their contributions remain small. In Tables 8.2 and 8.3 the weight of each component in the norm of the zero mode is tabulated for $\theta_H = 0.1$ and $m_{\rm KK} = 13$ TeV. It is seen that the zero mode sector is nearly chiral.

Table 8.2 Norm of the quark zero modes for $\theta_H = 0.1$ and $m_{\rm KK} = 13$ TeV. $N_{f_L} = \int_1^{z_L} dz |f_L(z)|^2$ etc. Wave functions are given in (8.30) and (8.36). In the table 10^{-20} implies $O(10^{-20})$. The zero modes have the chiral structure.

	N_{f_L}	N_{g_L}			N_{f_R}	N_{g_R}		
$u^{(0)}$	1.	10^{-20}			0.0025	0.9975		
$c^{(0)}$	1.	10^{-14}			0.0025	0.9975		
$t^{(0)}$	0.9998	0.0002			0.0017	0.9983		
	N_{f_L}	N_{g_L}	N_{h_L}	N_{k_L}	N_{f_R}	N_{g_R}	N_{h_R}	N_{k_R}
$d^{(0)}$	1.	10^{-23}	10^{-14}	10^{-14}	10^{-4}	0.0208	0.3870	0.5919
$s^{(0)}$	1.	10^{-18}	10^{-11}	10^{-11}	10^{-5}	0.0079	0.3710	0.6211
$b^{(0)}$	1.	10^{-11}	10^{-7}	10^{-7}	10^{-7}	0.0003	0.2653	0.7344

The wave functions of the KK excited modes are nearly vector-like. In Tables 8.4 and 8.5 the weight of each component in the norm of the KK excited modes ($n = 1 \sim 4$) in the first generation is tabulated for $\theta_H = 0.1$ and $m_{\rm KK} = 13$ TeV. In the quark sector $\{u^{(2\ell)}, d^{(2\ell)}\}$ and $\{u^{(2\ell-1)}, d^{(2\ell-1)}\}$

Table 8.3 Norm of the lepton zero modes for $\theta_H = 0.1$ and $m_{KK} = 13$ TeV. We have taken $m_{\nu_e} = m_{\nu_\mu} = m_{\nu_\tau} = 1$ meV and $M_e = M_\mu = M_\tau = 10^6$ GeV for evaluation. $N_{f_L} = \int_1^{z_L} dz |f_L(z)|^2$, $N_h = |h|^2$ etc. Wave functions are given in (8.38) and (8.41). In the table 10^{-26} implies $O(10^{-26})$. The zero modes have the chiral structure.

	N_{f_L}	N_{g_L}	N_{f_R}	N_{g_R}	
$e^{(0)}$	1.	10^{-26}	0.0025	0.9975	
$\mu^{(0)}$	1.	10^{-17}	0.0025	0.9975	
$\tau^{(0)}$	1.	10^{-12}	0.0025	0.9975	
	N_{f_L}	N_{g_L}	N_{f_R}	N_{g_R}	N_h
$\nu_e^{+(0)}$	1.	10^{-61}	10^{-20}	10^{-18}	10^{-18}
$\nu_\mu^{+(0)}$	1.	10^{-61}	10^{-25}	10^{-22}	10^{-18}
$\nu_\tau^{+(0)}$	1.	10^{-61}	10^{-27}	10^{-25}	10^{-18}

Table 8.4 Norm of the KK modes $\{u^{(n)}, d^{(n)}, D_d^{(n)}\}$ ($n = 1 \sim 4$) for $\theta_H = 0.1$ and $m_{KK} = 13$ TeV. Wave functions are given in (8.30) and (8.36).

	N_{f_L}	N_{g_L}			N_{f_R}	N_{g_R}		
$u^{(1)}$	10^{-19}	1.			10^{-11}	1.		
$u^{(2)}$	1.	10^{-19}			1.	10^{-11}		
$u^{(3)}$	10^{-18}	1.			10^{-11}	1.		
$u^{(4)}$	1.	10^{-18}			1.	10^{-11}		
	N_{f_L}	N_{g_L}	N_{h_L}	N_{k_L}	N_{f_R}	N_{g_R}	N_{h_R}	N_{k_R}
$d^{(1)}$	10^{-22}	1.	10^{-10}	10^{-11}	10^{-14}	1.	10^{-10}	10^{-10}
$d^{(2)}$	1.	10^{-21}	10^{-12}	10^{-12}	1.	10^{-14}	10^{-12}	10^{-12}
$d^{(3)}$	10^{-21}	1.	10^{-10}	10^{-10}	10^{-14}	1.	10^{-10}	10^{-10}
$d^{(4)}$	1.	10^{-21}	10^{-12}	10^{-12}	1.	10^{-14}	10^{-12}	10^{-12}
$D_d^{(1)}$	10^{-13}	10^{-10}	0.6971	0.3029	10^{-5}	0.0070	0.1246	0.8684
$D_d^{(2)}$	10^{-12}	10^{-10}	0.6334	0.3666	10^{-5}	0.0028	0.9517	0.0454
$D_d^{(3)}$	10^{-13}	10^{-10}	0.2843	0.7157	10^{-5}	0.0023	0.0337	0.9640
$D_d^{(4)}$	10^{-12}	10^{-10}	0.7808	0.2192	10^{-6}	0.0018	0.9733	0.0249

($\ell \geq 1$) are mostly in the (u, d) and (u', d') components of $\Psi^q_{(\mathbf{3},\mathbf{4})}$, respectively. Similarly in the lepton sector $\{e^{(2\ell)}, \nu_e^{\pm(2\ell)}\}$ and $\{e^{(2\ell-1)}, \nu_e^{\pm(2\ell-1)}\}$ ($\ell \geq 1$) are in the (e, ν_e) and (e', ν'_e) components of $\Psi^\ell_{(\mathbf{1},\mathbf{4})}$, respectively.

Table 8.5 Norm of the KK modes $\{e^{(n)}, \nu_e^{\pm(n)}\}$ $(n = 1 \sim 4)$ for $\theta_H = 0.1$ and $m_{\rm KK} = 13$ TeV. Wave functions are given in (8.38) and (8.41). The Majorana states $\nu_e^{+(n)}$ and $\nu_e^{-(n)}$ have almost degenerate masses.

	N_{f_L}	N_{g_L}	N_{f_R}	N_{g_R}	
$e^{(1)}$	10^{-25}	1.	10^{-14}	1.	
$e^{(2)}$	1.	10^{-25}	1.	10^{-14}	
$e^{(3)}$	10^{-25}	1.	10^{-14}	1.	
$e^{(4)}$	1.	10^{-24}	1.	10^{-14}	
	N_{f_L}	N_{g_L}	N_{f_R}	N_{g_R}	N_h
$\nu_e^{+(1)}$	10^{-28}	0.5	10^{-17}	0.5	10^{-15}
$\nu_e^{-(1)}$	10^{-28}	0.5	10^{-17}	0.5	10^{-15}
$\nu_e^{+(2)}$	0.5	10^{-28}	0.5	10^{-17}	10^{-18}
$\nu_e^{-(2)}$	0.5	10^{-27}	0.5	10^{-16}	10^{-17}
$\nu_e^{+(3)}$	10^{-27}	0.5	10^{-17}	0.5	10^{-15}
$\nu_e^{-(3)}$	10^{-27}	0.5	10^{-16}	0.5	10^{-14}
$\nu_e^{+(4)}$	0.5	10^{-25}	0.5	10^{-15}	10^{-15}
$\nu_e^{-(4)}$	0.5	10^{-26}	0.5	10^{-15}	10^{-16}

8.4 Gauge couplings

Gauge couplings are determined by inserting the KK expansions of gauge fields and fermion fields in the 5D action and integrating over the fifth-dimensional coordinate z. The evaluation of gauge and Higgs couplings is most conveniently done in the twisted gauge. We suppress the tilde ($\tilde{}$) sign in this and next subsections.

8.4.1 *Self-couplings of gauge fields*

The $SO(5)$ gauge fields are expanded as $A_M = \sum_{\alpha=1}^{10} A_M^\alpha T^\alpha$, (7.15). Making use of the relations in Eq. (A.18), one finds that

$$\begin{aligned}
-i[A_\mu, A_\nu] &= \varepsilon^{abc}\{A_\mu^{a_L} A_\nu^{b_L} T_L^c + A_\mu^{a_R} A_\nu^{b_R} T_R^c + \tfrac{1}{2} A_\mu^{\hat{a}} A_\nu^{\hat{b}} (T_L^c + T_R^c)\} \\
&\quad + \tfrac{1}{2}(A_\mu^{\hat{a}} A_\nu^{\hat{4}} - A_\mu^{\hat{4}} A_\nu^{\hat{a}})(T_L^a - T_R^a) \\
&\quad + \tfrac{1}{2}\varepsilon^{abc}\{(A_\mu^{a_L} + A_\mu^{a_R}) A_\nu^{\hat{b}} + A_\mu^{\hat{a}}(A_\nu^{b_L} + A_\nu^{b_R})\}\hat{T}^c \\
&\quad - \tfrac{1}{2}\{(A_\mu^{a_L} - A_\mu^{a_R}) A_\nu^{\hat{4}} - A_\mu^{\hat{4}}(A_\nu^{a_L} - A_\nu^{a_R})\}\hat{T}^a \\
&\quad + \tfrac{1}{2}\{(A_\mu^{a_L} - A_\mu^{a_R}) A_\nu^{\hat{a}} - A_\mu^{\hat{a}}(A_\nu^{a_L} - A_\nu^{a_R})\}\hat{T}^4 \, .
\end{aligned} \tag{8.54}$$

Let us focus on the couplings among the W, W_R, Z, Z_R and γ towers. A_μ and B_μ contain

$$\begin{aligned}
A_\mu \Rightarrow \sqrt{k}\Big\{&(\mathring{W}^L_\mu + \mathring{W}^L_{R\mu})T^-_L + (\mathring{W}^R_\mu + \mathring{W}^R_{R\mu})T^-_R + \hat{\mathring{W}}_\mu \hat{T}^- \\
&+ (\mathring{W}^{L\dagger}_\mu + \mathring{W}^{L\dagger}_{R\mu})T^+_L + (\mathring{W}^{R\dagger}_\mu + \mathring{W}^{R\dagger}_{R\mu})T^+_R + \hat{\mathring{W}}^\dagger_\mu \hat{T}^+ \\
&+ (\mathring{Z}^L_\mu + \mathring{\gamma}^L_\mu + \mathring{Z}^L_{R\mu})T^3_L + (\mathring{Z}^R_\mu + \mathring{\gamma}^R_\mu + \mathring{Z}^R_{R\mu})T^3_R + \hat{\mathring{Z}}_\mu \hat{T}^3\Big\}, \\
B_\mu \Rightarrow \sqrt{k}\Big\{&\mathring{Z}^B_\mu + \mathring{\gamma}^B_\mu + \mathring{Z}^B_{R\mu}\Big\},
\end{aligned} \tag{8.55}$$

where $T^\pm_L = (T^1_L \pm iT^2_L)/\sqrt{2}$ etc. By inserting the expansion Eq. (8.55) into the bilinear part of the action in Eq. (7.16) and making use of the orthonormality relations (8.18) and (8.25) it is easy to see that

$$\begin{aligned}
&\int \frac{dz}{kz}(-\tfrac{1}{4})\big(\mathrm{Tr}\, F_{\mu\nu}F_{\rho\sigma} + F^{(B)}_{\mu\nu}F^{(B)}_{\rho\sigma}\big)\eta^{\mu\rho}\eta^{\nu\sigma} \\
&\Rightarrow \sum_{n=0}^{\infty}\Big\{-\frac{1}{2}W^{(n)\dagger}_{\mu\nu}W^{(n)\mu\nu} - \frac{1}{4}Z^{(n)}_{\mu\nu}Z^{(n)\mu\nu} - \frac{1}{4}\gamma^{(n)}_{\mu\nu}\gamma^{(n)\mu\nu}\Big\} \\
&\quad + \sum_{n=1}^{\infty}\Big\{-\frac{1}{2}W^{(n)\dagger}_{R\mu\nu}W^{(n)\mu\nu}_R - \frac{1}{4}Z^{(n)}_{R\mu\nu}Z^{(n)\mu\nu}_R\Big\}
\end{aligned} \tag{8.56}$$

where $W^{(n)}_{\mu\nu} = \partial_\mu W^{(n)}_\nu - \partial_\nu W^{(n)}_\mu$ etc.

Cubic couplings among the W, W_R, Z, Z_R and γ towers are contained in the $\mathrm{Tr}\,(\partial_\mu A_\nu - \partial_\nu A_\mu)[A_\rho, A_\sigma]\eta^{\mu\rho}\eta^{\nu\sigma}$ part of the action. Inserting the expansion Eq. (8.55) into the cubic part of the action and using Eq. (8.54), one finds that

$$\begin{aligned}
&\int \frac{dz}{kz}\frac{ig_A}{2}\mathrm{Tr}\,(\partial_\mu A_\nu - \partial_\nu A_\mu)[A^\mu, A^\nu] \\
&\Rightarrow -\frac{ig_A\sqrt{k}}{2}\int\frac{dz}{z}\Big\{2K[\mathring{W}^{L\dagger} + \mathring{W}^{L\dagger}_R, \mathring{W}^L + \mathring{W}^L_R, \mathring{Z}^L + \mathring{\gamma}^L + \mathring{Z}^L_R] \\
&\quad + 2K[\mathring{W}^{R\dagger} + \mathring{W}^{R\dagger}_R, \mathring{W}^R + \mathring{W}^R_R, \mathring{Z}^R + \mathring{\gamma}^R + \mathring{Z}^R_R] \\
&\quad + K[\hat{\mathring{W}}^\dagger, \hat{\mathring{W}}, \mathring{Z}^L + \mathring{Z}^R + \mathring{\gamma}^L + \mathring{\gamma}^R + \mathring{Z}^L_R + \mathring{Z}^R_R] \\
&\quad + K[\hat{\mathring{W}}^\dagger, \mathring{W}^L + \mathring{W}^R + \mathring{W}^L_R + \mathring{W}^R_R, \hat{\mathring{Z}}] \\
&\quad + K[\mathring{W}^{L\dagger} + \mathring{W}^{R\dagger} + \mathring{W}^{L\dagger}_R + \mathring{W}^{R\dagger}_R, \hat{\mathring{W}}, \hat{\mathring{Z}}]\,,
\end{aligned} \tag{8.57}$$

where $K[A,B,C]$ is defined in Eq. (2.50). Inserting the expression in Eqs. (8.15) and (8.22) into (8.57), one obtains for the WWZ couplings

$$\begin{aligned}
\mathcal{L}_{WWZ} &= \sum_{m,n,\ell} -ig^{WWZ}_{nm\ell}\, K[W^{(n)\dagger}, W^{(m)}, Z^{(\ell)}],\\
g^{WWZ}_{nm\ell}\\
&= g_w\sqrt{kL}\int_1^{z_L}\frac{dz}{z}\Big\{h^{L*}_{W(n)}h^{L}_{W(m)}h^{L}_{Z(\ell)} + h^{R*}_{W(n)}h^{R}_{W(m)}h^{R}_{Z(\ell)}\\
&\quad + \hat{h}^{*}_{W(n)}\hat{h}_{W(m)}h^{V}_{Z(\ell)} + \hat{h}^{*}_{W(n)}h^{V}_{W(m)}\hat{h}_{Z(\ell)} + h^{V*}_{W(n)}\hat{h}_{W(m)}\hat{h}_{Z(\ell)}\Big\},\\
h^{V}_{W(n)} &= \tfrac{1}{2}(h^{L}_{W(n)} + h^{R}_{W(n)}),\quad h^{V}_{Z(\ell)} = \tfrac{1}{2}(h^{L}_{Z(\ell)} + h^{R}_{Z(\ell)})\ . \qquad (8.58)
\end{aligned}$$

For the $WW\gamma$ one finds

$$\begin{aligned}
\mathcal{L}_{WW\gamma} &= \sum_{m,n,\ell} -ig^{WW\gamma}_{nm\ell}\, K[W^{(n)\dagger}, W^{(m)}, \gamma^{(\ell)}],\\
g^{WW\gamma}_{nm\ell} &= g_w\sqrt{kL}\int_1^{z_L}\frac{dz}{z}\Big\{h^{L*}_{W(n)}h^{L}_{W(m)}\\
&\qquad + h^{R*}_{W(n)}h^{R}_{W(m)} + \hat{h}^{*}_{W(n)}\hat{h}_{W(m)}\Big\}\, h^{V}_{\gamma(\ell)},\\
h^{V}_{\gamma(\ell)} &= h^{L}_{\gamma(\ell)} = h^{R}_{\gamma(\ell)}\ . \qquad (8.59)
\end{aligned}$$

In particular $h^{V}_{\gamma(0)} = \sin\theta^0_W/\sqrt{kL}$ for the photon $\gamma^{(0)}$ so that

$$g^{WW\gamma}_{nm0} = g_w\sin\theta^0_W\delta_{nm} = e\,\delta_{nm}\ . \qquad (8.60)$$

The WWZ_R couplings are given by

$$\begin{aligned}
\mathcal{L}_{WWZ_R} &= \sum_{m,n,\ell} -ig^{WWZ_R}_{nm\ell}\, K[W^{(n)\dagger}, W^{(m)}, Z_R^{(\ell)}],\\
g^{WWZ_R}_{nm\ell} &= g_w\sqrt{kL}\int_1^{z_L}\frac{dz}{z}\Big\{h^{L*}_{W(n)}h^{L}_{W(m)}h^{L}_{Z_R^{(\ell)}} + h^{R*}_{W(n)}h^{R}_{W(m)}h^{R}_{Z_R^{(\ell)}}\\
&\qquad + \tfrac{1}{2}\hat{h}^{*}_{W(n)}\hat{h}_{W(m)}(h^{L}_{Z_R^{(\ell)}} + h^{R}_{Z_R^{(\ell)}})\Big\}. \qquad (8.61)
\end{aligned}$$

The WW_RZ and W_RWZ couplings are given by

$$\begin{aligned}
\mathcal{L}_{WW_RZ} &= \sum_{m,n,\ell} -ig^{WW_RZ}_{nm\ell}\, K[W^{(n)\dagger}, W_R^{(m)}, Z^{(\ell)}],\\
\mathcal{L}_{W_RWZ} &= \sum_{m,n,\ell} -ig^{W_RWZ}_{nm\ell}\, K[W_R^{(n)\dagger}, W^{(m)}, Z^{(\ell)}],\\
g^{WW_RZ}_{nm\ell} &= g_w\sqrt{kL}\int_1^{z_L}\frac{dz}{z}\Big\{h^{L*}_{W(n)}h^{L}_{W_R^{(m)}}h^{L}_{Z(\ell)} + h^{R*}_{W(n)}h^{R}_{W_R^{(m)}}h^{R}_{Z(\ell)}\\
&\qquad + \tfrac{1}{2}\hat{h}^{*}_{W(n)}(h^{L}_{W_R^{(m)}} + h^{R}_{W_R^{(m)}})\hat{h}_{Z(\ell)}\Big\},\\
g^{W_RWZ}_{nm\ell} &= (g^{WW_RZ}_{mn\ell})^*\ . \qquad (8.62)
\end{aligned}$$

Quartic couplings among the W, Z and γ towers are contained in the $\mathrm{Tr}\,[A_\mu, A_\nu][A^\mu, A^\nu]$ part of the action. Inserting the expression (8.55) there, one finds that

$$
\begin{aligned}
&-\int \frac{dz}{kz}\,\frac{(-ig_A)^2}{4}\,\mathrm{Tr}\,[A_\mu, A_\nu][A^\mu, A^\nu] \\
&\qquad \Rightarrow \mathcal{L}_{\gamma\gamma WW} + \mathcal{L}_{\gamma ZWW} + \mathcal{L}_{ZZWW} + \mathcal{L}_{WWWW}, \\
&\mathcal{L}_{\gamma\gamma WW} = -\sum_{\ell,p,n,m} \frac{1}{2} g^{\gamma\gamma WW}_{\ell pnm} M[\gamma^{(\ell)}, W^{(n)\dagger}, \gamma^{(p)}, W^{(m)}], \\
&\mathcal{L}_{\gamma ZWW} = -\sum_{\ell,p,n,m} \frac{1}{2} g^{\gamma ZWW}_{\ell pnm} \Big\{ M[\gamma^{(\ell)}, W^{(n)\dagger}, Z^{(p)}, W^{(m)}] \\
&\qquad\qquad + M[Z^{(p)}, W^{(n)\dagger}, \gamma^{(\ell)}, W^{(m)}] \Big\}, \\
&\mathcal{L}_{ZZWW} = -\sum_{\ell,p,n,m} \frac{1}{2} g^{ZZWW}_{\ell pnm} M[Z^{(\ell)}, W^{(n)\dagger}, Z^{(p)}, W^{(m)}], \\
&\mathcal{L}_{WWWW} = -\sum_{\ell,p,n,m} \frac{1}{4} g^{WWWW}_{\ell pnm} M[W^{(\ell)\dagger}, W^{(p)}, W^{(n)\dagger}, W^{(m)}],
\end{aligned}
\tag{8.63}
$$

where $M[A, B, C, D]$ is defined in Eq. (2.50) and the coupling constants are given by

$$
\begin{aligned}
g^{\gamma\gamma WW}_{\ell pnm} &= g_w^2 kL \int_1^{z_L} \frac{dz}{z}\, h^V_{\gamma(\ell)} h^V_{\gamma(p)} \\
&\quad \times \Big\{ h^{L*}_{W(n)} h^L_{W(m)} + h^{R*}_{W(n)} h^R_{W(m)} + \hat{h}^*_{W(n)} \hat{h}_{W(m)} \Big\},
\end{aligned}
\tag{8.64}
$$

$$
\begin{aligned}
g^{\gamma ZWW}_{\ell pnm} &= g_w^2 kL \int_1^{z_L} \frac{dz}{z}\, h^V_{\gamma(\ell)} \Big\{ h^L_{Z(p)} h^{L*}_{W(n)} h^L_{W(m)} \\
&\quad + h^R_{Z(p)} h^{R*}_{W(n)} h^R_{W(m)} + h^V_{Z(p)} \hat{h}^*_{W(n)} \hat{h}_{W(m)} \\
&\quad + \hat{h}_{Z(p)} h^{V*}_{W(n)} \hat{h}_{W(m)} + \hat{h}_{Z(p)} \hat{h}^*_{W(n)} h^V_{W(m)} \Big\},
\end{aligned}
\tag{8.65}
$$

$$
\begin{aligned}
g^{ZZWW}_{\ell pnm} &= g_w^2 kL \int_1^{z_L} \frac{dz}{z} \Big\{ h^L_{Z(\ell)} h^L_{Z(p)} h^{L*}_{W(n)} h^L_{W(m)} \\
&+ h^R_{Z(\ell)} h^R_{Z(p)} h^{R*}_{W(n)} h^R_{W(m)} \\
&+ \tfrac{1}{2} (h^L_{Z(\ell)} h^{L*}_{W(n)} + h^R_{Z(\ell)} h^{R*}_{W(n)}) \hat{h}_{Z(p)} \hat{h}_{W(m)} \\
&+ \tfrac{1}{2} \hat{h}_{Z(\ell)} \hat{h}^*_{W(n)} (h^L_{Z(p)} h^L_{W(m)} + h^R_{Z(p)} h^R_{W(m)})
\end{aligned}
$$

$$+ (h^V_{Z(\ell)} \hat{h}^*_{W(n)} + \hat{h}_{Z(\ell)} h^{V*}_{W(n)})(h^V_{Z(p)} \hat{h}_{W(m)} + \hat{h}_{Z(p)} h^V_{W(m)})$$

$$+ \tfrac{1}{2} \hat{h}_{Z(\ell)} \hat{h}_{Z(p)} \hat{h}^*_{W(n)} \hat{h}_{W(m)} \Big\}, \tag{8.66}$$

$$g^{WWWW}_{\ell pnm} = g_w^2 kL \int_1^{z_L} \frac{dz}{z} \Big\{ h^{L*}_{W(\ell)} h^L_{W(p)} h^{L*}_{W(n)} h^L_{W(m)}$$

$$+ h^{R*}_{W(\ell)} h^R_{W(p)} h^{R*}_{W(n)} h^R_{W(m)}$$

$$+ \tfrac{1}{2}(h^{L*}_{W(\ell)} h^L_{W(p)} + h^{R*}_{W(\ell)} h^R_{W(p)}) \hat{h}^*_{W(n)} \hat{h}_{W(m)}$$

$$+ \tfrac{1}{2} \hat{h}^*_{W(\ell)} \hat{h}_{W(p)} (h^{L*}_{W(n)} h^L_{W(m)} + h^{R*}_{W(n)} h^R_{W(m)})$$

$$+ \tfrac{1}{2}(h^{L*}_{W(\ell)} \hat{h}_{W(p)} + \hat{h}^*_{W(\ell)} h^L_{W(p)})(h^{R*}_{W(n)} \hat{h}_{W(m)} + \hat{h}^*_{W(n)} h^R_{W(m)})$$

$$+ \tfrac{1}{2}(h^{R*}_{W(\ell)} \hat{h}_{W(p)} + \hat{h}^*_{W(\ell)} h^R_{W(p)})(h^{L*}_{W(n)} \hat{h}_{W(m)} + \hat{h}^*_{W(n)} h^L_{W(m)})$$

$$+ \tfrac{1}{2} \hat{h}^*_{W(\ell)} \hat{h}_{W(p)} \hat{h}^*_{W(n)} \hat{h}_{W(m)} \Big\}. \tag{8.67}$$

Note that the photon ($\gamma^{(0)}$) couplings in $\mathcal{L}_{\gamma\gamma WW}$ and $\mathcal{L}_{\gamma ZWW}$ become, as $h^V_{\gamma^{(0)}} = \sin\theta^0_W/\sqrt{kL}$,

$$\begin{aligned} g^{\gamma\gamma WW}_{00nm} &= e^2\, \delta_{nm} \ , \\ g^{\gamma ZWW}_{0pnm} &= e\, g^{WWZ}_{nmp} \ . \end{aligned} \tag{8.68}$$

The deviation of the couplings among the zero modes of W and Z, particularly of the couplings g^{WWZ}_{000} and g^{ZZWW}_{0000}, from those in the SM is very small. In the SM $g^{\rm SM}_{WWZ} = g_w \cos\theta^{\rm SM}_W$, $g^{\rm SM}_{WWWW} = g_w^2$ and $g^{\rm SM}_{ZZWW} = g_w^2 \cos^2\theta^{\rm SM}_W$. In GHU, for $\theta_H = 0.1$, $m_{\rm KK} = 13\,{\rm TeV}$ and $\sin^2\theta^0_W = 0.2304$ for instance, one finds that

$$\begin{aligned} &\frac{g^{WWZ}_{000}}{g_w \cos\theta^0_W} - 1 = 2.7 \times 10^{-7} \ , \\ &\frac{g^{ZZWW}_{0000}}{g_w^2 \cos^2\theta^0_W} - 1 = 8.8 \times 10^{-7} \ , \\ &\frac{g^{WWWW}_{0000}}{g_w^2} - 1 = -0.00480 \ . \end{aligned} \tag{8.69}$$

Some of the couplings to KK excited modes are tabulated in Table 8.6. The couplings $g^{WWZ}_{00n}/g_w \cos\theta^0_W$, $g^{WWZ}_{0n0}/g_w \cos\theta^0_W$ and $g^{ZZWW}_{000n}/g_w^2 \cos^2\theta^0_W$ ($n \geq 1$) are very small, but $g^{WWZ}_{nn0}/g_w \cos\theta^0_W$, $g^{ZZWW}_{00nn}/g_w^2 \cos^2\theta^0_W$ and g^{WWWW}_{00nn}/g_w^2 ($n \geq 1$) are $O(1)$. Couplings among the KK modes such as $g^{WWZ}_{nn1}/g_w \cos\theta^0_W$ and g^{WWWW}_{11nn}/g_w^2 ($n \geq 1$) become large.

Table 8.6 Gauge couplings involving KK modes of W and Z for $\theta_H = 0.1$, $m_{\rm KK} = 13$ TeV and $\sin^2\theta_W^0 = 0.2304$ are listed.

n	$\dfrac{g^{WWZ}_{00n}}{g_w \cos\theta^0_W}$	$\dfrac{g^{WWZ}_{0n0}}{g_w \cos\theta^0_W}$	$\dfrac{g^{WWZ}_{nn0}}{g_w \cos\theta^0_W}$	$\dfrac{g^{WWZ}_{nn1}}{g_w \cos\theta^0_W}$	$\dfrac{g^{WWZ}_{nn3}}{g_w \cos\theta^0_W}$
1	−0.00021	−0.00040	0.99686	5.4301	1.7131
2	3×10^{-8}	-1×10^{-7}	0.35046	1.3813	−0.7648
3	-8×10^{-6}	−0.00001	0.99687	3.7231	−0.6589
4	-3×10^{-8}	-3×10^{-8}	0.35044	1.2554	−0.2391
n	$\dfrac{g^{ZZWW}_{00nn}}{g_w^2 \cos^2\theta^0_W}$	$\dfrac{g^{ZZWW}_{0n0n}}{g_w^2 \cos^2\theta^0_W}$	$\dfrac{g^{WWWW}_{00nn}}{g_w^2}$	$\dfrac{g^{WWWW}_{0n0n}}{g_w^2}$	$\dfrac{g^{WWWW}_{11nn}}{g_w^2}$
1	0.99436	0.99541	0.99583	0.99511	33.415
2	0.12487	0.56821	0.49892	0.00381	9.2802
3	0.99483	0.99612	0.99642	0.99516	20.138
4	0.12500	0.56833	0.49897	0.00386	9.6201

8.4.2 *Gauge couplings of fermions*

Gauge couplings of fermion fields are contained in the part of the action

$$\int d^4x \int_1^{z_L} \frac{dz}{k} \sum_J \bar{\check{\Psi}}^J \gamma^\mu (-i)\Big(g_A A_\mu + g_B Q_X B_\mu\Big)\check{\Psi}^J \tag{8.70}$$

where $\check{\Psi}^J = z^{-2}\Psi^J$. Inserting the KK expansion for A_μ, B_μ, $\check{\Psi}^J$ and $\bar{\check{\Psi}}^J$ into (8.70) and integrating over z, one finds gauge couplings of all fermion modes.

(i) W couplings

The W couplings are in

$$\mathcal{L}^W_{\rm int} = -ig_A \sum_J \int_1^{z_L} \frac{dz}{\sqrt{k}} \bar{\check{\Psi}}^J \gamma^\mu \Big(\mathring{W}^{L\dagger}_\mu T^+_L + \mathring{W}^{R\dagger}_\mu T^+_R + \hat{\mathring{W}}^\dagger_\mu \hat{T}^+\Big)\check{\Psi}^J. \tag{8.71}$$

For the first generation of the quark multiplet $\check{\Psi}^q_{(\mathbf{3},\mathbf{4})} = (\check{u}, \check{d}, \check{u}', \check{d}')^t$ we find

$$\begin{aligned}\mathcal{L}^{Wud} &= \frac{g_w}{\sqrt{2}} \sum_{\ell=0}^\infty W^{(\ell)\dagger}_\mu \\ &\quad \times \sum_{n,m=0}^\infty \Big\{ \hat{g}^{Wud}_{L\,\ell nm} \bar{u}^{(n)}_L \gamma^\mu d^{(m)}_L + \hat{g}^{Wud}_{R\,\ell nm} \bar{u}^{(n)}_R \gamma^\mu d^{(m)}_R \Big\},\end{aligned}$$

$$\hat{g}^{Wud}_{L/R\,\ell nm} = G_W[(h^L, h^R, \hat{h})_{W^{(\ell)}}; (f,g)^{u^{(n)}}_{L/R}, (f,g)^{d^{(m)}}_{L/R}], \tag{8.72}$$

where

$$G_W[(h^L, h^R, \hat{h})_\alpha; (f,g)_1, (f,g)_2]$$
$$= \sqrt{kL} \int_1^{z_L} dz \left\{ h_\alpha^{L*} f_1^* f_2 + h_\alpha^{R*} g_1^* g_2 + \frac{i}{\sqrt{2}} \hat{h}_\alpha^* (f_1^* g_2 - g_1^* f_2) \right\} \quad (8.73)$$

and u, d wave functions are given in Eqs. (8.30) and (8.36). Note that the $D^\pm$ components of the $d^{(m)}$ wave functions, $h_{L/R}^{d^{(m)}}$ and $k_{L/R}^{d^{(m)}}$, do not contribute to the W couplings, as $D^\pm$ fields are $SO(5)$ singlets. The couplings to the $D^{(m)}$ ($m \geq 1$) modes are given by replacing $d^{(m)}$ by $D^{(m)}$ in the above formula.

Similarly for the first generation of the lepton multiplet $\check{\Psi}^\ell_{(\mathbf{1},\mathbf{4})} = (\check{\nu}_e, \check{e}, \check{\nu}'_e, \check{e}')^t$ we find

$$\mathcal{L}^{W\nu_e e} = \frac{g_w}{\sqrt{2}} \sum_{\ell=0}^{\infty} W_\mu^{(\ell)\dagger}$$
$$\times \left[\sum_{n,m=0}^{\infty} \left\{ \hat{g}_{L\,\ell nm}^{W\nu_e^+ e}\, \bar{\nu}_{eL}^{+(n)} \gamma^\mu e_L^{(m)} + \hat{g}_{R\,\ell nm}^{W\nu_e^+ e}\, \bar{\nu}_{eR}^{+(n)} \gamma^\mu e_R^{(m)} \right\} \right.$$
$$\left. + \sum_{n=1}^{\infty} \sum_{m=0}^{\infty} \left\{ \hat{g}_{L\,\ell nm}^{W\nu_e^- e}\, \bar{\nu}_{eL}^{-(n)} \gamma^\mu e_L^{(m)} + \hat{g}_{R\,\ell nm}^{W\nu_e^- e}\, \bar{\nu}_{eR}^{-(n)} \gamma^\mu e_R^{(m)} \right\} \right],$$

$$\hat{g}_{L/R\,\ell nm}^{W\nu_e^\pm e} = G_W[(h^L, h^R, \hat{h})_{W^{(\ell)}}; (f,g)_{L/R}^{\nu_e^{\pm(n)}}, (f,g)_{L/R}^{e^{(m)}}]. \quad (8.74)$$

The values in the SM are $\hat{g}_{L\,000,\mathrm{SM}}^{Wud} = \hat{g}_{L\,000,\mathrm{SM}}^{W\nu_e^+ e} = 1$ and $\hat{g}_{R\,000,\mathrm{SM}}^{Wud} = \hat{g}_{R\,000,\mathrm{SM}}^{W\nu_e^+ e} = 0$. In GHU the couplings have matrix structure. The couplings $(\hat{g}_{L/R}^{W^{(0)}ud})_{nm} = \hat{g}_{L/R\,0nm}^{Wud}$ ($n, m = 0, \cdots, 4$) for $\theta_H = 0.1$ and $m_{\mathrm{KK}} = 13\,\mathrm{TeV}$ are given by

$$\hat{g}_L^{W^{(0)}ud} = \begin{pmatrix} 0.99765 & -0.02490 & 0.00002 & -0.00283 & 10^{-6} \\ -0.02490 & 0.00250 & 0.02839 & 10^{-7} & 0.00051 \\ 0.00002 & 0.02839 & 0.99762 & -0.02455 & 0.00002 \\ -0.00283 & 10^{-7} & -0.02455 & 0.00250 & 0.02702 \\ 10^{-6} & 0.00051 & 0.00002 & 0.02702 & 0.99762 \end{pmatrix},$$

$$\hat{g}_R^{W^{(0)}ud} = \begin{pmatrix} 10^{-12} & 10^{-7} & 10^{-7} & 10^{-9} & 10^{-7} \\ 10^{-8} & 0.00250 & 0.02415 & 10^{-8} & 10^{-6} \\ 10^{-7} & 0.02415 & 0.99763 & -0.02256 & 0.00002 \\ 10^{-10} & 10^{-8} & -0.02256 & 0.00250 & 0.02583 \\ 10^{-8} & 10^{-6} & 0.00002 & 0.02583 & 0.99763 \end{pmatrix}. \quad (8.75)$$

Here 10^{-7}, for instance, implies $O(10^{-7})$. The couplings for KK excited states are nearly vector-like. They have very small axial-vector components; $\hat{g}_A^{W^{(0)}ud} = \frac{1}{2}(\hat{g}_R^{W^{(0)}ud} - \hat{g}_L^{W^{(0)}ud})$ is $O(10^{-3})$ or less. The $\hat{g}_{L/R\,0nm}^{W\nu_e^{\pm}e}$ couplings are given by

$$\hat{g}_L^{W^{(0)}\nu_e^+ e} = \begin{pmatrix} 0.99765 & -0.02361 & 0.00002 & -0.00296 & 10^{-7} \\ -0.01669 & 0.00177 & 0.02009 & 10^{-8} & 0.00032 \\ -0.00001 & -0.02009 & -0.70542 & 0.01713 & 0.00002 \\ 0.00209 & 10^{-8} & 0.01712 & -0.00177 & -0.01915 \\ 10^{-7} & 0.00032 & 0.00002 & 0.01917 & 0.70538 \end{pmatrix}, \quad (8.76)$$

and

$$\hat{g}_{R\,0nm}^{W\nu_e^+ e} \begin{cases} \lesssim 10^{-17} & \text{for } n = 0, m = 0 \sim 4 \\ \lesssim 10^{-8} & \text{for } n = 1 \sim 4, m = 0 \\ \sim \hat{g}_{L\,0nm}^{W\nu_e^+ e} & \text{for } n, m \geq 1 \end{cases}$$

$$\hat{g}_{L/R\,0nm}^{W\nu_e^- e} \sim -\hat{g}_{L/R\,0nm}^{W\nu_e^+ e} \quad \text{for } n \geq 1, m \geq 0. \quad (8.77)$$

The couplings among the KK excited modes are almost vector-like. Note that the $\nu_e^{+(n)}$ and $\nu_e^{-(n)}$ $(n \geq 1)$ modes are nearly degenerate in mass.

The $W^{(0)}$ and $W^{(1)}$ couplings to leptons and quarks are tabulated in Table 8.7. The $g_L^{W^{(0)}}$ couplings to quark and lepton doublets except for (t, b) are almost universal with five digits accuracy, being smaller than the g_{000}^{WWZ} coupling by 0.24% for $\theta_H = 0.1$ and $m_{\rm KK} = 13$ TeV. The $W^{(1)}$ couplings to the left-handed quarks and leptons become large.

Table 8.7 The W and W_R couplings of quarks and leptons for $\theta_H = 0.1$, $m_{\rm KK} = 13$ TeV and $M = 10^3$ TeV. $\hat{g}_{R\,100}^{W\alpha\beta}$ is very small.

(α, β)	$\hat{g}_{L\,000}^{W\alpha\beta}$	$\hat{g}_{R\,000}^{W\alpha\beta}$	$\hat{g}_{L\,100}^{W\alpha\beta}$	$\hat{g}_{L\,300}^{W\alpha\beta}$	$\hat{g}_{L\,100}^{W_R\alpha\beta}$	$\hat{g}_{R\,100}^{W_R\alpha\beta}$
(ν_e, e)	0.997647	10^{-23}	5.7444	2.2599	0.0145	10^{-19}
(ν_μ, μ)	0.997644	10^{-21}	5.4697	1.7284	0.0139	10^{-17}
(ν_τ, τ)	0.997642	10^{-20}	5.2870	1.4132	0.0134	10^{-16}
(u, d)	0.997645	10^{-12}	5.5618	1.8986	0.0141	10^{-9}
(c, s)	0.997643	10^{-9}	5.3580	1.5322	0.0136	10^{-6}
(t, b)	0.997969	10^{-5}	4.4101	0.2608	0.0113	-0.0344

(ii) W_R couplings

The W_R couplings are in

$$\mathcal{L}_{\rm int}^{W_R} = -ig_A \sum_J \int_1^{z_L} \frac{dz}{\sqrt{k}} \bar{\check{\Psi}}^J \gamma^\mu \Big(\mathring{W}_{R\mu}^{L\dagger} T_L^+ + \mathring{W}_{R\mu}^{R\dagger} T_R^+ \Big) \check{\Psi}^J \tag{8.78}$$

where

$$\begin{pmatrix} \mathring{W}_{R\mu}^L \\ \mathring{W}_{R\mu}^R \end{pmatrix} = \sum_{n=1}^{\infty} W_{R\mu}^{(n)}(x) \begin{pmatrix} h_{W_R^{(n)}}^L(z) \\ h_{W_R^{(n)}}^R(z) \end{pmatrix} . \tag{8.79}$$

It follows immediately, with the definition of couplings similar to (8.72) and (8.74), that

$$\hat{g}_{L/R\,\ell nm}^{W_R ud} = G_W[(h^L, h^R, 0)_{W_R^{(\ell)}}; (f,g)_{L/R}^{u^{(n)}}, (f,g)_{L/R}^{d^{(m)}}] \tag{8.80}$$

etc. $(h^L, h^R, 0)_{W_R^{(\ell)}}$ in the above means that the $\hat{h}$ component is zero. The $W_R^{(1)}$ couplings to quarks and leptons are tabulated in Table 8.7.

(iii) γ couplings

The γ couplings are in

$$\mathcal{L}_{\rm int}^{\gamma} = -ig_A \sum_J \int_1^{z_L} \frac{dz}{\sqrt{k}} \bar{\check{\Psi}}^J \gamma^\mu \Big(\mathring{\gamma}_\mu^L T_L^3 + \mathring{\gamma}_\mu^R T_R^3 + \frac{g_B}{g_A} \mathring{\gamma}_\mu^B Q_X \Big) \check{\Psi}^J . \tag{8.81}$$

Making use of $g_B/g_A = \sin\theta_W^0/(1 - 2\sin^2\theta_W^0)^{1/2}$, $Q_{\rm EM} = T_L^3 + T_R^3 + Q_X$ and the expression for $h_{\gamma^{(\ell)}}$ in Eq. (8.23), $\mathcal{L}_{\rm int}^{\gamma}$ becomes

$$\mathcal{L}_{\rm int}^{\gamma} = -ie \sum_J \sum_{\ell=0}^{\infty} \gamma_\mu^{(\ell)}(x) \sqrt{\frac{L}{k}} \int_1^{z_L} dz\, h_{\gamma^{(\ell)}} \bar{\check{\Psi}}^J \gamma^\mu Q_{\rm EM} \check{\Psi}^J . \tag{8.82}$$

Hence for the first generation of quarks, for instance,

$$\begin{aligned} \mathcal{L}_{\rm int}^{\gamma\, ud} &= -ie \sum_{\ell,n,m=0}^{\infty} \gamma_\mu^{(\ell)} \Big\{ \hat{Q}_{L\,\ell nm}^u \bar{u}_L^{(n)} \gamma^\mu u_L^{(m)} + \hat{Q}_{R\,\ell nm}^u \bar{u}_R^{(n)} \gamma^\mu u_R^{(m)} \\ &\qquad + \hat{Q}_{L\,\ell nm}^d \bar{d}_L^{(n)} \gamma^\mu d_L^{(m)} + \hat{Q}_{R\,\ell nm}^d \bar{d}_R^{(n)} \gamma^\mu d_R^{(m)} \Big\}, \\ \hat{Q}_{L/R\,\ell nm}^u &= Q_u\, G_\gamma^u[h_{\gamma^{(\ell)}}; (f,g)_{L/R}^{u^{(n)}}, (f,g)_{L/R}^{u^{(m)}}], \\ \hat{Q}_{L/R\,\ell nm}^d &= Q_d\, G_\gamma^d[h_{\gamma^{(\ell)}}; (f,g,h,k)_{L/R}^{d^{(n)}}, (f,g,h,k)_{L/R}^{d^{(m)}}], \end{aligned} \tag{8.83}$$

where $Q_u = \frac{2}{3}$, $Q_d = -\frac{1}{3}$ and

$$
\begin{aligned}
G^u_\gamma[h_\gamma; (f,g)_1, (f,g)_2] &= \sqrt{kL} \int_1^{z_L} dz\, h_\gamma \big(f_1^* f_2 + g_1^* g_2 \big), \\
G^d_\gamma[h_\gamma; (f,g,h,k)_1, (f,g,h,k)_2] & \\
&= G^u_\gamma[h_\gamma; (f,g)_1, (f,g)_2] + G^u_\gamma[h_\gamma; (h,k)_1, (h,k)_2] \\
&= \sqrt{kL} \int_1^{z_L} dz\, h_\gamma \big(f_1^* f_2 + g_1^* g_2 + h_1^* h_2 + k_1^* k_2 \big). \qquad (8.84)
\end{aligned}
$$

Note that $\hat{Q}^{u/d}_{L/R\,0nm} = Q_{u/d}\, \delta_{nm}$. Similar expression holds for leptons.

(iv) Z couplings

The Z couplings are in

$$
\mathcal{L}^Z_{\rm int} = -i g_A \sum_J \int_1^{z_L} \frac{dz}{\sqrt{k}}
$$

$$
\bar{\check{\Psi}}^J \gamma^\mu \Big(\mathring{Z}^L_\mu T^3_L + \mathring{Z}^R_\mu T^3_R + \hat{\mathring{Z}}_\mu \hat{T}^3 + \frac{g_B}{g_A} \mathring{Z}^B_\mu Q_X \Big) \check{\Psi}^J. \qquad (8.85)
$$

By making use of the decomposition of the Z wave functions into the h^{su2} and h^{em} parts in Eq. (8.23), we find that for the (u,d) towers

$$
\begin{aligned}
\mathcal{L}^{Z,u,d} &= -i \frac{g_w}{\cos\theta^0_W} \sum_{\ell=0}^\infty Z^{(\ell)}_\mu \\
&\times \sum_{n,m=0}^\infty \Big\{ \hat{g}^{Zuu}_{L\,\ell nm} \bar{u}^{(n)}_L \gamma^\mu u^{(m)}_L + \hat{g}^{Zuu}_{R\,\ell nm} \bar{u}^{(n)}_R \gamma^\mu u^{(m)}_R \\
&\qquad + \hat{g}^{Zdd}_{L\,\ell nm} \bar{d}^{(n)}_L \gamma^\mu d^{(m)}_L + \hat{g}^{Zdd}_{R\,\ell nm} \bar{d}^{(n)}_R \gamma^\mu d^{(m)}_R \Big\}, \qquad (8.86)
\end{aligned}
$$

where

$$
\begin{aligned}
\hat{g}^{Zuu}_{L/R\,\ell nm} &= \hat{g}^{Zuu,su2}_{L/R\,\ell nm} - \sin^2\theta^0_W\, \hat{g}^{Zuu,em}_{L/R\,\ell nm}, \\
\hat{g}^{Zuu,su2}_{L/R\,\ell nm} &= \cos\theta^0_W T^3_u\, G_W[(h^L, h^R, \hat{h})^{su2}_{Z^{(\ell)}}; (f,g)^{u^{(n)}}_{L/R}, (f,g)^{u^{(m)}}_{L/R}], \\
\hat{g}^{Zuu,em}_{L/R\,\ell nm} &= \cos\theta^0_W Q_u\, G^u_\gamma[h^{em}_{Z^{(\ell)}}; (f,g)^{u^{(n)}}_{L/R}, (f,g)^{u^{(m)}}_{L/R}], \\
\hat{g}^{Zdd}_{L/R\,\ell nm} &= \hat{g}^{Zdd,su2}_{L/R\,\ell nm} - \sin^2\theta^0_W\, \hat{g}^{Zdd,em}_{L/R\,\ell nm}, \\
\hat{g}^{Zdd,su2}_{L/R\,\ell nm} &= \cos\theta^0_W T^3_d\, G_W[(h^L, h^R, \hat{h})^{su2}_{Z^{(\ell)}}; (f,g,h,k)^{d^{(n)}}_{L/R}, (f,g,h,k)^{d^{(m)}}_{L/R}], \\
\hat{g}^{Zdd,em}_{L/R\,\ell nm} &= \cos\theta^0_W Q_d\, G^d_\gamma[h^{em}_{Z^{(\ell)}}; (f,g,h,k)^{d^{(n)}}_{L/R}, (f,g,h,k)^{d^{(m)}}_{L/R}], \qquad (8.87)
\end{aligned}
$$

and $(T_u^3, T_d^3) = (\frac{1}{2}, -\frac{1}{2})$. The Z couplings to $d^{(n)}D^{(m)}$ and $D^{(n)}D^{(m)}$ are very small, and have been omitted in the above. In the SM $\hat{g}_{\rm SM}^Z = T_L^3 - \sin^2\theta_W^{\rm SM}\, Q_{\rm EM}$. In GHU, for $\theta_H = 0.1$ and $m_{\rm KK} = 13$ TeV, one finds that

$$\begin{pmatrix} \hat{g}_{L\,000}^{Zuu,su2} \\ \hat{g}_{R\,000}^{Zuu,su2} \end{pmatrix} = \begin{pmatrix} 0.498844 \\ 10^{-11} \end{pmatrix}, \quad \begin{pmatrix} \hat{g}_{L\,000}^{Zuu,em} \\ \hat{g}_{R\,000}^{Zuu,em} \end{pmatrix} = \begin{pmatrix} 0.666791 \\ 0.666725 \end{pmatrix},$$

$$\begin{pmatrix} \hat{g}_{L\,000}^{Zdd,su2} \\ \hat{g}_{R\,000}^{Zdd,su2} \end{pmatrix} = \begin{pmatrix} -0.498844 \\ 10^{-13} \end{pmatrix}, \quad \begin{pmatrix} \hat{g}_{L\,000}^{Zdd,em} \\ \hat{g}_{R\,000}^{Zdd,em} \end{pmatrix} = \begin{pmatrix} -0.333395 \\ -0.333372 \end{pmatrix}. \tag{8.88}$$

Similarly for leptons one finds that

$$\begin{aligned} \mathcal{L}^{Z,\nu_e^\pm,e} &= -i\frac{g_w}{\cos\theta_W^0}\sum_{\ell=0}^{\infty} Z_\mu^{(\ell)} \\ &\times \sum_{n,m=0}^{\infty}\Big\{ \hat{g}_{L\,\ell nm}^{Zee}\bar{e}_L^{(n)}\gamma^\mu e_L^{(m)} + \hat{g}_{R\,\ell nm}^{Zee}\bar{e}_R^{(n)}\gamma^\mu e_R^{(m)} \\ &+ \sum_{a,b=\pm}\Big(\hat{g}_{L\,\ell nm}^{Z\nu_e^a\nu_e^b}\bar{\nu}_{eL}^{a(n)}\gamma^\mu\nu_{eL}^{b(m)} + \hat{g}_{R\,\ell nm}^{Z\nu_e^a\nu_e^b}\bar{\nu}_{eR}^{a(n)}\gamma^\mu\nu_{eR}^{b(m)}\Big)\Big\}, \end{aligned} \tag{8.89}$$

where

$$\begin{aligned} \hat{g}_{L/R\,\ell nm}^{Zee} &= \hat{g}_{L/R\,\ell nm}^{Zee,su2} - \sin^2\theta_W^0\,\hat{g}_{L/R\,\ell nm}^{Zee,em}, \\ \hat{g}_{L/R\,\ell nm}^{Zee,su2} &= \cos\theta_W^0 T_e^3\, G_W[(h^L,h^R,\hat{h})_{Z^{(\ell)}}^{su2};(f,g)_{L/R}^{e^{(n)}},(f,g)_{L/R}^{e^{(m)}}], \\ \hat{g}_{L/R\,\ell nm}^{Zee,em} &= \cos\theta_W^0 Q_e\, G_\gamma^u[h_{Z^{(\ell)}}^{em};(f,g)_{L/R}^{e^{(n)}},(f,g)_{L/R}^{e^{(m)}}], \\ \hat{g}_{L/R\,\ell nm}^{Z\nu_e^a\nu_e^b} &= \hat{g}_{L/R\,\ell nm}^{Z\nu_e^a\nu_e^b,su2} \\ &= \cos\theta_W^0 T_{\nu_e}^3\, G_W[(h^L,h^R,\hat{h})_{Z^{(\ell)}}^{su2};(f,g)_{L/R}^{\nu_e^{a(n)}},(f,g)_{L/R}^{\nu_e^{b(m)}}], \end{aligned} \tag{8.90}$$

and $(T_{\nu_e}^3, T_e^3) = (\frac{1}{2}, -\frac{1}{2})$. Here $\hat{g}_{L/R\,\ell 0m}^{Z\nu_e^-\nu_e^\pm} = \hat{g}_{L/R\,\ell n0}^{Z\nu_e^\pm\nu_e^-} = 0$ has been understood.

The $Z^{(0)}$ couplings among KK excited states of both (u,d) and (ν_e, e) towers are nearly vector-like. Further $\hat{g}_{L/R\,0nm}^{Z\nu_e^+\nu_e^+} \sim \hat{g}_{L/R\,0nm}^{Z\nu_e^-\nu_e^-} \sim -\hat{g}_{L/R\,0nm}^{Z\nu_e^+\nu_e^-} \sim -\hat{g}_{L/R\,0nm}^{Z\nu_e^-\nu_e^+}$ for $n,m \geq 1$. The $Z^{(0)}$ couplings to leptons and quarks are tabulated in Table 8.8. One sees that the couplings normalized by the $W^{(0)}\nu_e e$ coupling, $\hat{g}_{L/R\,000}^{Z\alpha\alpha}/\hat{g}_{L\,000}^{W\nu_e e}$, are very close to the SM values.

Table 8.8 The $Z^{(0)}$ couplings of leptons and quarks for $\theta_H = 0.1$, $m_{\rm KK} = 13$ TeV, $\sin^2\theta_W^0 = 0.2304$ and $M = 10^3$ TeV. The SM values $T_L^3 - \sin^2\theta_W^{\rm SM} Q_{\rm EM}$ with $\sin^2\theta_W^{\rm SM} = 0.2312$ are listed as well. The ratio $\hat{g}^{Z\alpha\alpha}_{L/R\,000}/\hat{g}^{W\nu_e e}_{L\,000}$ should be compared with the SM value.

α	$\hat{g}^{Z\alpha\alpha}_{L\,000}$	$\hat{g}^{Z\alpha\alpha}_{R\,000}$	$\dfrac{\hat{g}^{Z\alpha\alpha}_{L\,000}}{\hat{g}^{W\nu_e e}_{L\,000}}$	$\dfrac{\hat{g}^{Z\alpha\alpha}_{R\,000}}{\hat{g}^{W\nu_e e}_{L\,000}}$	$\hat{g}^{\rm SM}_L$	$\hat{g}^{\rm SM}_R$
ν_e	0.498845	10^{-32}	0.500022	10^{-32}		
ν_μ	0.498843	10^{-32}	0.500020	10^{-32}	$\frac{1}{2}$	0
ν_τ	0.498842	10^{-32}	0.500019	10^{-32}		
e	−0.268450	0.230372	−0.269083	0.230915		
μ	−0.268449	0.230372	−0.269082	0.230915	−0.2688	0.2312
τ	−0.268449	0.230371	−0.269082	0.230915		
u	0.345248	−0.153581	0.346062	−0.153943		
c	0.345247	−0.153581	0.346061	−0.153943	0.3459	−0.1541
t	0.345578	−0.153251	0.346393	−0.153612		
d	−0.422046	0.076793	−0.423041	0.076974		
s	−0.422045	0.076793	−0.423040	0.076974	−0.4229	0.0771
b	−0.422039	0.076795	−0.423034	0.076977		

(v) Z_R couplings

The Z_R couplings are in

$$\mathcal{L}^{Z_R}_{\rm int} = -ig_A \sum_J \int_1^{z_L} \frac{dz}{\sqrt{k}}\, \bar{\check{\Psi}}^J \gamma^\mu \Big(\mathring{Z}^L_{R\mu} T^3_L + \mathring{Z}^R_{R\mu} T^3_R + \frac{g_B}{g_A} \mathring{Z}^B_{R\mu} Q_X \Big) \check{\Psi}^J ,$$

$$\begin{pmatrix} \mathring{Z}^L_{R\mu} \\ \mathring{Z}^R_{R\mu} \\ \mathring{Z}^B_{R\mu} \end{pmatrix} = \sum_{n=1}^\infty Z^{(n)}_{R\mu}(x) \begin{pmatrix} h^L_{Z_R^{(n)}}(z) \\ h^R_{Z_R^{(n)}}(z) \\ h^B_{Z_R^{(n)}}(z) \end{pmatrix} . \tag{8.91}$$

It follows that for (u,d)

$$\begin{aligned} \mathcal{L}^{Z_R,u,d} &= -i \frac{g_w}{\cos\theta_W^0} \sum_{\ell=1}^\infty Z^{(\ell)}_{R\mu} \\ &\times \sum_{n,m=0}^\infty \Big\{ \hat{g}^{Z_R uu}_{L\,\ell nm} \bar{u}^{(n)}_L \gamma^\mu u^{(m)}_L + \hat{g}^{Z_R uu}_{R\,\ell nm} \bar{u}^{(n)}_R \gamma^\mu u^{(m)}_R \\ &\quad + \hat{g}^{Z_R dd}_{L\,\ell nm} \bar{d}^{(n)}_L \gamma^\mu d^{(m)}_L + \hat{g}^{Z_R dd}_{R\,\ell nm} \bar{d}^{(n)}_R \gamma^\mu d^{(m)}_R \Big\}, \end{aligned} \tag{8.92}$$

where

$$\hat{g}^{Z_R uu}_{L/R\,\ell nm} = \cos\theta^0_W \Big\{ T^3_u\, G_W[(h^L, h^R, 0)_{Z_R^{(\ell)}}; (f,g)^{u^{(n)}}_{L/R}, (f,g)^{u^{(m)}}_{L/R}]$$

$$+ \frac{\sin\theta^0_W}{\sqrt{1 - 2\sin^2\theta^0_W}} Q^u_X G^u_\gamma [h^B_{Z_R^{(\ell)}}; (f,g)^{u^{(n)}}_{L/R}, (f,g)^{u^{(m)}}_{L/R}] \Big\},$$

$$\hat{g}^{Z_R dd}_{L/R\,\ell nm} = \cos\theta^0_W \Big\{ T^3_d\, G_W[(h^L, h^R, 0)_{Z_R^{(\ell)}}; (f,g)^{d^{(n)}}_{L/R}, (f,g)^{d^{(m)}}_{L/R}]$$

$$+ \frac{\sin\theta^0_W}{\sqrt{1 - 2\sin^2\theta^0_W}} \Big(Q^d_X G^u_\gamma [h^B_{Z_R^{(\ell)}}; (f,g)^{d^{(n)}}_{L/R}, (f,g)^{d^{(m)}}_{L/R}]$$

$$+ Q^D_X G^u_\gamma [h^B_{Z_R^{(\ell)}}; (h,k)^{d^{(n)}}_{L/R}, (h,k)^{d^{(m)}}_{L/R}] \Big) \Big\}, \tag{8.93}$$

$Q^u_X = Q^d_X = \frac{1}{6}$ and $Q^D_X = -\frac{1}{3}$. Similar expression holds for leptons.

The couplings of $Z^{(1)}$, $Z^{(1)}_R$ and $\gamma^{(1)}$ to the left-handed components of leptons and quarks become large, which leads to distinct signals in e^+e^- collisions at high energies as explained in the next chapter. Those couplings in units of $g_w/\cos\theta^0_W$ are tabulated in Table 8.9.

Table 8.9 The $Z^{(1)}$, $Z^{(1)}_R$ and $\gamma^{(1)}$ couplings of leptons and quarks in units of $g_w/\cos\theta^0_W$ for $\theta_H = 0.1$, $m_{\rm KK} = 13$ TeV, $\sin^2\theta^0_W = 0.2304$ and $M = 10^3$ TeV.

α	$\hat{g}^{Z\alpha\alpha}_{L\,100}$	$\hat{g}^{Z\alpha\alpha}_{R\,100}$	$\hat{g}^{Z_R\alpha\alpha}_{L\,100}$	$\hat{g}^{Z_R\alpha\alpha}_{R\,100}$	$\hat{g}^{\gamma\alpha\alpha}_{L\,100}$	$\hat{g}^{\gamma\alpha\alpha}_{R\,100}$
ν_e	2.87438	10^{-31}	−0.90388	10^{-28}	0	0
ν_μ	2.73697	10^{-31}	−0.86274	10^{-29}	0	0
ν_τ	2.64554	10^{-31}	−0.83534	10^{-29}	0	0
e	−1.54683	−0.05121	−0.91457	10^{-11}	−2.41860	0.09391
μ	−1.47288	−0.05089	−0.87294	10^{-7}	−2.30295	0.09332
τ	−1.42368	−0.05116	−0.84521	0.00005	−2.22599	0.09381
u	1.92614	0.03386	0.29909	10^{-8}	1.56115	−0.06210
c	1.85555	0.03399	0.28866	−0.00001	1.50392	−0.06234
t	1.52723	−0.28634	0.23996	−0.64903	1.23798	0.52748
d	−2.35459	0.10181	0.28873	−0.15776	−0.78058	−0.18533
s	−2.26830	0.12797	0.27866	−0.19240	−0.75196	−0.23299
b	−1.86701	0.25410	0.23173	−0.35866	−0.61889	−0.46285

8.5 Higgs couplings

The 4D Higgs field $H(x) = H^{(0)}(x)$ is contained in A_z.

$$A_z \Rightarrow \frac{1}{\sqrt{k}} \mathring{H} \hat{T}^4 \ , \quad \mathring{H} = \sum_{\ell=0}^{\infty} H^{(\ell)}(x) u_{H^{(\ell)}}(z) \ , \tag{8.94}$$

where $u_{H^{(\ell)}}(z)$ is given in Eq. (8.27). Only $H^{(0)}$ is physical.

8.5.1 *Gauge-Higgs couplings*

Higgs couplings of W and Z fields are contained in the part $(-\frac{1}{4})\mathrm{Tr}\, F_{MN}F^{MN}$ of the action. First note that

$$\begin{aligned} -i\,[A_\mu, A_z] \Rightarrow \tfrac{1}{2}\mathring{H}\Big\{ & \hat{\mathring{W}}_\mu^\dagger (T_L^+ - T_R^+) - (\mathring{W}_\mu^{L\dagger} - \mathring{W}_\mu^{R\dagger})\hat{T}^+ \\ & + \hat{\mathring{W}}_\mu (T_L^- - T_R^-) - (\mathring{W}_\mu^L - \mathring{W}_\mu^R)\hat{T}^- \\ & + \hat{\mathring{Z}}_\mu (T_L^3 - T_R^3) - (\mathring{Z}_\mu^L - \mathring{Z}_\mu^R)\hat{T}^3 \Big\} \ . \end{aligned} \tag{8.95}$$

WWH and ZZH couplings are found to be

$$\begin{aligned} & -ig_A \int \frac{kdz}{z} \mathrm{Tr}\,\partial_z A_\mu [A^\mu, A_z] \Rightarrow \\ & \sum_{n,m,\ell=0}^{\infty} \Big\{ g_{nm\ell}^{WWH} W_\mu^{(n)\dagger} W^{(m)\mu} H^{(\ell)} + \frac{1}{2} g_{nm\ell}^{ZZH} Z_\mu^{(n)} Z^{(m)\mu} H^{(\ell)} \Big\}, \end{aligned} \tag{8.96}$$

where

$$\begin{aligned} g_{nm\ell}^{WWH} &= \frac{g_w k\sqrt{kL}}{2} \int_1^{z_L} \frac{dz}{z} u_{H^{(\ell)}} \\ & \times \Big\{ \partial_z \big(h_{W^{(n)}}^{L*} - h_{W^{(n)}}^{R*}\big)\hat{h}_{W^{(m)}} + \hat{h}_{W^{(n)}}^* \partial_z \big(h_{W^{(m)}}^L - h_{W^{(m)}}^R\big) \\ & - \partial_z \hat{h}_{W^{(n)}}^* \big(h_{W^{(m)}}^L - h_{W^{(m)}}^R\big) - \big(h_{W^{(n)}}^{L*} - h_{W^{(n)}}^{R*}\big)\partial_z \hat{h}_{W^{(m)}} \Big\}, \\ g_{nm\ell}^{ZZH} &= \frac{g_w k\sqrt{kL}}{2} \int_1^{z_L} \frac{dz}{z} u_{H^{(\ell)}} \\ & \times \Big\{ \partial_z \big(h_{Z^{(n)}}^L - h_{Z^{(n)}}^R\big)\hat{h}_{Z^{(m)}} + \hat{h}_{Z^{(n)}} \partial_z \big(h_{Z^{(m)}}^L - h_{Z^{(m)}}^R\big) \\ & - \partial_z \hat{h}_{Z^{(n)}} \big(h_{Z^{(m)}}^L - h_{Z^{(m)}}^R\big) - \big(h_{Z^{(n)}}^L - h_{Z^{(n)}}^R\big)\partial_z \hat{h}_{Z^{(m)}} \Big\}. \end{aligned} \tag{8.97}$$

The couplings $g_{nn\ell}^{WWH}$ and $g_{nn\ell}^{ZZH}$ are simplified, by inserting the wave functions in Eqs. (8.16), (8.23) and (8.27) into Eq. (8.97) and making use of the relations in Eqs. (7.82) and (C.15), to

$$\begin{aligned} g_{nn\ell}^{WWH} &= -\delta_{\ell 0}\, g_w s_H c_H m_{W^{(n)}} \frac{\sqrt{kL(z_L^2-1)}}{r_{W^{(n)}}} \frac{C(1;\lambda_{W^{(n)}})}{S(1;\lambda_{W^{(n)}})}, \\ g_{nn\ell}^{ZZH} &= -\delta_{\ell 0}\, g_w s_H c_H m_{Z^{(n)}} \frac{\sqrt{kL(z_L^2-1)}}{r_{Z^{(n)}}} \frac{C(1;\lambda_{Z^{(n)}})}{S(1;\lambda_{Z^{(n)}})}. \end{aligned} \tag{8.98}$$

We have seen in Section 8.2 from the viewpoint of effective interactions that $g_{WWH} \sim g_w m_W \cos\theta_H$ and $g_{ZZH} \sim g_w m_Z \cos\theta_H / \cos\theta_W^0$. Let us define

$$\begin{aligned} I_{W^{(n)}} &= \frac{g_{nn0}^{WWH}}{g_w m_{W^{(n)}} \cos\theta_H}, \\ I_{Z^{(n)}} &= \frac{g_{nn0}^{ZZH}}{g_w m_{Z^{(n)}} \cos\theta_H / \cos\theta_W^0}. \end{aligned} \tag{8.99}$$

For $\theta_H = 0.1$ and $m_{\rm KK} = 13$ TeV, one finds that $I_{W^{(0)}} = 1.00021$ and $I_{Z^{(0)}} = 1.00027$. It is confirmed that the estimate in Eq. (8.12) is good. The behavior of $I_{W^{(n)}}$ and $I_{Z^{(n)}}$ for $n \geq 1$ is striking too. As shown in Table 8.10, the sign of $I_{W^{(n)}}$ and $I_{Z^{(n)}}$ alternates as n increases. This behavior of $I_{W^{(n)}}$ becomes important in discussing the decay rate of the Higgs boson in the next chapter.

$WWHH$ and $ZZHH$ couplings are found to be

$$\begin{aligned} &\frac{1}{2} k g_A^2 \int_1^{z_L} \frac{dz}{z}\, \mathrm{Tr}\,[A_\mu, A_z]\,[A^\mu, A_z] \\ &\Rightarrow - \sum_{n,m,\ell,p=0}^{\infty} \bigg\{ \frac{1}{4} g_{nm\ell p}^{WWHH} W_\mu^{(n)\dagger} W^{(m)\mu} H^{(\ell)} H^{(p)} \\ &\qquad\qquad + \frac{1}{8} g_{nm\ell p}^{ZZHH} Z_\mu^{(n)} Z^{(m)\mu} H^{(\ell)} H^{(p)} \bigg\}, \end{aligned} \tag{8.100}$$

where

$$\begin{aligned} g_{nm\ell p}^{WWHH} &= g_w^2\, kL \int_1^{z_L} \frac{dz}{z} u_{H^{(\ell)}} u_{H^{(p)}} \\ &\quad \times \Big\{ \big(h_{W^{(n)}}^{L*} - h_{W^{(n)}}^{R*}\big)\big(h_{W^{(m)}}^{L} - h_{W^{(m)}}^{R}\big) + 2\hat{h}_{W^{(n)}}^* \hat{h}_{W^{(m)}} \Big\}, \\ g_{nm\ell p}^{ZZHH} &= g_w^2\, kL \int_1^{z_L} \frac{dz}{z} u_{H^{(\ell)}} u_{H^{(p)}} \\ &\quad \times \Big\{ \big(h_{Z^{(n)}}^{L} - h_{Z^{(n)}}^{R}\big)\big(h_{Z^{(m)}}^{L} - h_{Z^{(m)}}^{R}\big) + 2\hat{h}_{Z^{(n)}} \hat{h}_{Z^{(m)}} \Big\}. \end{aligned} \tag{8.101}$$

The values g_{nn00}^{WWHH}/g_w^2 and $g_{nn00}^{ZZHH}/(g_w^2/\cos^2\theta_W^0)$ for $\theta_H = 0.1$ and $m_{\rm KK} =$ 13 TeV are tabulated in Table 8.10. One sees that $g_{0000}^{WWHH} = 0.9936\, g_w^2$ and $g_{0000}^{ZZHH} = 0.9937\, g_w^2/\cos^2\theta_W^0$, both of which are very close to the values in the SM. However, the couplings to the KK excited modes, g_{nn00}^{WWHH} and g_{nn00}^{ZZHH} $(n \geq 1)$, become large.

8.5.2 *Yukawa couplings*

The Yukawa couplings of fermions are contained in the part $(-ig_A)\bar{\check{\Psi}}^J\gamma^5 {e_5}^z A_z\check{\Psi}^J$ of the action. Noting ${e_5}^z = kz$, one finds that for the up-quark tower

$$\begin{aligned}
&\int_1^{z_L} \frac{dz}{kz}\,(-ig_A)\bar{\check{\Psi}}^J\gamma^5 {e_5}^z A_z\check{\Psi}^J \\
&\Rightarrow -\frac{i}{\sqrt{2}}\sum_{\ell,n,m=0}^{\infty} H^{(\ell)}\Big\{y_{\ell nm}^{Huu}\, u_L^{(n)\dagger} u_R^{(m)} - y_{\ell nm}^{Huu\,*}\, u_R^{(m)\dagger} u_L^{(n)}\Big\}, \\
&y_{\ell nm}^{Huu} = -ig_w\frac{\sqrt{kL}}{2}\int_1^{z_L} dz\, u_{H^{(\ell)}}\big\{f_L^{u^{(n)}*} g_R^{u^{(m)}} + g_L^{u^{(n)}*} f_R^{u^{(m)}}\big\}. \quad (8.102)
\end{aligned}$$

By inserting the wave functions in Eqs. (8.27) and (8.30) into (8.102) and making use of the relation $C_LC_R - S_LS_R = 1$ in Eq. (C.17), the diagonal element $y_{\ell nn}^{Huu}$ is simplified to be

$$y_{\ell nn}^{Huu} = -\delta_{\ell 0}\, g_w\, \frac{\bar{s}_H\bar{c}_H\sqrt{kL(z_L^2-1)}}{2\sqrt{2}\, r_{u^{(n)}}}\, \frac{C_L(1;\lambda_{u^{(n)}},c_u)}{S_L(1;\lambda_{u^{(n)}},c_u)}\,. \quad (8.103)$$

In the discussion of the effective interactions of the Higgs field in Section 8.2 we have seen that the Yukawa coupling of the $u = u^{(0)}$ quark is about $y_u^{\rm GHU,eff} = y_u^{\rm SM}\cos^2\frac{1}{2}\theta_H$ where the SM value is $y_u^{\rm SM} = m_u/f_H\sin\theta_H$. One finds that

$$\frac{y_{000}^{Huu}}{y_u^{\rm SM}\cos^2\frac{1}{2}\theta_H} - 1 = 1.0\times 10^{-9}\,. \quad (8.104)$$

The formulas for the Yukawa couplings of other quarks and leptons take the same form as in Eqs. (8.102) and (8.103) where $u^{(n)}$ is replaced by the respective field. The Yukawa couplings of quarks and leptons are suppressed by a factor $\cos^2\frac{1}{2}\theta_H$ compared with the SM values.

Let us define

$$I_{u^{(n)}} = \frac{y_{0nn}^{Huu}}{y_u^{\rm SM}\cos^2\frac{1}{2}\theta_H}\,, \quad I_{t^{(n)}} = \frac{y_{0nn}^{Htt}}{y_t^{\rm SM}\cos^2\frac{1}{2}\theta_H}\,. \quad (8.105)$$

Table 8.10 Higgs couplings to the W, Z boson towers and u, t quark towers ($0 \leq n \leq 20$) are tabulated for $\theta_H = 0.1$, $m_{\rm KK} = 13$ TeV and $\sin^2 \theta_W^0 = 0.2304$. $I_{W^{(n)}}, I_{Z^{(n)}}, I_{u^{(n)}}$ and $I_{t^{(n)}}$ are defined in Eqs. (8.99) and (8.105). $J_{W^{(n)}} = g_{nn00}^{WWHH}/g_w^2$ and $J_{Z^{(n)}} = g_{nn00}^{ZZHH}/(g_w^2/\cos^2\theta_W^0)$.

n	$I_{W^{(n)}}$	$I_{Z^{(n)}}$	$I_{u^{(n)}}$	$I_{t^{(n)}}$	$J_{W^{(n)}}$	$J_{Z^{(n)}}$
0	1.0002	1.0003	1.	1.0006	0.9936	0.9937
1	−0.0313	−0.0355	−0.00001	−0.1097	29.068	29.111
2	0.0330	0.0375	0.00001	0.1104	35.210	27.371
3	−0.0326	−0.0369	−0.00002	−0.1059	19.773	19.802
4	0.0332	0.0376	0.00002	0.1061	35.210	27.370
⋮	⋮	⋮	⋮	⋮	⋮	⋮
17	−0.0348	−0.0394	−0.00005	−0.1029	17.612	17.637
18	0.0349	0.0395	0.00005	0.1029	35.210	27.370
19	−0.0349	−0.0396	−0.00005	−0.1028	17.593	17.619
20	0.0350	0.0397	0.00005	0.1028	35.210	27.370

The values of $I_{u^{(n)}}$ and $I_{t^{(n)}}$ are tabulated in Table 8.10. The sign of $I_{t^{(n)}}$ alternates; $I_{t^{(n)}}/|I_{t^{(n)}}| = (-1)^n$, and its absolute value $|I_{t^{(n)}}|$ tends to approach a constant.

The Higgs couplings g_{000}^{WWH}, g_{000}^{ZZH}, g_{0000}^{WWHH}, g_{0000}^{ZZHH} and Yukawa couplings of quarks and leptons are very close to those in the SM. However, the couplings g_{nn00}^{WWHH}, g_{nn00}^{ZZHH}, g_{nn1}^{WWZ}, g_{nn11}^{WWWW} ($n \geq 1$) etc. become large. This fact indicates that the perturbative description based on the 4D KK modes is not appropriate at an energy scale $E > 2\,m_{\rm KK}$. A fully five-dimensional description is necessary there.

Chapter 9

Predictions

We have seen in the previous chapter that the gauge couplings among the SM fields and particles are nearly the same as in the SM. The Higgs couplings to W and Z and the Yukawa couplings of quarks and leptons are suppressed by a factor $\cos\theta_H$ or $\cos^2\frac{1}{2}\theta_H$ compared with the SM couplings so that the deviation from the SM is small for $|\theta_H| \lesssim 0.1$.

We shall see that there appears large deviation from the SM in the Higgs cubic and quartic self-couplings. Quarks and leptons have distinct couplings to the first excited KK modes of gauge bosons with large parity violation, which can be seen in future collider experiments. Further the W boson mass is predicted to be larger than the value in the SM. In this chapter we clarify how to confirm the GHU scenario in the coming experiments.

9.1 The Fermi constant and W boson mass

In the SM the mass of the W boson, m_W, is predicted once the electromagnetic fine structure constant α, the Fermi constant G_F in the muon (μ) decay, the masses of the Z boson, Higgs boson, quarks and leptons, and the strong coupling constant α_s in QCD are given. By taking higher-order radiative corrections into account m_W is evaluated, which is to be compared with the observed value of m_W.

The W boson mass is one of the output quantities in GHU as well. The predicted value, however, slightly differs from that in the SM, which, therefore, serves as one of the tests for the GHU scenario. In this section we evaluate m_W in GHU.

There are two additional relevant parameters in GHU, namely θ_H and $m_{\rm KK}$. These two are not completely independent. Once $m_{\rm KK}$ is specified, for instance, the allowed range of θ_H is restricted as seen below. In the previous chapters the value of $\sin^2\theta_W^0$ in (7.72) also needs to be specified in order to evaluate the KK mass spectra and gauge couplings. $\sin^2\theta_W^0$ is related to G_F. We are going to see that $\sin^2\theta_W^0$ and m_W are simultaneously determined in a self-consistent manner.

In the SM the μ decay proceeds by emitting the W boson at the tree level. (Fig. 9.1) The Fermi constant G_F is written in the SM as

$$\frac{G_F}{\sqrt{2}} = \frac{\pi\alpha}{2s_W^2}\,\frac{1}{m_W^2}\,(1+\Delta r_{\rm SM}^{\rm loop})\,, \tag{9.1}$$

$$s_W^2 = 1 - \frac{m_W^2}{m_Z^2}\,, \tag{9.2}$$

where $e = g_w s_W$, $\alpha^{-1} = 4\pi/e^2 = 137.035999084(21)$, $G_F = 1.1663788(6)\times 10^{-5}\,{\rm GeV}^{-2}$ and $m_Z = 91.1876(21)$ GeV. [67] $\Delta r_{\rm SM}^{\rm loop}$ represents the sum of all loop corrections, which depends on α, α_s, m_W, m_Z, m_H, and masses of quarks and leptons. s_W^2 is the weak mixing angle in the on-shell scheme in the SM. Combining Eqs. (9.1) and (9.2), one can write the W boson mass in the SM as

$$m_W^{\rm SM} = \frac{m_Z}{\sqrt{2}}\left[1+\sqrt{1-\frac{4\pi\alpha(1+\Delta r_{\rm SM}^{\rm loop})}{\sqrt{2}\,G_F m_Z^2}}\,\right]^{1/2}. \tag{9.3}$$

At the tree level $\Delta r_{\rm SM}^{\rm loop} = 0$ so that $m_W^{\rm SM}|_{\rm tree} = 80.9387$ GeV, which is larger than the observed W mass. The loop correction $\Delta r_{\rm SM}^{\rm loop}$ has been estimated to be $\Delta r_{\rm SM}^{\rm loop} \simeq 0.0383\pm 0.0004$, which leads to $m_W^{\rm SM} \simeq 80.354\pm 0.007$ GeV. [68–70]

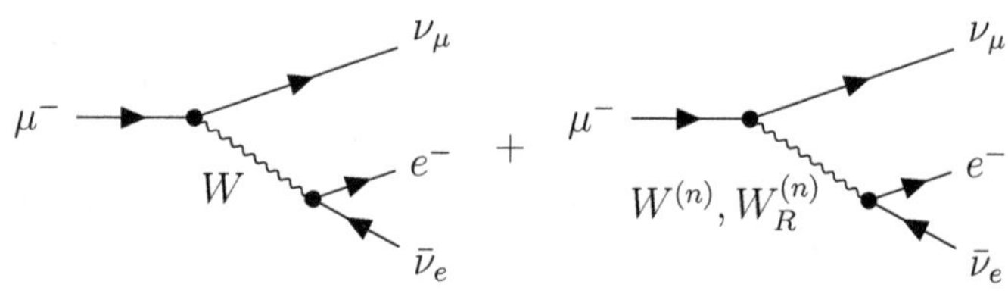

Fig. 9.1 The muon decay $\mu^- \to e^-\bar\nu_e\nu_\mu$ at the tree level. In the SM the decay proceeds by exchanging W. In GHU it proceeds by exchanging not only $W = W^{(0)}$, but also $W^{(n)}$ and $W_R^{(n)}$ $(n\geq 1)$.

In GHU the μ decay proceeds by emitting not only $W = W^{(0)}$, but also $W^{(n)}$ and $W_R^{(n)}$ $(n \geq 1)$. (Fig. 9.1) The relation (9.1) is replaced by

$$\frac{G_F}{\sqrt{2}} = \frac{\pi\alpha}{2\sin^2\theta_W^{\rm shell}} \frac{\hat{g}_{L\,000}^{W\nu_\mu^+\mu}\,\hat{g}_{L\,000}^{W\nu_e^+e}}{m_{W^{(0)}}^2}(1+\Delta r_G)\,(1+\Delta r_{\rm GHU}^{\rm loop}),$$

$$\Delta r_G = \frac{1}{\hat{g}_{L\,000}^{W\nu_\mu^+\mu}\,\hat{g}_{L\,000}^{W\nu_e^+e}} \sum_{n=1}^{\infty} \left\{ \hat{g}_{L\,n00}^{W\nu_\mu^+\mu}\,\hat{g}_{L\,n00}^{W\nu_e^+e} \left[\frac{m_{W^{(0)}}}{m_{W^{(n)}}}\right]^2 \right.$$

$$\left. + \hat{g}_{L\,n00}^{W_R\nu_\mu^+\mu}\,\hat{g}_{L\,n00}^{W_R\nu_e^+e} \left[\frac{m_{W^{(0)}}}{m_{W_R^{(n)}}}\right]^2 \right\}, \tag{9.4}$$

where $e = g_w \sin\theta_W^{\rm shell}$. The right-handed couplings are very small, and have been omitted in the expression above. Δr_G represents the contributions from the exchange of $W^{(n)}$ and $W_R^{(n)}$ ($n \geq 1$) at the tree level. $\Delta r_{\rm GHU}^{\rm loop}$ represents the sum of loop corrections.

Comparing (9.1) and (9.4), one sees that $\sin^2\theta_W^{\rm shell}$ in GHU corresponds to s_W^2 in the SM, but its relation to the ratio m_W^2/m_Z^2 is modified. At the tree level the masses $(m_W, m_Z) = (k\lambda_{W^{(0)}}, k\lambda_{Z^{(0)}})$ are determined by Eqs. (7.62) and (7.70).

$$2S(1;\lambda_{W^{(0)}}, z_L)C'(1;\lambda_{W^{(0)}}, z_L) + \lambda_{W^{(0)}}\sin^2\theta_H = 0\ , \tag{9.5}$$

$$2S(1;\lambda_{Z^{(0)}}, z_L)C'(1;\lambda_{Z^{(0)}}, z_L) + \frac{\lambda_{Z^{(0)}}\sin^2\theta_H}{1-\sin^2\theta_W^{\rm shell}} = 0\ . \tag{9.6}$$

The bare weak mixing angle $\sin^2\theta_W^0$ in Eq. (7.72) has been determined from the ratio of the two gauge couplings g_B/g_A. At the tree level $\sin^2\theta_W^{\rm shell}$ in (9.4) and (9.6) and $\sin^2\theta_W^0$ in (7.72) are the same, but need to be distinguished from each other at the loop level. The relations (9.5) and (9.6) have been derived from the boundary conditions satisfied by the gauge fields, which remain intact in the original gauge even at higher orders. With this observation the relations (9.5) and (9.6) are promoted to all orders with physical m_W and m_Z. In other words, $\sin^2\theta_W^{\rm shell}$ is the weak mixing angle in the on-shell scheme in GHU just as s_W^2 is in the SM. $\sin^2\theta_W^0$ corresponds to the weak mixing angle in the $\overline{\rm MS}$ scheme in the SM, which depends on the energy scale under consideration.

In GHU the W boson mass $m_W = m_{W^{(0)}}$ is determined by solving (9.4) and (9.5) simultaneously. For this end the value of $\sin^2\theta_W^{\rm shell}$ must be determined self-consistently. We take the following procedure for the given values of θ_H and $m_{\rm KK}$. One picks a tentative value of $\sin^2\theta_W^{\rm shell}$. Then the value of k and z_L are fixed from (9.6) and $m_{\rm KK} = \pi k/(z_L - 1)$ in order to reproduce the observed value of $m_Z = k\lambda_{Z^{(0)}}$. With these k and z_L, $m_W = k\lambda_{W^{(0)}} \equiv m_W^a$ is determined from (9.5). On the other hand, with

$\sin^2\theta_W^{\rm shell}$ given, all gauge couplings and mass spectra appearing in (9.4) are determined. Further for the loop corrections only the SM particles are relevant at the m_μ scale and their gauge couplings are very close to those in the SM as shown in the previous chapter. One can therefore make an approximation $\Delta r_{\rm GHU}^{\rm loop} \sim \Delta r_{\rm SM}^{\rm loop}$. Now one evaluates $m_{W^{(0)}} \equiv m_W^b$ from (9.4) with the values of G_F and α inserted. The two values m_W^a and m_W^b generally differ from each other. One adjusts $\sin^2\theta_W^{\rm shell}$ such that these two values for m_W coincide. In this manner m_W and $\sin^2\theta_W^{\rm shell}$ are determined to satisfy (9.4) and (9.5) simultaneously. [71]

For $m_{\rm KK} = 13$ TeV and $\theta_H = 0.10$, for instance, one finds that $m_W = 80.396$ GeV, $\sin^2\theta_W^{\rm shell} = 0.22266$, $\hat{g}_{L\,000}^{W\nu_\mu^+\mu}\hat{g}_{L\,000}^{W\nu_e^+e} = 0.99530$ and $\Delta r_G = 0.0020$. In the infinite sum in Δr_G in (9.4) only the $W^{(1)}$ term gives an appreciable contribution (about 0.0019). The sum rapidly converges. In Fig. 9.2 the predicted values for m_W are plotted for various values of $m_{\rm KK}$ and θ_H. In the SM $m_W^{\rm SM} \simeq 80.354 \pm 0.007$ GeV. In 2022 the CDF Collaboration reported on the W mass, $m_W^{\rm CDF} = 80.4335 \pm 0.0094$ GeV. [72] The ATLAS Collaboration also reanalyzed the data in 2011 to obtain $m_W^{\rm ATLAS} = 80.360 \pm 0.016$ GeV. [73] The experimental situation has not been settled. We see that the predicted m_W in GHU for 13 TeV $\leq m_{\rm KK} \leq$ 20 TeV lies between $m_W^{\rm SM}$ and $m_W^{\rm CDF}$. In Fig. 9.3 the predicted values for $\sin^2\theta_W^{\rm shell}$ are plotted for various values of $m_{\rm KK}$ and θ_H. $s_W^2|_{\rm SM} = 1 - (m_W^2/m_Z^2) = 0.22339 \pm 0.00010$ is the value in the on-shell scheme in the SM.

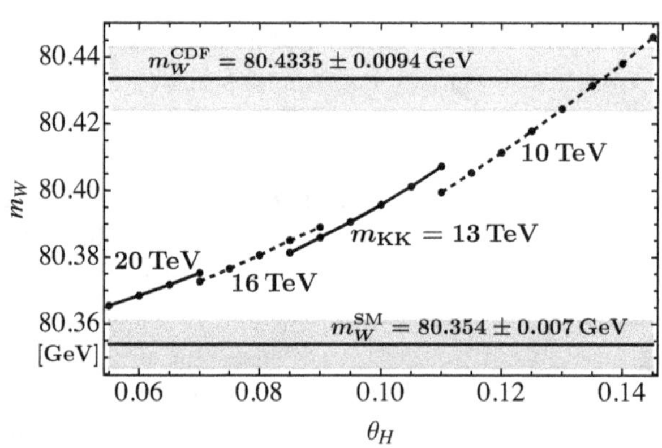

Fig. 9.2 The W boson mass m_W in GHU is plotted as a function of θ_H with various $m_{\rm KK}$. The constraint $m_{\rm KK} \gtrsim 13$ TeV is obtained from the experimental data at LHC. [74] The predicted m_W in GHU for 13 TeV $\leq m_{\rm KK} \leq$ 20 TeV lies between $m_W^{\rm SM}$ and $m_W^{\rm CDF}$.

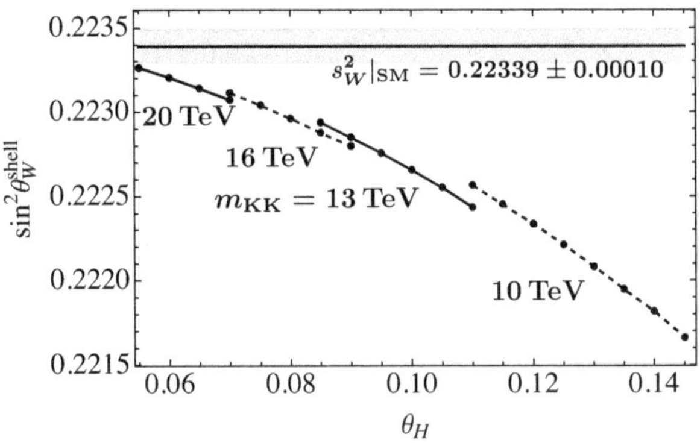

Fig. 9.3 $\sin^2\theta_W^{\rm shell}$ in the on-shell scheme in GHU is plotted as a function of θ_H with various $m_{\rm KK}$. $s^2_W|_{\rm SM} = 1 - (m^2_W/m^2_Z) = 0.22339 \pm 0.00010$ is the value in the on-shell scheme in the SM, listed in Table 10.2 of Ref. [67].

It is seen that the predicted values of m_W and $\sin^2\theta_W^{\rm shell}$ are mostly determined by the value of θ_H. Once the value of $m_{\rm KK}$ is specified, the range of θ_H is restricted to $\theta_H^{\rm min} \le \theta_H \le \theta_H^{\rm max}$. For $\theta_H < \theta_H^{\rm min}$ the top quark mass m_t cannot be reproduced. The top quark mass $m_t = k\lambda_{t^{(0)}}$ is determined by Eq. (7.89), or

$$S_L S_R(1; \lambda_{t^{(0)}}, c_t; z_L) + \sin^2\frac{\theta_H}{2} = 0 . \tag{9.7}$$

With given θ_H, z_L and $\lambda_{t^{(0)}}$, a solution for $|c_t|$ exists only for $\theta_H \ge \theta_H^{\rm min}$. ($c_t = 0$ for $\theta_H = \theta_H^{\rm min}$.) The value of θ_H is also constrained by $\theta_H < \theta_H^{\rm max}$. For $\theta_H > \theta_H^{\rm max}$ the μ-e universality is spoiled particularly in the Z couplings of the right-handed e and μ at the m_Z scale, and the AdS curvature ($\Lambda = -6k^2$) becomes larger, exceeding the Planck mass scale.

To understand how m_W in GHU becomes larger than $m_W^{\rm SM}$, let us rewrite the formula (9.4) in the form

$$\begin{aligned} \frac{G_F}{\sqrt{2}} &= C\,\frac{\pi\alpha}{2s^2_{W,\rm GHU}}\,\frac{1}{m^2_{W^{(0)}}}\,(1 + \Delta r^{\rm loop}_{\rm GHU}) , \\ C &= \frac{s^2_{W,\rm GHU}}{\sin^2\theta_W^{\rm shell}}\,\hat g^{W^{(0)}}_{\mu\nu_\mu,L}\hat g^{W^{(0)}}_{e\nu_e,L}\,(1+\Delta r_G) , \\ s^2_{W,\rm GHU} &= 1 - \frac{m^2_{W^{(0)}}}{m^2_Z} . \end{aligned} \tag{9.8}$$

There are three factors in C. For $m_{\rm KK} = 13$ TeV and $\theta_H = 0.10$ one finds $s^2_{W,\rm GHU}/\sin^2\theta_W^{\rm shell} = 1.00014$, $\hat g^{W^{(0)}}_{\mu\nu_\mu,L}\hat g^{W^{(0)}}_{e\nu_e,L} = 0.99530$, $1 + \Delta r_G = 1.0020$

and therefore $C = 0.99743 < 1$. In other words the effective $\Delta r^{\rm eff}_{\rm GHU}$ defined by $1 + \Delta r^{\rm eff}_{\rm GHU} = C(1 + \Delta r^{\rm loop}_{\rm GHU})$ becomes smaller than $\Delta r^{\rm loop}_{\rm GHU} \sim \Delta r^{\rm loop}_{\rm SM}$, which in turn makes $m^{(0)}_W$ larger than $m^{\rm SM}_W$. The most dominant factor is $\hat{g}^{W^{(0)}}_{\mu\nu_\mu,L}\hat{g}^{W^{(0)}}_{e\nu_e,L} < 1$, which is related to the fact that the bulk mass parameters c_e and c_μ are negative.

Once θ_H and $m_{\rm KK}$ are given, the weak mixing angle $\sin^2\theta^{\rm shell}_W$ in the on-shell scheme in GHU is determined. In the SM, $s^2_W = 0.22339$ in the on-shell scheme corresponds to $\hat{s}^2_Z = \sin^2\hat{\theta}_W(m_Z) = 0.23122$ in the $\overline{\rm MS}$ scheme in the SM. [67] To derive GHU predictions for the observables in e^-e^+ collisions at the m_Z pole, for instance, one needs to know $\sin^2\theta^0_W(m_Z)$ at the m_Z scale. It is reasonable to take

$$\sin^2\theta^0_W(m_Z) \sim \frac{\hat{s}^2_Z}{s^2_W}\bigg|_{\rm SM} \sin^2\theta^{\rm shell}_W \sim 1.0351\,\sin^2\theta^{\rm shell}_W\,. \tag{9.9}$$

For $\theta_H = 0.1$ and $m_{\rm KK} = 13$ TeV, one finds $\sin^2\theta^0_W(m_Z) = 0.23035$.

With $\sin^2\theta^0_W(m_Z)$ one evaluates the Z couplings $\hat{g}^{Zee}_{L/R\,000}$ and $\hat{g}^{Z\mu\mu}_{L/R\,000}$. The forward-backward asymmetry in the $e^-e^+ \to \mu^-\mu^+$ process at the m_Z pole is given by

$$\begin{aligned} A^\mu_{FB} &= \frac{3}{4}\,A_e A_\mu\,, \\ A_f &= \frac{\big(\hat{g}^{Zff}_{L\,000}\big)^2 - \big(\hat{g}^{Zff}_{R\,000}\big)^2}{\big(\hat{g}^{Zff}_{L\,000}\big)^2 + \big(\hat{g}^{Zff}_{R\,000}\big)^2}\,. \end{aligned} \tag{9.10}$$

A^μ_{FB} is plotted as a function of θ_H with various $m_{\rm KK}$ in Fig. 9.4. It is seen that the predicted values in GHU are consistent with the experimental data $A^\mu_{\rm FB}(m_Z)^{\rm exp} = 0.0169 \pm 0.0013$ for $13\,{\rm TeV} \le m_{\rm KK} \le 20\,{\rm TeV}$.

It is important to determine m_W by experiments. The W^+W^- pair production process in the e^+e^- collisions near the threshold will give indispensable information on m_W. Once m_W is determined, the values of $m_{\rm KK}$ and θ_H in GHU is severely constrained.

9.2 Decay of the Higgs boson

The Higgs boson was discovered at LHC in 2012 in the decay mode $H \to \gamma\gamma$, which takes place at the one loop level. The Higgs boson is neutral, and does not couple to photons directly. In the SM the Higgs boson first become $t\bar{t}$ and W^+W^- pairs, and those pairs emit two photons and annihilate. (Fig. 9.5) As a result the Higgs boson decays into two photons. Although the branching ratio of the $H \to \gamma\gamma$ process is small, it gives clean events

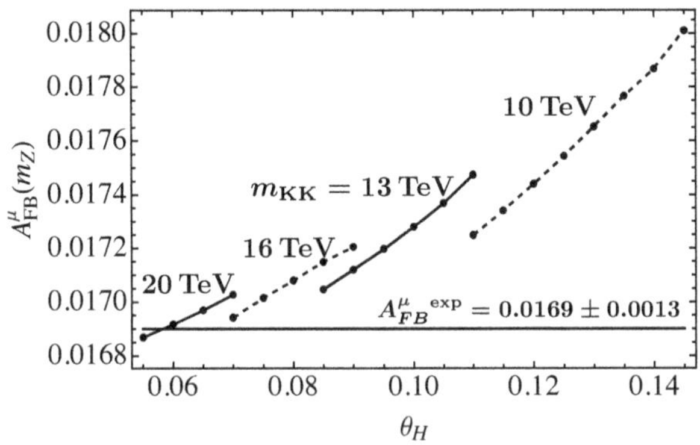

Fig. 9.4 The forward-backward asymmetry in the process $e^-e^+ \to \mu^-\mu^+$ at the Z pole $A^{\mu}_{\rm FB}(m_Z)$ is plotted as a function of θ_H with various $m_{\rm KK}$. $A^{\mu}_{\rm FB}(m_Z)$ for $13\,{\rm TeV} \leq m_{\rm KK} \leq 20\,{\rm TeV}$ is consistent with the observed value $A^{\mu}_{\rm FB}(m_Z)^{\rm exp} = 0.0169 \pm 0.0013$.

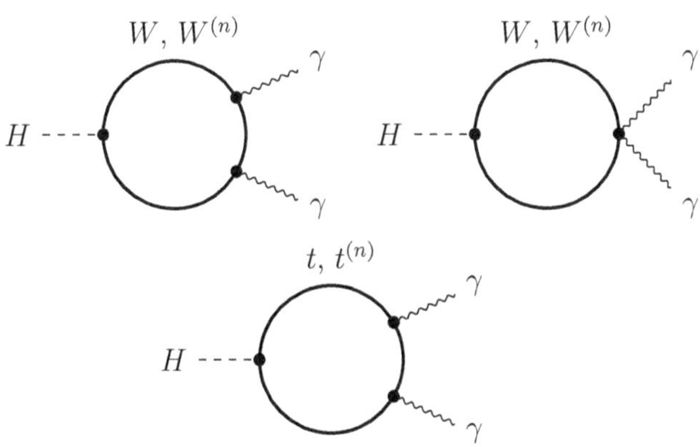

Fig. 9.5 The decay of the Higgs boson $H \to \gamma\gamma$ takes place at the one loop level. In GHU virtual pairs of $W^{(n)}$ and $t^{(n)}$ $(n \geq 1)$ also contribute.

with well-controlled background in experiments. Note that other quarks and leptons give negligible contributions to $H \to \gamma\gamma$ as their Higgs couplings are small.

In GHU, in addition to $W = W^{(0)}$ and $t = t^{(0)}$, their KK excited modes $W^{(n)}$ and $t^{(n)}$ $(n \geq 1)$ also contribute. Contributions from other fermions are negligible. The decay width $\Gamma(H \to \gamma\gamma)$ is given, at the one loop level, by [54, 75]

$$\Gamma(H \to \gamma\gamma) = \frac{\alpha^2 g_w^2}{1024\pi^3} \frac{m_H^3}{m_W^2} \left|\mathcal{F}_{\rm total}\right|^2 ,$$

$$\mathcal{F}_{\rm total} = \mathcal{F}_W + \mathcal{F}_t \ ,$$

$$\mathcal{F}_W = \cos\theta_H \sum_{n=0}^{\infty} I_{W^{(n)}} \frac{m_W}{m_{W^{(n)}}} F_1(\tau_{W^{(n)}}) \ ,$$

$$\mathcal{F}_t = \frac{4}{3} \cos^2 \frac{\theta_H}{2} \sum_{n=0}^{\infty} I_{t^{(n)}} \frac{m_t}{m_{t^{(n)}}} F_{1/2}(\tau_{t^{(n)}}) \ ,$$

$$\tau_j = \frac{4m_j^2}{m_H^2} \ , \tag{9.11}$$

where $I_{W^{(n)}}$ and $I_{t^{(n)}}$ are defined in Eqs. (8.99) and (8.105), and the functions $F_1(\tau)$ and $F_{1/2}(\tau)$ are given by

$$F_1(\tau) = 2 + 3\tau + 3\tau(2-\tau) f(\tau) \ ,$$

$$F_{1/2}(\tau) = -2\tau[1 + (1-\tau) f(\tau)] \ ,$$

$$f(\tau) = \begin{cases} \left[\sin^{-1}\left(\sqrt{1/\tau}\right)\right]^2 & \text{for } \tau \geq 1 \ , \\ -\frac{1}{4}\left[\ln \dfrac{1+\sqrt{1-\tau}}{1-\sqrt{1-\tau}} - i\pi\right]^2 & \text{for } \tau < 1 \ . \end{cases} \tag{9.12}$$

In the $\tau \to \infty$ limit, $F_1 \to 7$ and $F_{1/2} \to -\frac{4}{3}$. As has been observed in Section 8.5, Table 8.10, $I_{W^{(n)}}$ and $I_{t^{(n)}}$ behave as $(-1)^n \cdot$ (const) for large n. Therefore the infinite sums in $\mathcal{F}_W$ and $\mathcal{F}_t$ in Eq. (9.11) rapidly converge as $\sum(-1)^n/n$.

For $\theta_H = 0.1$, $m_{\rm KK} = 13$ TeV and $\sin^2\theta_W^0 = \sin^2\theta_W^{\rm shell} = 0.22266$, one finds that

$$\begin{pmatrix} \mathcal{F}_{\rm total} \\ \mathcal{F}_W \\ \mathcal{F}_t \end{pmatrix} = \begin{pmatrix} 6.4553 \\ 8.2860 \\ -1.8307 \end{pmatrix} , \quad \begin{pmatrix} \mathcal{F}_{\rm total}^{\rm SM} \\ \mathcal{F}_W^{\rm SM} \\ \mathcal{F}_t^{\rm SM} \end{pmatrix} = \begin{pmatrix} 6.4916 \\ 8.3268 \\ -1.8352 \end{pmatrix} , \tag{9.13}$$

where the values in the SM have been written on the right. The suppression relative to the SM values comes mostly from the factors $\cos\theta_H$ and $\cos^2\frac{1}{2}\theta_H$ in $\mathcal{F}_W$ and $\mathcal{F}_t$. Contributions from the KK excited modes $W^{(n)}$ and $t^{(n)}$ $(n \geq 1)$ are 1.1×10^{-4} to $\mathcal{F}_{\rm total}$, -8.9×10^{-4} to $\mathcal{F}_W$, and 1.01×10^{-3} to $\mathcal{F}_t$. The resultant decay width is

$$\Gamma(H \to \gamma\gamma) = 8.72\,{\rm keV} \ . \tag{9.14}$$

As the cubic couplings among the Higgs boson and SM particles are like $g_{000}^{WWH} \sim g_{\rm SM}^{WWH} \cos\theta_H$ and $y_{000}^{Htt} \sim y_t^{\rm SM} \cos^2 \frac{1}{2}\theta_H$, the width of each decay mode of the Higgs boson $\Gamma(H \to j)$ is suppressed by a factor $\cos^2\theta_H$ or $\cos^4 \frac{1}{2}\theta_H$ compared to the SM value $\Gamma(H \to j)^{\rm SM}$. The branching fraction $B(H \to j) = \Gamma(H \to j)/\Gamma(H \to {\rm all})$ becomes very close to that in the SM for small $\theta_H \lesssim 0.1$. Similarly, the production rate of the Higgs boson in pp collision experiments is suppressed by a factor $\cos^2\theta_H$ or $\cos^4 \frac{1}{2}\theta_H$ compared to that in the SM.

9.3 Self-couplings of the Higgs boson

It is of great interest to know the θ_H and $m_{\rm KK}$ dependence of the cubic and quartic self-couplings, λ_3 and λ_4 given in Eq. (8.1), of the Higgs boson. As shown in Section 8.1, although the effective potential $V_{\rm eff}(\theta)$ depends on θ_H, $m_{\rm KK}$ and other parameters in the dark fermion sector, λ_3 and λ_4 are determined solely by θ_H and $m_{\rm KK}$ to high accuracy. [58]

The result is depicted in Fig. 9.6. In the SM $\lambda_3^{\rm SM} = 31.8$ GeV and $\lambda_4^{\rm SM} = 0.0323$ at the tree level. In GHU λ_3 is 12% smaller than $\lambda_3^{\rm SM}$, and λ_4 is 39% smaller than $\lambda_4^{\rm SM}$.

9.4 Production of W' and Z' bosons

KK excited modes of the various fields appear as new particles. They have large masses of $O(m_{\rm KK})$. They are produced in collider experiments at high energies. At the Large Hadron Collider (LHC) in Switzerland, for instance, proton beams are accelerated in two opposite directions along the circular ring, and make head-on collisions with center-of-mass energies up to 14 TeV. The rate of the conversion of quarks to their KK excited modes via W or Z exchange is very small. Further KK excited modes of quarks/antiquarks must be produced in a pair, which requires at least the collision energy $\sim 1.8\, m_{\rm KK}$.

The KK excited modes of gauge bosons can be produced as single particles. The mass of $W^{(1)}$ or $Z^{(1)}$ is about $0.8\, m_{\rm KK}$. $W^{(1)}$ and $W_R^{(1)}$ are called W' bosons, and $Z^{(1)}$, $Z_R^{(1)}$ and $\gamma^{(1)}$ are called Z' bosons. These W' and Z' bosons are produced at LHC in the processes depicted in Fig. 9.7. Quarks and anti-quarks inside protons pair-annihilate to become a Z' or W' boson, which subsequently decays into a lepton or quark pair. The energy carried by the quark or anti-quark is much less than the energy of the proton so that the Z' or W' boson is generated as a virtual particle. Nevertheless the

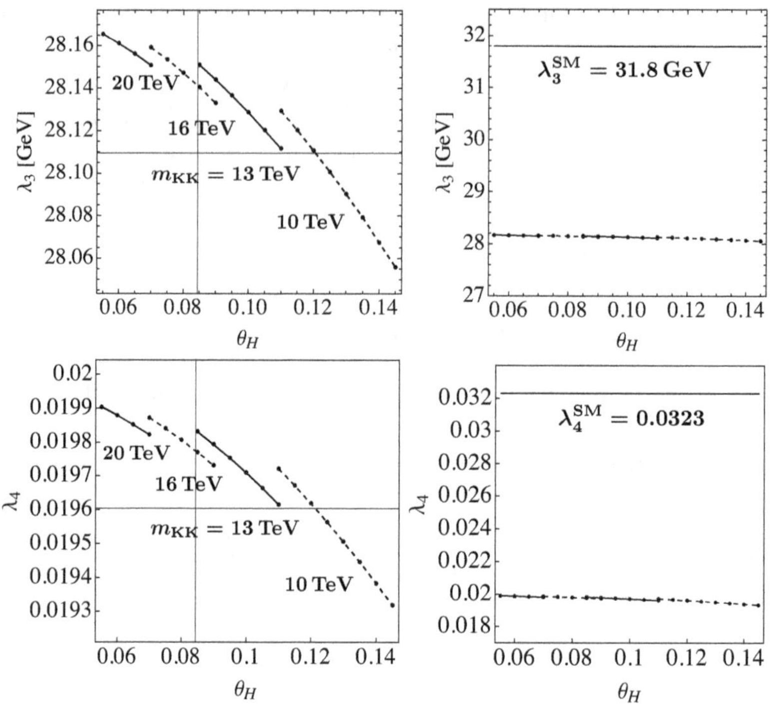

Fig. 9.6 The cubic and quartic self-couplings, λ_3 and λ_4, of the Higgs boson. In the figures on the right the values $\lambda_3^{\rm SM}$ and $\lambda_4^{\rm SM}$ in the SM are also shown.

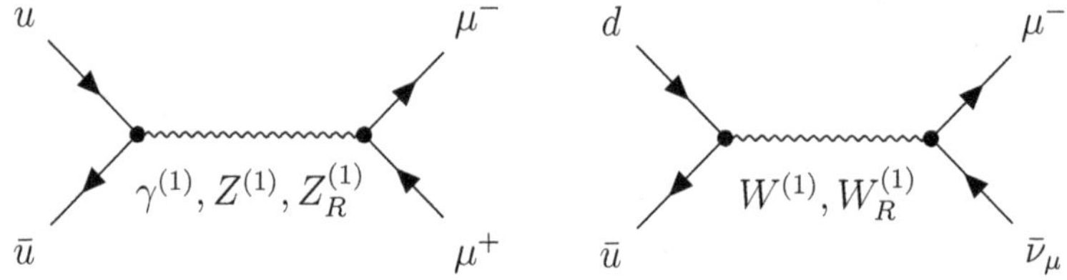

Fig. 9.7 The production of Z' and W' bosons at LHC. Virtual Z' and W' bosons are produced by pair annihilation of quarks and antiquarks inside protons.

event $pp \to \mu^-\mu^+$, for instance, serves as a clean signal for Z' bosons with small background. The couplings of $\gamma^{(1)}, Z^{(1)}, Z_R^{(1)}$ to the left-handed components of quarks and leptons are large. Their decay widths become very large so that events of Z' bosons appear as a broad peak in pp collisions. Although $m_{Z^{(1)}}, m_{Z_R^{(1)}}, m_{\gamma^{(1)}} \sim 10$ TeV for $\theta_H = 0.1$ and $m_{\rm KK} = 13$ TeV, the existence of Z' bosons can be explored at the High Luminosity LHC. At the moment no convincing signals for Z' have been found, which gives a constraint $m_{\rm KK} \gtrsim 13$ TeV. [74]

9.5 Z' interference effect in e^-e^+ collisions

In electron-positron (e^-e^+) collisions at high energies, say, in the 250 GeV–1 TeV range, Z' bosons are virtually produced, leading to striking interference with γ and Z events in GHU. In e^-e^+ collisions there occurs fermion pair production $e^-e^+ \to f\bar{f}$ where the final state $f\bar{f}$ includes e^-e^+, $\mu^-\mu^+$, $u\bar{u}$, $d\bar{d}$, $c\bar{c}$, $b\bar{b}$ and $t\bar{t}$. As shown in Fig. 9.7, diagrams for the final states other than e^-e^+ are classified in two categories: diagrams containing γ and Z in the intermediate state and diagrams containing $\gamma^{(1)}$, $Z^{(1)}$ and $Z_R^{(1)}$ in the intermediate state. The interference terms of these two types of amplitudes gives rise to distinct signals of Z' bosons even for $\sqrt{s} \ll m_{\rm KK}$. [76–80]

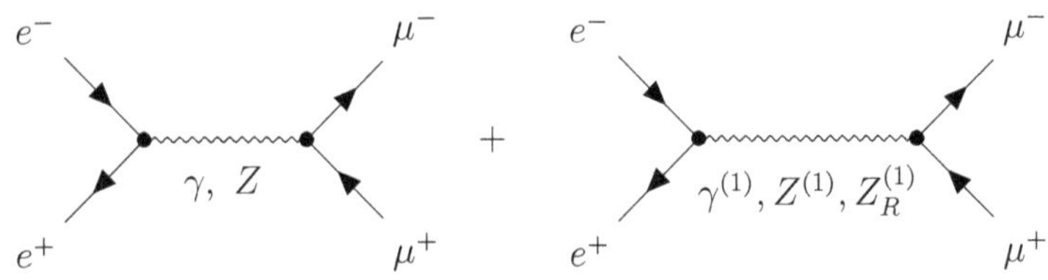

Fig. 9.8 Main diagrams at the tree level in the process $e^-e^+ \to \mu^-\mu^+$.

As shown in Section 8.4.2 and tabulated in Table 8.9, the left-handed components of leptons and quarks have rather large gauge couplings to $\gamma^{(1)}$, $Z^{(1)}$ and $Z_R^{(1)}$, whereas the couplings of the right-handed components are small. There is large parity violation in the gauge couplings of $\gamma^{(1)}$, $Z^{(1)}$ and $Z_R^{(1)}$. Wave functions of $\gamma^{(n)}$, $Z^{(n)}$ and $Z_R^{(n)}$ ($n \geq 1$) are localized near the TeV brane at $z = z_L$ ($y = L$). In the GUT inspired GHU the bulk mass parameters of quark-lepton multiplets, $\Psi^{q,\alpha}_{\mathbf{(3,4)}}$ and $\Psi^{\ell,\alpha}_{\mathbf{(1,4)}}$, are negative so that the left-handed components of the zero modes, namely quarks and leptons, are localized near the TeV brane whereas the right-handed components are localized near the UV brane at $z = 1$ ($y = 0$). This is why the left-handed components have large couplings to the Z' bosons. If the bulk mass parameters were positive, the right-handed components would have large couplings.

The scattering amplitude for the process $e^-(p)e^+(p') \to \mu^-(k)\mu^+(k')$

depicted in Fig. 9.8 is given, in the $\xi = 1$ gauge, by

$$\begin{aligned}\mathcal{M} &= \bar{v}(p')\gamma^\mu u(p)\,\frac{e^2}{s}\,\bar{v}(k')\gamma_\mu u(k)\\ &+ \frac{g_w^2}{\cos^2\theta_W^0}\sum_V{}'\,\bar{v}(p')\gamma^\mu\left\{\hat{g}_{eL}^V\frac{1-\gamma^5}{2}+\hat{g}_{eR}^V\frac{1+\gamma^5}{2}\right\}u(p)\\ &\times\frac{1}{s-m_V^2+im_V\Gamma_V}\bar{v}(k')\gamma_\mu\left\{\hat{g}_{\mu L}^V\frac{1-\gamma^5}{2}+\hat{g}_{\mu R}^V\frac{1+\gamma^5}{2}\right\}u(k)\ , \end{aligned} \tag{9.15}$$

where $s = -(p+p')^2$, $V = Z, \gamma^{(1)}, Z^{(1)}, Z_R^{(1)}$ in the sum $\sum'$, and $\hat{g}_{eL}^{Z^{(1)}} = \hat{g}_{L\,100}^{Zee}$ etc. Γ_V is the decay width of the V boson. Let $\mathcal{M}_0$ ($\mathcal{M}_{Z'}$) be the amplitude mediated by γ and Z ($\gamma^{(1)}, Z^{(1)}$ and $Z_R^{(1)}$), and therefore $\mathcal{M} = \mathcal{M}_0 + \mathcal{M}_{Z'}$.

At e^-e^+ linear colliders such as ILC, the initial electron and positron beams can be polarized with polarization P_{e^-} and P_{e^+}. $P_{e^\pm} = +1$ (-1) corresponds to purely right-handed (left-handed) $e^\pm$. The differential cross section in the center-of-mass frame becomes

$$\begin{aligned}&\frac{d\sigma}{d\cos\theta}(P_{e^-}, P_{e^+}, \cos\theta)\\ &\quad= \frac{1-P_{e^-}P_{e^+}}{4}\left\{(1-P_{\rm eff})\frac{d\sigma_{LR}}{d\cos\theta}+(1+P_{\rm eff})\frac{d\sigma_{RL}}{d\cos\theta}\right\},\\ &P_{\rm eff} = \frac{P_{e^-}-P_{e^+}}{1-P_{e^-}P_{e^+}}\ . \end{aligned} \tag{9.16}$$

$d\sigma_{LR}/d\cos\theta$ and $d\sigma_{RL}/d\cos\theta$ are differential cross sections for $e_L^-e_R^+ \to \mu^-\mu^+$ and $e_R^-e_L^+ \to \mu^-\mu^+$, respectively. For $\sqrt{s} \gg 2m_\mu$ they are given by

$$\begin{aligned}\frac{d\sigma_{LR}}{d\cos\theta} &\simeq \frac{s}{32\pi}\Big\{(1+\cos\theta)^2\,|Q_{e_L\mu_L}|^2+(1-\cos\theta)^2\,|Q_{e_L\mu_R}|^2\Big\},\\ \frac{d\sigma_{RL}}{d\cos\theta} &\simeq \frac{s}{32\pi}\Big\{(1+\cos\theta)^2\,|Q_{e_R\mu_R}|^2+(1-\cos\theta)^2\,|Q_{e_R\mu_L}|^2\Big\}. \end{aligned} \tag{9.17}$$

Here $Q_{e_a\mu_b}$ $(a, b = L, R)$ is given by

$$Q_{e_a\mu_b} = \Big(\frac{g_w}{\cos\theta_W^0}\Big)^2\sum_V\frac{\hat{g}_{e\,a}^V\hat{g}_{\mu\,b}^V}{s-m_V^2+im_V\Gamma_V}, \tag{9.18}$$

where the sum extends over $V = \gamma, Z, \gamma^{(1)}, Z^{(1)}$ and $Z_R^{(1)}$. For a photon $m_\gamma = \Gamma_\gamma = 0$ and $\hat{g}_{eL/R}^\gamma = \hat{g}_{\mu L/R}^\gamma = -\sin\theta_W^0\cos\theta_W^0$. The total cross section is

$$\sigma_{\rm tot} = \tfrac{1}{4}(1 - P_{e^-}P_{e^+})\{(1 - P_{\rm eff})\sigma_{LR} + (1 + P_{\rm eff})\sigma_{RL}\},$$

$$\sigma_{LR} = \frac{s}{12\pi}\big(|Q_{e_L\mu_L}|^2 + |Q_{e_L\mu_R}|^2\big),$$

$$\sigma_{RL} = \frac{s}{12\pi}\big(|Q_{e_R\mu_R}|^2 + |Q_{e_R\mu_L}|^2\big). \tag{9.19}$$

Let $Q^0_{e_a\mu_b}$ and $Q^{Z'}_{e_a\mu_b}$ be the contributions to $Q_{e_a\mu_b}$ from γ and Z and from $\gamma^{(1)}$, $Z^{(1)}$ and $Z_R^{(1)}$, respectively. Then

$$|Q_{e_a\mu_b}|^2 = |Q^0_{e_a\mu_b}|^2 + (Q^0_{e_a\mu_b}Q^{Z'*}_{e_a\mu_b} + Q^{0*}_{e_a\mu_b}Q^{Z'}_{e_a\mu_b}) + |Q^{Z'}_{e_a\mu_b}|^2\,. \tag{9.20}$$

The first term is the contribution from γ and Z, and is very close to the one in the SM, $|Q^{\rm SM}_{e_a\mu_b}|^2$. The third term represents the contribution due only to the production of Z' particles. The second term is the interference term characteristic in quantum theory. For $\sqrt{s} \ll m_{Z'} \equiv \frac{1}{3}(m_{\gamma^{(1)}} + m_{Z^{(1)}} + m_{Z_R^{(1)}})$ the contribution of the third term is negligible, but the effect of the second term can be explored by precision measurements in e^-e^+ collider experiments.

To estimate the magnitude of the second term for $m_Z^2 \ll s \ll m_{Z'}^2$, note that

$$Q^0_{e_a\mu_b} \sim \frac{g_w^2}{\cos^2\theta_W^0}\frac{\kappa^0_{ab}}{s}\,, \quad \kappa^0_{ab} = \sin^2\theta^0_W\cos^2\theta^0_W + \hat{g}^Z_{e\,a}\hat{g}^Z_{\mu\,b}\,,$$

$$Q^{Z'}_{e_a\mu_b} \sim -\frac{g_w^2}{\cos^2\theta^0_W}\frac{\kappa^{Z'}_{ab}}{m^2_{Z'}}\,,$$

$$\kappa^{Z'}_{ab} = \hat{g}^{\gamma^{(1)}}_{e\,a}\hat{g}^{\gamma^{(1)}}_{\mu\,b} + \hat{g}^{Z^{(1)}}_{e\,a}\hat{g}^{Z^{(1)}}_{\mu\,b} + \hat{g}^{Z_R^{(1)}}_{e\,a}\hat{g}^{Z_R^{(1)}}_{\mu\,b}\,, \tag{9.21}$$

so that

$$\frac{Q^0_{e_a\mu_b}Q^{Z'*}_{e_a\mu_b} + Q^{0*}_{e_a\mu_b}Q^{Z'}_{e_a\mu_b}}{|Q^0_{e_a\mu_b}|^2} \sim -\frac{2\kappa^{Z'}_{ab}}{\kappa^0_{ab}}\frac{s}{m^2_{Z'}}\,. \tag{9.22}$$

$\kappa^{Z'}_{ab}$, κ^0_{ab} for $\theta_H = 0.1$ and $m_{\rm KK} = 13$ TeV and $\kappa^{\rm SM}_{ab}$ in the SM are

$$\begin{matrix} & & LL & LR & RL & RR \end{matrix}$$
$$\begin{pmatrix}\kappa^{Z'}_{ab}\\ \kappa^0_{ab}\\ \kappa^{\rm SM}_{ab}\end{pmatrix} = \begin{pmatrix} 8.6466 & -0.1470 & -0.1408 & 0.0114\\ 0.2494 & 0.1154 & 0.1154 & 0.2304\\ 0.25 & 0.1156 & 0.1156 & 0.2312\end{pmatrix}. \tag{9.23}$$

It is seen that $\kappa^{Z'}_{LL} \gg \kappa^0_{LL}$. $m_{Z'} \sim 10$ TeV so that even for $\sqrt{s} = 250$ GeV, the ratio for the LL component in (9.22) becomes ~ -0.043. The effect of

Z' bosons can be clearly seen in the process where the $Q_{e_L \mu_L}$ term becomes important. For instance, the total cross section for unpolarized e^- and e^+ beams ($P_{e^\pm} = 0$) becomes, with $\sigma^0_{\rm tot}$ denoting the contribution from γ and Z only,

$$\frac{\sigma_{\rm tot} - \sigma^0_{\rm tot}}{\sigma^0_{\rm tot}} \sim -2 \frac{\kappa^0_{LL}\kappa^{Z'}_{LL} + \kappa^0_{LR}\kappa^{Z'}_{LR} + \kappa^0_{RL}\kappa^{Z'}_{RL} + \kappa^0_{RR}\kappa^{Z'}_{RR}}{(\kappa^0_{LL})^2 + (\kappa^0_{LR})^2 + (\kappa^0_{RL})^2 + (\kappa^0_{RR})^2} \frac{s}{m^2_{Z'}}$$

$$\sim -29.96 \frac{s}{m^2_{Z'}} , \tag{9.24}$$

which amounts to -0.019 for $\sqrt{s} = 250$ GeV.

The total cross section $\sigma_{\rm tot}(e^-e^+ \to \mu^-\mu^+)$ is plotted in Fig. 9.9 (the left figure) for $\theta_H = 0.1$ and $m_{\rm KK} = 13$ TeV. $(m_{\gamma^{(1)}}, m_{Z^{(1)}}, m_{Z_R^{(1)}}) = (10.2, 10.2, 10.0)$ TeV and $(\Gamma_{\gamma^{(1)}}, \Gamma_{Z^{(1)}}, \Gamma_{Z_R^{(1)}}) = (3.3, 7.8, 0.8)$ TeV. As $\gamma^{(1)}$ and $Z^{(1)}$ have large decay widths, the signal of the Z' boson appears as a broad resonance around $m_{Z'} \sim 10$ TeV. In the right figure in Fig. 9.9

$$\Delta_{\sigma_{\rm tot}}(s) = \frac{\sigma^{\rm GHU}_{\rm tot} - \sigma^{\rm SM}_{\rm tot}}{\sigma^{\rm SM}_{\rm tot}} \tag{9.25}$$

is plotted for $100\,{\rm GeV} \le \sqrt{s} \le 500\,{\rm GeV}$ with the polarization $(P_{e^-}, P_{e^+}) = (0, 0)$ and $(\pm 0.8, 0)$. Note that even for $\sqrt{s} = 250$ GeV, $\Delta_{\sigma_{\rm tot}} =$

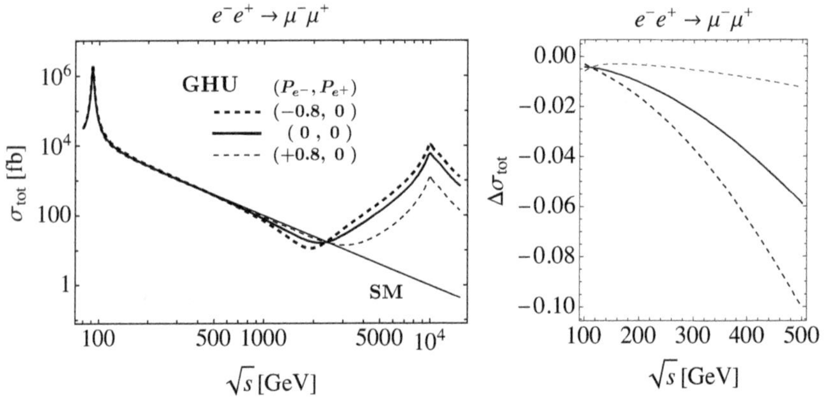

Fig. 9.9 In the left figure the total cross section $\sigma_{\rm tot}(e^-e^+ \to \mu^-\mu^+)$ in GHU with unpolarized and polarized e^- beams for $\theta_H = 0.1$ and $m_{\rm KK} = 13$ TeV are shown in the range $80\,{\rm GeV} < \sqrt{s} < 15\,{\rm TeV}$. The total cross section in the SM with unpolarized beams is also shown. In the right figure $\Delta_{\sigma_{\rm tot}}(s) = \sigma^{\rm GHU}_{\rm tot}/\sigma^{\rm SM}_{\rm tot} - 1$ is plotted for $\theta_H = 0.1$ and $m_{\rm KK} = 13$ TeV with unpolarized and polarized e^- beams in the range $100\,{\rm GeV} < \sqrt{s} < 500\,{\rm GeV}$.

$-0.026, -0.016, -0.004$ for $(P_{e^-}, P_{e^+}) = (-0.8, 0), (0, 0), (+0.8, 0)$, respectively, which can be confirmed in precision experiments at e^-e^+ colliders in future.

As is seen in Fig. 9.9, the total cross section has distinct dependence on the polarization of the incident $e^\mp$ beams. The polarization dependence resulting from the large parity violation in the Z' couplings becomes more evident in the forward-backward asymmetry in the fermion pair production processes. The forward-backward asymmetry A_{FB} is given by

$$A_{FB} = \frac{\sigma_F - \sigma_B}{\sigma_F + \sigma_B} ,$$

$$\sigma_F = \int_0^1 d\cos\theta \, \frac{d\sigma}{d\cos\theta} , \quad \sigma_B = \int_{-1}^0 d\cos\theta \, \frac{d\sigma}{d\cos\theta} . \tag{9.26}$$

It follows from Eqs. (9.16) and (9.17) that in the $e^-e^+ \to \mu^-\mu^+$ process, for $\sqrt{s} \gg m_\mu$,

$$A^\mu_{FB} = \frac{3}{4}\frac{B_1 - B_2}{B_1 + B_2} ,$$

$$B_1 = (1 - P_{\text{eff}})|Q_{e_L\mu_L}|^2 + (1 + P_{\text{eff}})|Q_{e_R\mu_R}|^2,$$

$$B_2 = (1 - P_{\text{eff}})|Q_{e_L\mu_R}|^2 + (1 + P_{\text{eff}})|Q_{e_R\mu_L}|^2, \tag{9.27}$$

where the effective polarization P_{eff} is defined in (9.16).

In the left figure in Fig. 9.10, A^μ_{FB} is plotted as a function of $\sqrt{s}$ with $P_{\text{eff}} = -0.8, 0$ and $+0.8$. At high energies $\sqrt{s} > 500\,\text{GeV}$ large deviation from the SM is seen. In the right figure the P_{eff}-dependence of A^μ_{FB} is plotted for $\sqrt{s} = 250$ GeV and 800 GeV. The deviation from the SM is large for $P_{\text{eff}} \sim -0.8$, while it becomes small for $P_{\text{eff}} \sim +0.8$. This behavior can be confirmed even at $\sqrt{s} = 250$ GeV at ILC.

Similarly one can explore the left-right (LR) asymmetry, just by switching P_{e^-}. Consider the LR asymmetry in the $e^-e^+ \to \mu^-\mu^+$ process with unpolarized positron beams ($P_{e^+} = 0$).

$$\begin{aligned} A^\mu_{LR}(s, P_{e^-}) &= \frac{\sigma_{\text{tot}}(s, P_{e^-}) - \sigma_{\text{tot}}(s, -P_{e^-})}{\sigma_{\text{tot}}(s, P_{e^-}) + \sigma_{\text{tot}}(s, -P_{e^-})} \\ &= -P_{e^-}\frac{\sigma_{LR} - \sigma_{RL}}{\sigma_{LR} + \sigma_{RL}} \\ &= -P_{e^-}\frac{C_1 - C_2}{C_1 + C_2} , \\ C_1(s) &= |Q_{e_L\mu_L}|^2 + |Q_{e_L\mu_R}|^2 , \\ C_2(s) &= |Q_{e_R\mu_R}|^2 + |Q_{e_R\mu_L}|^2 . \end{aligned} \tag{9.28}$$

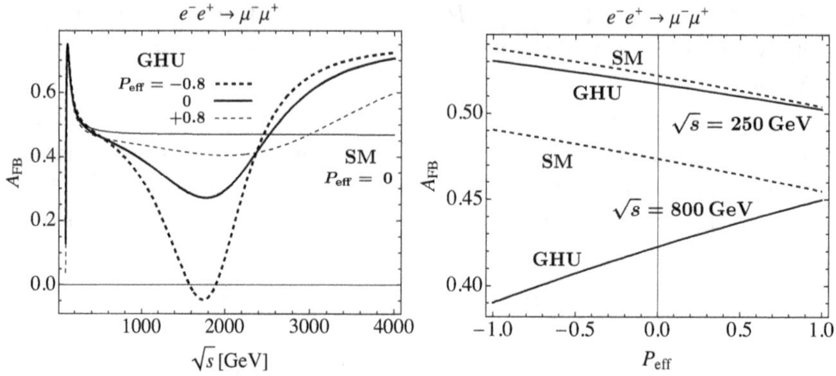

Fig. 9.10 The forward-backward asymmetry A^{μ}_{FB} is plotted as a function of $\sqrt{s}$ for $\theta_H = 0.1$ and $m_{\rm KK} = 13$ TeV. In the right figure A^{μ}_{FB} is plotted as a function of $P_{\rm eff}$ for $\sqrt{s} = 250$ GeV and 800 GeV.

In Fig. 9.11 the $\sqrt{s}$-dependence of $-A^{\mu}_{LR}/P_{e^-}$ is plotted for $\theta_H = 0.1$ and $m_{\rm KK} = 13$ TeV. One recognizes that with polarized electron beams large deviation from the SM can be seen even at $\sqrt{s} = 150\,{\rm GeV} \sim 250\,{\rm GeV}$. $A^{\mu}_{LR} = -0.089\,P_{e^-}$ $(-0.060\,P_{e^-})$ at $\sqrt{s} = 160$ GeV (250 GeV) in GHU, whereas $A^{\mu}_{LR} = -0.093\,P_{e^-}$ $(-0.073\,P_{e^-})$ in the SM.

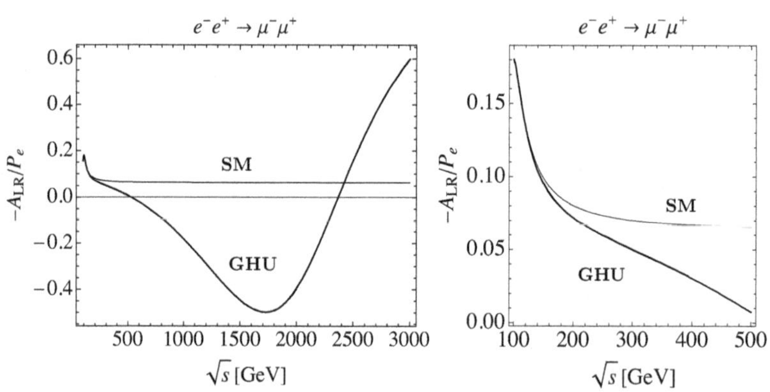

Fig. 9.11 The left-right asymmetry A^{μ}_{LR} divided by $-P_{e^-}$ in Eq. (9.28) is plotted as a function of $\sqrt{s}$ for $\theta_H = 0.1$ and $m_{\rm KK} = 13$ TeV. Large deviation from the SM is seen even at $\sqrt{s} = 160$ GeV.

9.6 Coupling sum rules

In the SM the gauge couplings of quarks, leptons, and W, Z bosons are determined by the representation to which each field belongs in the gauge group. There are many relations among the gauge couplings which become crucial to guarantee unitarity in scattering processes and the finiteness of physical quantities. In GHU the 5D gauge couplings of each 5D multiplet are determined by the representation to which the field belongs in the gauge group. However, when a 5D field is expanded in 4D KK modes, the gauge couplings among the 4D fields become nontrivial, depending on the AB phase θ_H, the spacetime parameters k, z_L and additional field parameters such as the bulk mass parameters c of quarks and leptons. The gauge couplings among the SM particles in GHU slightly deviate from those in the SM, which immediately leads to a question whether or not the unitarity and finiteness mentioned above are maintained in GHU.

In this section we shall see that the contributions from KK modes are important for having the unitarity in scattering processes and the finiteness of physical quantities. There appear many nontrivial relations among the gauge couplings in GHU, which we call coupling sum rules.

9.6.1 *Unitarity at the tree level*

Consider the process $e^- e^+ \to W^- W^+$. Diagrams contributing to this process at the tree level are shown in Fig. 9.12(a). It is known that the high energy ($\sqrt{s} \gg m_W$) behavior of this process is well controlled in the SM as a result of the cancellation among the amplitudes of the diagrams. [81–83]

In the SM $e = g_w \sin\theta_W^{\rm SM}$, $g_{WWZ}^{\rm SM} = g_w \cos\theta_W^{\rm SM}$, $g_{We\nu_e\,L}^{\rm SM} = g_w/\sqrt{2}$, $g_{We\nu_e\,R}^{\rm SM} = 0$, $g_{Zee\,L}^{\rm SM} = (g_w/\cos\theta_W^{\rm SM})(-\frac{1}{2} + \sin^2\theta_W^{\rm SM})$, and $g_{Zee\,R}^{\rm SM} = (g_w/\cos\theta_W^{\rm SM}) \sin^2\theta_W^{\rm SM}$. The part of the total amplitude increasing as s vanishes in the SM as a result of the identity

$$\begin{aligned} e_L^- :& \quad (-e)(+e) + g_{Zee\,L}^{\rm SM}\, g_{WWZ}^{\rm SM} + (g_{We\nu_e\,L}^{\rm SM})^2 = 0 \,, \\ e_R^- :& \quad (-e)(+e) + g_{Zee\,R}^{\rm SM}\, g_{WWZ}^{\rm SM} + (g_{We\nu_e\,R}^{\rm SM})^2 = 0 \,. \end{aligned} \tag{9.29}$$

In GHU the various KK modes contribute in the intermediate states, and the part of the total amplitude increasing as s vanishes provided [84, 85]

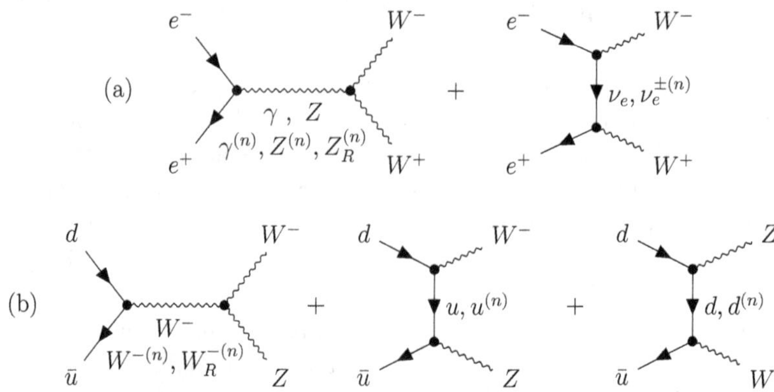

Fig. 9.12 (a) Diagrams for $e^-e^+ \to W^-W^+$. (b) Diagrams for $d\,\bar{u} \to W^-Z$. In GHU the various $n \geq 1$ KK modes also contribute as shown in the figures.

$$
\begin{aligned}
&F^s_{L/R} + F^t_{L/R} = 0 \, , \\
&g_w^2 F^s_{L/R} = \sum_{n=0}^{\infty} e\,\hat{Q}^e_{L/R,n00}\, g_{00n}^{WW\gamma} + \sum_{n=0}^{\infty} \frac{g_w}{\cos\theta_W^0}\, \hat{g}^{Zee}_{L/R,n00}\, g_{00n}^{WWZ} \\
&\qquad\qquad + \sum_{n=1}^{\infty} \frac{g_w}{\cos\theta_W^0}\, \hat{g}^{Z_R ee}_{L/R,n00}\, g_{00n}^{WWZ_R} , \\
&g_w^2 F^t_{L/R} = \sum_{n=0}^{\infty} \left(\frac{g_w}{\sqrt{2}}\, \hat{g}^{W\nu_e^+ e}_{L/R,0n0} \right)^2 + \sum_{n=1}^{\infty} \left(\frac{g_w}{\sqrt{2}}\, \hat{g}^{W\nu_e^- e}_{L/R,0n0} \right)^2 .
\end{aligned}
\tag{9.30}
$$

The coupling sum rule (9.30) is confirmed numerically. For $\theta_H = 0.1$, $m_{\rm KK} = 13$ TeV and $\sin^2\theta_W^0 = 0.230352$, one finds that

$$
\begin{aligned}
&F^s_L = -0.497934 \, , \quad F^t_L = 0.497933 \, , \quad F^s_L + F^t_L = -2 \times 10^{-7} \, , \\
&F^s_R = 1 \times 10^{-7} \, , \quad F^t_R = 8 \times 10^{-17} \, , \quad F^s_R + F^t_R = 1 \times 10^{-7} \, .
\end{aligned}
\tag{9.31}
$$

We stress that the coupling sum rule is valid only if one takes into account the contributions from $\gamma^{(n)}$, $Z^{(n)}$, $Z_R^{(n)}$ and $\nu_e^{\pm(n)}$ ($n \geq 1$). If contributions only from the zero modes ($\gamma^{(0)}, Z^{(0)}, \nu_e^{+(0)}$) were taken into account, $F^s_L + F^t_L$ would become -0.001152. For the left-handed electron e_L^- each component in F^s_L and F^t_L is given by

$$
\begin{aligned}
&F_L^{s,\gamma} = -0.229531 \, , \quad F_L^{s,Z} = -0.268113 \, , \quad F_L^{s,Z_R} = -0.000289 \, , \\
&F_L^{t,\nu_e^+} = 0.497792 \, , \quad F_L^{t,\nu_e^-} = 0.000142 \, .
\end{aligned}
\tag{9.32}
$$

As a consequence of the coupling sum rule the unitarity at the tree level is guaranteed in the process $e^- e^+ \to W^- W^+$ in GHU.

Similar arguments apply for the process $d\,\bar{u} \to W^- Z$ as well. Diagrams contributing to this process are shown in Fig. 9.12(b). (Contributions from $D^{(n)}$ in the intermediate state in the last diagram are numerically negligible.) In the SM the high energy behavior of the amplitude is controlled by the identities

$$g^{\rm SM}_{Wud\,L} g^{\rm SM}_{WWZ} - g^{\rm SM}_{Wud\,L} g^{\rm SM}_{Zuu\,L} + g^{\rm SM}_{Wud\,L} g^{\rm SM}_{Zdd\,L} = 0 \,, \tag{9.33}$$

and $g^{\rm SM}_{Wud\,R} = 0$. In GHU the part of the total amplitude increasing as s vanishes provided [84, 85]

$$\begin{aligned}
&G^s_{L/R} - G^t_{L/R} + G^u_{L/R} = 0 \,, \\
&\frac{g_w^2 G^s_{L/R}}{\sqrt{2}\cos\theta^0_W} = \sum_{n=0}^{\infty} \frac{g_w}{\sqrt{2}} \hat{g}^{Wud}_{L/R,n00}\, g^{WWZ}_{0n0} + \sum_{n=1}^{\infty} \frac{g_w}{\sqrt{2}} \hat{g}^{W_R ud}_{L/R,n00}\, g^{WW_R Z}_{0n0} \,, \\
&\frac{g_w^2 G^t_{L/R}}{\sqrt{2}\cos\theta^0_W} = \sum_{n=0}^{\infty} \frac{g_w}{\sqrt{2}} \hat{g}^{Wud}_{L/R,0n0} \frac{g_w}{\cos\theta^0_W} \hat{g}^{Zuu}_{L/R,0n0} \,, \\
&\frac{g_w^2 G^u_{L/R}}{\sqrt{2}\cos\theta^0_W} = \sum_{n=0}^{\infty} \frac{g_w}{\sqrt{2}} \hat{g}^{Wud}_{L/R,00n} \frac{g_w}{\cos\theta^0_W} \hat{g}^{Zdd}_{L/R,00n} \,.
\end{aligned} \tag{9.34}$$

The coupling sum rule (9.34) is confirmed numerically. For $\theta_H = 0.1$, $m_{\rm KK} = 13$ TeV and $\sin^2\theta^0_W = 0.230352$, one finds that

$$\begin{aligned}
&G^s_L = 0.766112 \,, \quad G^t_L = 0.344749 \,, \quad G^u_L = -0.421367 \,, \\
&G^s_L - G^t_L + G^u_L = 4.6 \times 10^{-7} \,, \\
&(G^s_R, G^t_R, G^u_R) = (-3.4, -4.7, 1.1) \times 10^{-13} \,.
\end{aligned} \tag{9.35}$$

The contributions from the KK modes are important. If only contributions from the zero modes ($W^{(0)}$, $u^{(0)}$, $d^{(0)}$) were taken into account, $G^s_L - G^t_L + G^u_L$ would become 0.002350.

The arguments can be easily generalized to any processes $f\bar{f}' \to W^- W^+$, $W^\pm Z$ or ZZ, leading to coupling sum rules similar to (9.30) or (9.34). The five-dimensional gauge invariance guarantees the unitarity at the tree level in these processes.

9.6.2 *Oblique corrections at the 1-loop level*

Let us consider quantum corrections to the propagators of the gauge bosons γ, W and Z, particularly fermion one-loop corrections as depicted in

Fig. 9.13. The diagram is generated for gauge fields X_μ and Y_μ coupled to fermions ψ_j and ψ_k by

$$X_\mu \bar{\psi}_n \gamma^\mu (g_{n\ell}^{X,V} - g_{n\ell}^{X,A} \gamma^5) \psi_\ell + Y_\mu \bar{\psi}_\ell \gamma^\mu (g_{\ell n}^{Y,V} - g_{\ell n}^{Y,A} \gamma^5) \psi_n \, . \tag{9.36}$$

The vacuum polarization $\Pi_{XY}^{\mu\nu}(p) = \Pi_{XY}(p^2)\, \eta^{\mu\nu} - \Sigma_{XY}(p^2)\, p^\mu p^\nu$ is divergent in four dimensions ($d = 4$). The divergent part of $\Pi_{XY}(p^2)$ is found, in the dimensional regularization scheme, to be

$$\begin{aligned} \Pi_{XY}(p^2)^{\rm div} &= \frac{1}{4\pi^2} \bigg[(g_{n\ell}^{X,V} g_{\ell n}^{Y,V} + g_{n\ell}^{X,A} g_{\ell n}^{Y,A}) \{ \tfrac{1}{3} p^2 - \tfrac{1}{2}(m_n^2 + m_\ell^2) \} \\ &\qquad + (g_{n\ell}^{X,V} g_{\ell n}^{Y,V} - g_{n\ell}^{X,A} g_{\ell n}^{Y,A})\, m_n m_\ell \bigg] \hat{E} \, , \\ \hat{E} &= -\frac{2}{4-d} + \gamma_E - \ln 4\pi\mu^2 \, , \end{aligned} \tag{9.37}$$

where m_n and m_ℓ are masses of ψ_n and ψ_ℓ, respectively.

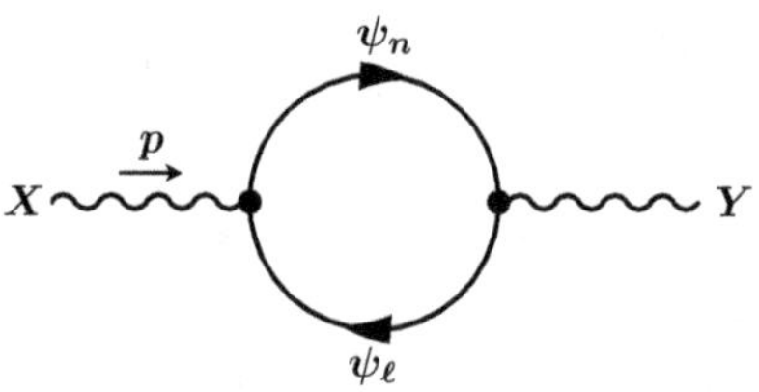

Fig. 9.13 Vacuum polarization diagram. $X, Y = \gamma, Z, W, W^\dagger$. Along the internal loop various fermion KK modes propagate.

The contributions from the (u, d) multiplet to $\Pi_{XY}(p^2)^{\rm div}$ $(X, Y = \gamma, Z, W, W^\dagger)$ are given by

$$\begin{aligned} \Pi_{ZZ}^{ud}(p^2)^{\rm div} &= \frac{N_C g_w^2}{4\pi^2 \cos^2\theta_W^0} \\ &\qquad \times \Big\{ \frac{1}{3} (A_0^{ud} - 2\sin^2\theta_W^0 B^{ud} + \sin^4\theta_W^0 C^{ud}) p^2 - \frac{1}{2} A_1^{ud} \Big\} \hat{E} \, , \\ \Pi_{Z\gamma}^{ud}(p^2)^{\rm div} &= \frac{N_C g_w^2 \sin\theta_W^0}{4\pi^2 \cos\theta_W^0} \frac{1}{3} (B^{ud} - \sin^2\theta_W^0 C^{ud}) p^2 \, \hat{E} \, , \\ \Pi_{\gamma\gamma}^{ud}(p^2)^{\rm div} &= \frac{N_C g_w^2 \sin^2\theta_W^0}{4\pi^2} \frac{1}{3} C^{ud} p^2 \, \hat{E} \, , \\ \Pi_{WW^\dagger}^{ud}(p^2)^{\rm div} &= \frac{N_C g_w^2}{8\pi^2} \Big\{ \frac{1}{3} D_0^{ud} p^2 - \frac{1}{2} D_1^{ud} \Big\} \hat{E} \, , \end{aligned} \tag{9.38}$$

where $N_C = 3$ and

$$
\begin{aligned}
A_0^{ud} &= \mathrm{Tr}\left\{\hat{g}_V^{W^3u}\hat{g}_V^{W^3u} + \hat{g}_A^{Zu}\hat{g}_A^{Zu} + \hat{g}_V^{W^3d}\hat{g}_V^{W^3d} + \hat{g}_A^{Zd}\hat{g}_A^{Zd}\right\}, \\
A_1^{ud} &= \sum_{n,\ell}\Big\{\hat{g}_{V,n\ell}^{W^3u}\hat{g}_{V,\ell n}^{W^3u}(m_{u^{(n)}} - m_{u^{(\ell)}})^2 + \hat{g}_{A,n\ell}^{Zu}\hat{g}_{A,\ell n}^{Zu}(m_{u^{(n)}} + m_{u^{(\ell)}})^2 \\
&\qquad + \hat{g}_{V,n\ell}^{W^3d}\hat{g}_{V,\ell n}^{W^3d}(m_{d^{(n)}} - m_{d^{(\ell)}})^2 + \hat{g}_{A,n\ell}^{Zd}\hat{g}_{A,\ell n}^{Zd}(m_{d^{(n)}} + m_{d^{(\ell)}})^2\Big\}, \\
B^{ud} &= Q_u \mathrm{Tr}\, \hat{g}_V^{W^3u} + Q_d \mathrm{Tr}\, \hat{g}_V^{W^3d}, \\
C^{ud} &= Q_u^2 \mathrm{Tr}\, I + Q_d^2 \mathrm{Tr}\, I, \\
D_0^{ud} &= \mathrm{Tr}\left\{\hat{g}_V^{Wud}\hat{g}_V^{W^\dagger du} + \hat{g}_A^{Wud}\hat{g}_A^{W^\dagger du}\right\}, \\
D_1^{ud} &= \sum_{n,\ell}\Big\{\hat{g}_{V,n\ell}^{Wud}\hat{g}_{V,\ell n}^{W^\dagger du}(m_{u^{(n)}} - m_{d^{(\ell)}})^2 \\
&\qquad + \hat{g}_{A,n\ell}^{Wud}\hat{g}_{A,\ell n}^{W^\dagger du}(m_{u^{(n)}} + m_{d^{(\ell)}})^2\Big\}. \qquad (9.39)
\end{aligned}
$$

Here Tr in $Q_u^2 \mathrm{Tr}\, I$ implies the trace over the $u^{(n)}$ states, and the couplings are denoted, with $\hat{g}_V = \frac{1}{2}(\hat{g}_R + \hat{g}_L)$ and $\hat{g}_A = \frac{1}{2}(\hat{g}_R - \hat{g}_L)$, as

$$
\begin{aligned}
&\hat{g}_{V,n\ell}^{W^3u} = \hat{g}_{V\,0n\ell}^{Zuu} + \sin^2\theta_W^0\, Q_u \delta_{n\ell}, \quad \hat{g}_{A,n\ell}^{Zu} = \hat{g}_{A\,0n\ell}^{Zuu}, \\
&\hat{g}_{V,n\ell}^{W^3d} = \hat{g}_{V\,0n\ell}^{Zdd} + \sin^2\theta_W^0\, Q_d \delta_{n\ell}, \quad \hat{g}_{A,n\ell}^{Zd} = \hat{g}_{A\,0n\ell}^{Zdd}, \\
&\hat{g}_{V/A,n\ell}^{Wud} = \hat{g}_{V/A\,0n\ell}^{Wud}, \quad \hat{g}_{V/A,\ell n}^{W^\dagger du} = (\hat{g}_{V/A,n\ell}^{Wud})^*. \qquad (9.40)
\end{aligned}
$$

There are six divergent terms in (9.38). In the SM three divergent coefficients in those terms are absorbed in the renormalization constants. Further, one can form three linearly independent combinations of $\Pi_{XY}(p^2)$ which are finite. Indeed $A_0 = B = \frac{1}{2}D_0 = \frac{1}{4}$, $A_1^{ud} = \frac{1}{2}D_1^{ud} = \frac{1}{4}(m_u^2 + m_d^2)$ in the SM. Those three combinations correspond to the S, T, U parameters of Peskin-Takeuchi, which serve as an indispensable tool for exploring physics beyond the SM. [83, 86–88]

In GHU as well three equality relations hold among A_0, A_1, B, D_0 and D_1. [89] They are

$$
\begin{cases}
A_0^{ud} = h^{ud} B^{ud} & \text{for } S \\
D_1^{ud} = 2A_1^{ud} & \text{for } T, \\
D_0^{ud} = 2A_0^{ud} & \text{for } U
\end{cases}
$$

$$
h^{ud} = \hat{g}_{L\,000}^{Zuu,su2} - \hat{g}_{L\,000}^{Zdd,su2}, \qquad (9.41)
$$

where $\hat{g}_{L\,000}^{Zuu,su2}$ and $\hat{g}_{L\,000}^{Zdd,su2}$ are given in Eq. (8.88). Note that each KK level gives rise to finite contributions to $A_0 \sim D_1$ in (9.39). The relations are obtained for each doublet of quarks and leptons. The coupling sum rules in (9.41) are confirmed numerically by summing the KK mode indices n, ℓ in (9.39) from 0 to $n_{\rm max}$, and we examine

$$\Delta_S = \frac{A_0 - hB}{A_0} \,, \quad \Delta_T = \frac{A_1 - \frac{1}{2}D_1}{A_1} \,, \quad \Delta_U = \frac{A_0 - \frac{1}{2}D_0}{A_0} \,. \tag{9.42}$$

The result is tabulated for (ν_e, e) and (u, d) doublets in Table 9.1 for $\theta_H = 0.1$, $m_{\rm KK} = 13\,{\rm TeV}$ and $n_{\rm max} = 12$. It is seen that the three coupling sum rules in (9.41) hold with 5- to 7-digit accuracy. If the KK mode sum is increased to $n_{\rm max} = 16$, for instance, A_0 and A_1 for the (ν_e, e) doublet change as $A_0 : 3.24 \to 4.24$ and $A_1/m_Z^2 : 44.1 \to 57.9$. Nevertheless the accuracy of the sum rules remains intact, $\Delta_S : 3.6 \times 10^{-5} \to 7.4 \times 10^{-5}$, $\Delta_T : 5.34 \times 10^{-5} \to 5.34 \times 10^{-5}$, $\Delta_U : -3.44 \times 10^{-7} \to -3.41 \times 10^{-7}$.

Table 9.1 The couplings sum rules (9.41). The values of A_0, A_1, Δ_S, Δ_T, Δ_U and h in (9.39) and (9.42) are tabulated for (ν_e, e) and (u, d) doublets for $\theta_H = 0.1$, $m_{\rm KK} = 13$ TeV and $M = 10^3$ TeV. The numerical values are evaluated by including the contributions coming from the KK towers of fermions up to the $n = 12$ level.

	A_0	A_1/m_Z^2	Δ_S	Δ_T	Δ_U	h
(ν_e, e)	3.24204	44.1410	3.6×10^{-5}	5.3×10^{-5}	-3.4×10^{-7}	0.997690
(u, d)	3.24210	43.7063	-4.0×10^{-6}	5.5×10^{-5}	-3.4×10^{-7}	0.997688

With the sum rules (9.41) one can define the oblique parameters in GHU which are finite and free from the divergence. The appearance of the h factor in the relation $A_0 = hB$ is related to the Z couplings of quarks and leptons. In the SM the Z couplings are given by $(g_w/\cos\theta_W^{\rm SM})(T_L^3 - \sin^2\theta_W^{\rm SM} Q_{\rm EM})$. In GHU the Z couplings of the zero modes are approximately given by

$$\frac{g_w}{\cos\theta_W^0}\Big(h^\beta\, T_L^3 - \sin^2\theta_W^0\, Q_{\rm EM}\Big) \,, \tag{9.43}$$

where $h^\beta = \hat{g}_{L\,000}^{Zuu,su2} - \hat{g}_{L\,000}^{Zdd,su2}$ for $\beta = ud$. The relevant quantity in determining the forward-backward asymmetry A_{FB} in the $e^- e^+ \to f\bar{f}$ process discussed in Section 9.4 is $\sin^2\theta_W^0/h^\beta$.

9.6.3 *Gauge anomalies*

Another type of coupling sum rules emerges in relation to gauge anomalies. In gauge theory in four dimensions gauge anomalies arise when the gauge couplings of fermions are chiral. [90–93] Anomalies are generated at the quantum level through triangular loop diagrams. The magnitude of anomalies is determined by the gauge coupling constants and the representation in the gauge group to which fermions belong to. In the SM gauge anomalies are cancelled in each generation of quarks and leptons. [94,95] It is known that the cancellation of gauge anomalies is crucial and necessary for a theory to be consistent in four dimensions.

In GHU the situation is more involved. There are infinitely many KK fermion modes which contribute to gauge anomalies. Furthermore, the gauge couplings of the zero modes of fermions, namely quarks and leptons, slightly deviate from those in the SM and the amount of the deviation depends on the fermion species as shown in the previous chapter. This could spoil the cancellation of gauge anomalies in each generation at $\theta_H \neq 0$.

It turns out that special relations hold among gauge anomalies arising from left-handed and right-handed components of fermions after all contributions from the KK modes inside the loop are summed over. The magnitude of the anomaly changes and flows with θ_H. The total anomaly stemming from each 5D fermion multiplet turns out independent of the bulk mass parameter. [96,97] We are going to investigate the details in the next section.

9.7 Anomaly flow

Gauge anomalies have many special features in GHU. In order to simplify the argument and make the essence clear, let us consider $SU(2)$ gauge theory in the RS space. Orbifold boundary conditions are imposed such that $SU(2)$ is reduced to $U(1)$. $SU(2)$ doublets of fermion multiplets are introduced with various types of boundary conditions.

The action is given by

$$I = \int d^5x \sqrt{-g} \Big\{ -\frac{1}{2} \text{Tr}\, F_{MN} F^{MN} + \overline{\Psi} \mathcal{D}(c) \Psi \Big\} ,$$

$$\mathcal{D}(c) = \gamma^a {e_a}^M \Big(\partial_M - i g_A A_M + \frac{1}{8} \omega_{Mbc} [\gamma^b, \gamma^c] \Big) - c\, \sigma'(y) , \quad (9.44)$$

where $A_M = \sum_{j=1}^3 \frac{1}{2} A_M^j \tau^j$ and $F_{MN} = \partial_M A_N - \partial_N A_M - i g_A [A_M, A_N]$. Orbifold boundary conditions in the y-coordinate system are given, with

$(y_0, y_1) = (0, L)$, by

$$\begin{pmatrix} A_\mu \\ A_y \end{pmatrix}(x, y_j - y) = P_j \begin{pmatrix} A_\mu \\ -A_y \end{pmatrix}(x, y_j + y) P_j^{-1} ,$$

$$\Psi(x, y_j - y) = \begin{cases} +P_j \gamma^5 \Psi(x, y_j + y) & \text{type 1A} \\ -P_j \gamma^5 \Psi(x, y_j + y) & \text{type 1B} \\ (-1)^j P_j \gamma^5 \Psi(x, y_j + y) & \text{type 2A} \\ (-1)^{j+1} P_j \gamma^5 \Psi(x, y_j + y) & \text{type 2B} \end{cases},$$

$$P_0 = P_1 = \tau^3 . \tag{9.45}$$

The $SU(2)$ symmetry is broken to $U(1)$ by the boundary conditions (9.45). $A_\mu^3, A_y^{1,2}$ are parity even at both y_0 and y_1, and have zero modes. The zero mode of A_μ^3 is the 4D $U(1)$ gauge field, and the 4D gauge coupling is given by $g_4 = g_A/\sqrt{L}$.

Without loss of generality one may assume that $\langle A_y^1 \rangle = 0$. The AB phase θ_H along the fifth dimension is

$$\theta_H = \frac{\langle A_z^{2(0)} \rangle}{f_H} , \quad f_H = \frac{1}{g_4} \sqrt{\frac{2k}{L(z_L^2 - 1)}} . \tag{9.46}$$

The twisted gauge is related to the original gauge by a large gauge transformation

$$\Omega(z) = e^{i\theta(z)\tau^2/2} , \quad \theta(z) = \theta_H \frac{z_L^2 - z^2}{z_L^2 - 1} , \tag{9.47}$$

as in the $SO(5)$ theory discussed in Section 7.8. The AB phase θ_H intertwines A_μ^1 and A_μ^3 fields. The KK expansions in the twisted gauge in the region $1 \le z \le z_L$ are written as

$$\begin{pmatrix} \tilde{A}_\mu^1(x, z) \\ \tilde{A}_\mu^3(x, z) \end{pmatrix} = \frac{1}{\sqrt{L}} \sum_{n=0}^{\infty} Z_\mu^{(n)}(x) \begin{pmatrix} \tilde{h}_n(z) \\ \tilde{k}_n(z) \end{pmatrix} . \tag{9.48}$$

The mass spectrum $\{m_n(\theta_H) = k\lambda_n(\theta_H)\}$ $(\lambda_0 < \lambda_1 < \lambda_2 < \cdots)$ of $Z_\mu^{(n)}$ is given by

$$SC'(1; \lambda_n) + \lambda_n \sin^2 \theta_H = 0 \tag{9.49}$$

and the wave functions are given by

$$\begin{pmatrix} \tilde{h}_n(z) \\ \tilde{k}_n(z) \end{pmatrix} = \frac{1}{\sqrt{r_n}} \begin{pmatrix} -s_H \check{S}(z; \lambda_n) \\ c_H C(z; \lambda_n) \end{pmatrix} ,$$

$$r_n = \frac{1}{kL} \int_1^{z_L} \frac{dz}{z} \Big\{ s_H^2 \check{S}(z; \lambda_n)^2 + c_H^2 C(z; \lambda_n)^2 \Big\}, \tag{9.50}$$

where $(s_H, c_H) = (\sin\theta_H, \cos\theta_H)$. The normalization factor of the wave functions in (9.48) and (9.50) differs from the previous one by a factor of $\sqrt{kL}$. This normalization turns out to be convenient when discussing anomaly formulas below.

Let us denote the doublet fermion as $\Psi = (u, d)^t$. The AB phase θ_H intertwines u and d. The KK expansion of the fermion field $\check{\Psi} = z^{-2}\Psi$ in the twisted gauge in the region $1 \le z \le z_L$ is expressed as

$$\begin{pmatrix} \tilde{\check{u}}_R(x,z) \\ \tilde{\check{d}}_R(x,z) \end{pmatrix} = \sqrt{k}\sum_{n=0}^{\infty} \chi_R^{(n)}(x) \begin{pmatrix} \tilde{f}_{Rn}(z) \\ \tilde{g}_{Rn}(z) \end{pmatrix},$$

$$\begin{pmatrix} \tilde{\check{u}}_L(x,z) \\ \tilde{\check{d}}_L(x,z) \end{pmatrix} = \sqrt{k}\sum_{n=0}^{\infty} \chi_L^{(n)}(x) \begin{pmatrix} \tilde{f}_{Ln}(z) \\ \tilde{g}_{Ln}(z) \end{pmatrix}. \tag{9.51}$$

The spectrum of the KK modes $\chi^{(n)}(x)$ is determined by

$$\begin{cases} S_L S_R(1;\lambda_n, c) + \sin^2 \frac{1}{2}\theta_H = 0 & \text{for type 1A/B} \\ S_L S_R(1;\lambda_n, c) + \cos^2 \frac{1}{2}\theta_H = 0 & \text{for type 2A/B} \end{cases}, \tag{9.52}$$

and the wave functions for type 1A are given by

$$\begin{pmatrix} \tilde{f}_{Rn}(z) \\ \tilde{g}_{Rn}(z) \end{pmatrix} = \frac{1}{\sqrt{r_n}} \begin{pmatrix} \bar{c}_H C_R(z;\lambda_n, c) \\ -\bar{s}_H \check{S}_R(z;\lambda_n, c) \end{pmatrix},$$

$$\begin{pmatrix} \tilde{f}_{Ln}(z) \\ \tilde{g}_{Ln}(z) \end{pmatrix} = \frac{1}{\sqrt{r_n}} \begin{pmatrix} \bar{s}_H \check{S}_L(z;\lambda_n, c) \\ \bar{c}_H C_L(z;\lambda_n, c) \end{pmatrix},$$

$$\begin{aligned} r_n &= \int_1^{z_L} dz \left\{ \bar{c}_H^2 C_R(z;\lambda_n, c)^2 + \bar{s}_H^2 \check{S}_R(z;\lambda_n, c)^2 \right\} \\ &= \int_1^{z_L} dz \left\{ \bar{s}_H^2 \check{S}_L(z;\lambda_n, c)^2 + \bar{c}_H^2 C_L(z;\lambda_n, c)^2 \right\}, \end{aligned} \tag{9.53}$$

where $(\bar{s}_H, \bar{c}_H) = (\sin\frac{1}{2}\theta_H, \cos\frac{1}{2}\theta_H)$.

The spectrum $m_n/m_{\rm KK} = \lambda_n(z_L - 1)/\pi$ of $Z_\mu^{(n)}$ and $\chi^{(n)}$ (type 1A) is shown for $z_L = 10$ and $c = 0.25$ in Fig. 9.14. Notice that the behavior of the spectrum is similar to that of the Z boson and top quark in Fig. 7.1 in the GHU model. There is no level crossing in the RS space. The $Z_\mu^{(0)}$ field becomes a massless gauge boson at $\theta_H = \pi$, whereas the $\chi^{(0)}$ mode is massive at $\theta_H = \pi$.

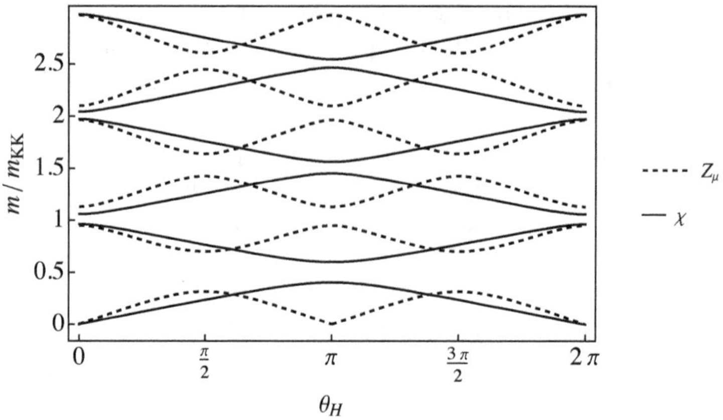

Fig. 9.14 The spectrum $m_n/m_{\rm KK} = \lambda_n(z_L - 1)/\pi$ of $Z_\mu^{(n)}$ and $\chi^{(n)}$ (type 1A) in the $SU(2)$ model is shown for $z_L = 10$ and $c = 0.25$.

9.7.1 *Anomaly coefficients*

The gauge coupling of the $\chi^{(0)}$ mode of type 1A is purely chiral at $\theta_H = 0$, but it becomes purely vectorlike at $\theta_H = \pi$. The chiral anomaly is generated at $\theta_H = 0$ at the one loop level. However the $\chi^{(0)}$ mode becomes massive with purely vectorlike coupling at $\theta_H = \pi$ so that the anomaly must vanish there. As is seen in Fig. 9.14, everything changes smoothly in the RS space as θ_H varies from 0 to π. The magnitude of the anomaly must change smoothly from the nonzero value to zero. How does it happen?

Chiral anomalies arise from triangular loop diagrams as depicted in Fig. 9.15. As in Eqs. (8.72) and (8.86), gauge couplings are written as

$$\begin{aligned}\mathcal{L}_{\rm int}^Z &= -ig_A \int_1^{z_L} \frac{dz}{k}\, \bar{\tilde{\Psi}} \gamma^\mu \tfrac{1}{2}\Big(\tilde{A}_\mu^1 \tau^1 + \tilde{A}_\mu^3 \tau^3\Big) \tilde{\Psi} \\ &= -i\frac{g_4}{2} \sum_{\ell=0}^\infty \sum_{n=0}^\infty \sum_{m=0}^\infty Z_\mu^{(\ell)} \\ &\qquad \times \Big\{ \hat{t}_{\ell nm}^R \bar{\chi}_R^{(n)} \gamma^\mu \chi_R^{(m)} + \hat{t}_{\ell nm}^L \bar{\chi}_L^{(n)} \gamma^\mu \chi_L^{(m)} \Big\}, \end{aligned} \tag{9.54}$$

where

$$\begin{aligned}\hat{t}_{\ell nm}^{R/L} = \int_1^{z_L} dz \Big\{ &\tilde{k}_\ell \big(\tilde{f}_{R/Ln}^* \tilde{f}_{R/Lm} - \tilde{g}_{R/Ln}^* \tilde{g}_{R/Lm} \big) \\ &+ \tilde{h}_\ell \big(\tilde{f}_{R/Ln}^* \tilde{g}_{R/Lm} + \tilde{g}_{R/Ln}^* \tilde{f}_{R/Lm} \big) \Big\}. \end{aligned} \tag{9.55}$$

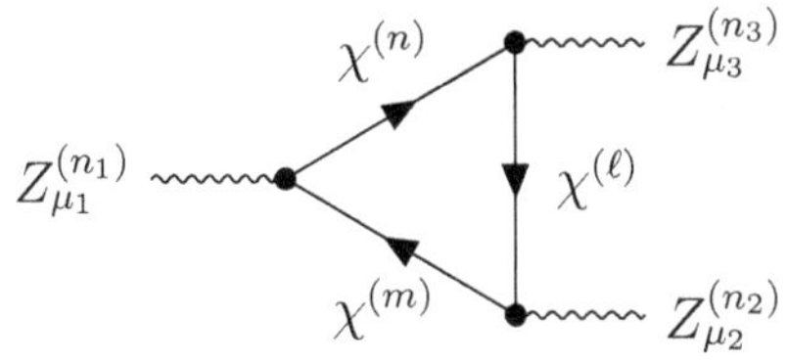

Fig. 9.15 Gauge anomalies in the $SU(2)$ model are generated by triangular loop diagrams with three legs of gauge fields.

Note that $\hat{t}_{\ell nm}^{R/L} = \hat{t}_{\ell nm}^{R/L}(\theta_H, z_L, c)$. The anomaly coefficient associated with the three legs of $Z_{\mu_1}^{(n_1)} Z_{\mu_2}^{(n_2)} Z_{\mu_3}^{(n_3)}$ in Fig. 9.15 is given by

$$\begin{aligned} a_{n_1 n_2 n_3} &= a_{n_1 n_2 n_3}^R - a_{n_1 n_2 n_3}^L , \\ a_{n_1 n_2 n_3}^R &= \sum_{n,m,\ell=0}^{\infty} \hat{t}_{n_1 nm}^R \hat{t}_{n_2 m\ell}^R \hat{t}_{n_3 \ell n}^R , \\ a_{n_1 n_2 n_3}^L &= \sum_{n,m,\ell=0}^{\infty} \hat{t}_{n_1 nm}^L \hat{t}_{n_2 m\ell}^L \hat{t}_{n_3 \ell n}^L . \end{aligned} \tag{9.56}$$

The coefficients $a_{n_1 n_2 n_3}^R$ and $a_{n_1 n_2 n_3}^L$ depend on θ_H, z_L and c. We are going to show that the total anomaly coefficient $a_{n_1 n_2 n_3}$ becomes independent of the bulk mass parameter c of the fermion field.

The gauge couplings are gauge invariant. In the original gauge the wave functions are given, in the fundamental region $0 \le y \le L$ $(1 \le z \le z_L)$, by

$$\begin{aligned} \begin{pmatrix} h_n(y) \\ k_n(y) \end{pmatrix} &= \begin{pmatrix} \cos\theta(z) & \sin\theta(z) \\ -\sin\theta(z) & \cos\theta(z) \end{pmatrix} \begin{pmatrix} \tilde{h}_n(z) \\ \tilde{k}_n(z) \end{pmatrix} , \\ \begin{pmatrix} f_{R/Ln}(y) \\ g_{R/Ln}(y) \end{pmatrix} &= \begin{pmatrix} \cos\frac{1}{2}\theta(z) & -\sin\frac{1}{2}\theta(z) \\ \sin\frac{1}{2}\theta(z) & \cos\frac{1}{2}\theta(z) \end{pmatrix} \begin{pmatrix} \tilde{f}_{R/Ln}(z) \\ \tilde{g}_{R/Ln}(z) \end{pmatrix} . \end{aligned} \tag{9.57}$$

Outside the fundamental region they are given by

$$\begin{aligned} \begin{pmatrix} h_n(y) \\ k_n(y) \end{pmatrix} &= \begin{pmatrix} -h_n(-y) \\ k_n(-y) \end{pmatrix} = \begin{pmatrix} h_n(y+2L) \\ k_n(y+2L) \end{pmatrix} , \\ \begin{pmatrix} f_{Rn}(y) \\ g_{Rn}(y) \end{pmatrix} &= S \begin{pmatrix} f_{Rn}(-y) \\ -g_{Rn}(-y) \end{pmatrix} = T \begin{pmatrix} f_{Rn}(y+2L) \\ g_{Rn}(y+2L) \end{pmatrix} , \\ \begin{pmatrix} f_{Ln}(y) \\ g_{Ln}(y) \end{pmatrix} &= S \begin{pmatrix} -f_{Ln}(-y) \\ g_{Ln}(-y) \end{pmatrix} = T \begin{pmatrix} f_{Ln}(y+2L) \\ g_{Ln}(y+2L) \end{pmatrix} , \\ (S,T) &= \begin{pmatrix} (+,+) & (-,+) \\ (+,-) & (-,-) \end{pmatrix} \quad \text{for type} \begin{pmatrix} \text{1A} & \text{1B} \\ \text{2A} & \text{2B} \end{pmatrix} . \end{aligned} \tag{9.58}$$

The coupling $\hat{t}^{R/L}_{\ell nm}$ in Eq. (9.55) can be expressed as

$$\hat{t}^{R/L}_{\ell nm} = \frac{k}{2} \int_{-b}^{2L-b} dy\, e^{\sigma(y)} \Big\{ k_\ell \big(f^*_{R/Ln} f_{R/Lm} - g^*_{R/Ln} g_{R/Lm} \big) + h_\ell \big(f^*_{R/Ln} g_{R/Lm} + g^*_{R/Ln} f_{R/Lm} \big) \Big\}. \tag{9.59}$$

Here the constant b in the integration range is arbitrary.

9.7.2 *Anomaly flow*

The equations of motion following from the action I in (9.44) are, at the classical level,

$$\begin{aligned} &\frac{1}{\sqrt{-g}} \partial_M \big(\sqrt{-g}\, F^{MN} \big) - i g_A [A_M, F^{MN}] + J^N = 0 \ , \\ &\mathcal{D}(c)\, \Psi = 0 \ , \\ &J^N = (J^N)^k \frac{T^k}{2} \ , \quad (J^N)^k = -i g_A \overline{\Psi} \gamma^a e_a{}^N \frac{T^k}{2} \Psi \ . \end{aligned} \tag{9.60}$$

The current in five dimensions is covariantly conserved:

$$\frac{1}{\sqrt{-g}} \partial_N \big(\sqrt{-g}\, J^N \big) - i g_A [A_N, J^N] = 0 \ . \tag{9.61}$$

The derivative term in the fifth coordinate generates mass terms in four dimensions when expanded in the KK modes. At the quantum level there arises an anomaly term on the righthand side of Eq. (9.61). The four-dimensional current $j^\mu_{(\ell)}(x)$ which couples with $Z^{(\ell)}_\mu(x)$ is

$$j^\mu_{(\ell)}(x) = \int_0^L dy\, \sqrt{-g}\, \big\{ h_n(y) (J^\mu)^1 + k_n(y) (J^\mu)^3 \big\} \ . \tag{9.62}$$

The divergence $\partial_\mu j^\mu_{(\ell)}$ picks up an anomalous term $j^{\rm anomaly}_{(\ell)}$ given by

$$j^{\rm anomaly}_{(\ell)} = -\frac{1}{16\pi^2} \Big(\frac{g_4}{2} \Big)^3 \sum_{n=0}^\infty \sum_{m=0}^\infty a_{\ell nm}\, Z^{(n)}_{\mu\nu} \tilde{Z}^{(m)\mu\nu} \tag{9.63}$$

where $a_{\ell nm}$ is given by Eq. (9.56), $Z^{(n)}_{\mu\nu} = \partial_\mu Z^{(n)}_\nu - \partial_\nu Z^{(n)}_\mu$ and $\tilde{Z}^{(m)\mu\nu} = \frac{1}{2} \epsilon^{\mu\nu\rho\sigma} Z^{(m)}_{\rho\sigma}$.

The coefficient $a_{\ell nm}$ is evaluated in two ways. In the first method one evaluates the gauge couplings $\hat{t}^R_{\ell nm}$ and $\hat{t}^L_{\ell nm}$, and then does the infinite sums in (9.56) to find $a^R_{\ell nm}$ and $a^L_{\ell nm}$, finally getting $a_{\ell nm} = a^R_{\ell nm} - a^L_{\ell nm}$. The θ_H dependence of a^R_{000}, a^L_{000} and a_{000} is shown for $z_L = 10$ in the presence of a fermion doublet of type 1A with $c = 0.25$ and 0.8 in Fig. 9.16. First the

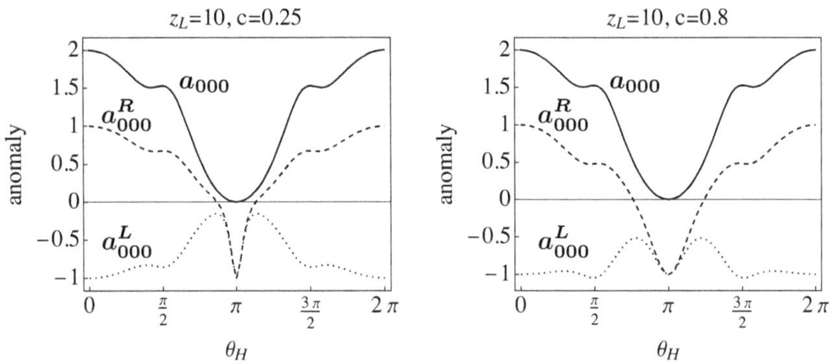

Fig. 9.16 Anomaly coefficients a_{000}, a^R_{000}, a^L_{000} are plotted as functions of θ_H in the presence of a fermion doublet of type 1A with $c = 0.25$ and 0.8 for $z_L = 10$. Although $a^R_{000}(\theta_H)$ and $a^L_{000}(\theta_H)$ depend on the bulk mass parameter c of the fermion, $a_{000}(\theta_H)$ is independent of c.

anomaly flow is observed, that is, the anomalies vary with θ_H. Secondly, although each of a^L_{000} and a^R_{000} depends on the bulk mass parameter c of the fermion field, the total anomaly $a_{000} = a^R_{000} - a^L_{000}$ is independent of c. The c-independence of $a_{\ell nm}(\theta_H)$ is confirmed as well. [96] The $Z^{(0)}_\mu$ field is massless and becomes a gauge field at $\theta_H = 0$ and π. The anomaly coefficient a_{000} takes a quantized value there.

9.7.3 *Anomaly formula*

The c-independence of $a_{n_1 n_2 n_3}$ observed above indicates the existence of an universal formula. There is an alternative method for evaluating $a_{n_1 n_2 n_3}$. We insert the couplings $\hat{t}^{R/L}_{\ell nm}$ in Eq. (9.59) into the expression for $a_{n_1 n_2 n_3}$ in Eq. (9.56), and do the infinite sum over n, m, ℓ first to find

$$\begin{aligned}
a_{n_1 n_2 n_3} &= \Big(\frac{k}{2}\Big)^3 \int\int\int_b^{2L-b} dy_1 dy_2 dy_3 \, e^{\sigma(y_1)+\sigma(y_2)+\sigma(y_3)} \\
&\times \Big[k_1 k_2 k_3 \big\{ A_R(1,2) A_R(2,3) A_R(3,1) - B_R(1,2) B_R(2,3) B_R(3,1) \\
&\qquad - A_L(1,2) A_L(2,3) A_L(3,1) + B_L(1,2) B_L(2,3) B_L(3,1) \big\} \\
&+ k_1 h_2 h_3 \{\cdots\} + h_1 k_2 h_3 \{\cdots\} + h_1 h_2 k_3 \{\cdots\} \Big] , \\
k_j &= k_{n_j}(y_j) \, , \quad h_j = h_{n_j}(y_j) \, , \\
\begin{pmatrix} A_{R/L}(j,k) \\ B_{R/L}(j,k) \end{pmatrix} &= \sum_{n=0}^{\infty} \begin{pmatrix} f_{R/Ln}(y_j) f^*_{R/Ln}(y_k) \\ g_{R/Ln}(y_j) g^*_{R/Ln}(y_k) \end{pmatrix} .
\end{aligned} \tag{9.64}$$

Here we have made use of $\sum_{n=0}^{\infty} f_{R/Ln}(y) g^*_{R/Ln}(y') = 0$. The explicit form of $A_{R/L}(j,k) \equiv A_{R/L}(y_j, y_k)$ and $B_{R/L}$ is found for $c = 0$.

$$\underline{\text{type 1A,}\ c=0}$$

$$A_R(y,y') = B_L(y,y') = \frac{e^{-\sigma(y)}}{k}\{\delta_{2L}(y-y') + \delta_{2L}(y+y')\} ,$$

$$B_R(y,y') = A_L(y,y') = \frac{e^{-\sigma(y)}}{k}\{\delta_{2L}(y-y') - \delta_{2L}(y+y')\} ,$$

$$\underline{\text{type 2A,}\ c=0}$$

$$A_R(y,y') = B_L(y,y') = \frac{e^{-\sigma(y)}}{k}\{\hat{\delta}_{2L}(y-y') + \hat{\delta}_{2L}(y+y')\} ,$$

$$B_R(y,y') = A_L(y,y') = \frac{e^{-\sigma(y)}}{k}\{\hat{\delta}_{2L}(y-y') - \hat{\delta}_{2L}(y+y')\} ,$$

$$\delta_L(x) = \sum_{n=-\infty}^{\infty} \delta(x - nL) , \quad \hat{\delta}_{2L}(y) = \delta_{4L}(y) - \delta_{4L}(y - 2L) . \tag{9.65}$$

Formulas for type 1B/2B are obtained by interchanging R and L in the formulas for type 1A/2A. When one inserts (9.65) into (9.64), there appear the products of three delta functions in the integrand. With $0 < b < L$ the products of delta functions reduce to

$$\left.\begin{array}{l} \delta_{2L}(y_1-y_2)\delta_{2L}(y_2-y_3)\delta_{2L}(y_3+y_1) \\ \delta_{2L}(y_1+y_2)\delta_{2L}(y_2+y_3)\delta_{2L}(y_3+y_1) \end{array}\right\}$$
$$\Rightarrow \frac{1}{2}\Big\{\delta(y_1)\delta(y_2)\delta(y_3) + \delta(y_1-L)\delta(y_2-L)\delta(y_3-L)\Big\} ,$$
$$\left.\begin{array}{l} \hat{\delta}_{2L}(y_1-y_2)\hat{\delta}_{2L}(y_2-y_3)\hat{\delta}_{2L}(y_3+y_1) \\ \hat{\delta}_{2L}(y_1+y_2)\hat{\delta}_{2L}(y_2+y_3)\hat{\delta}_{2L}(y_3+y_1) \end{array}\right\}$$
$$\Rightarrow \frac{1}{2}\Big\{\delta(y_1)\delta(y_2)\delta(y_3) - \delta(y_1-L)\delta(y_2-L)\delta(y_3-L)\Big\} . \tag{9.66}$$

Furthermore, as $h_n(0) = h_n(L) = 0$, only the terms proportional to $k_1k_2k_3$ in (9.64) survive.

We arrive at the following formula for the anomaly coefficients;

$$a_{n\ell m}(\theta_H, z_L) = Q_0 k_n(0) k_\ell(0) k_m(0) + Q_1 k_n(L) k_\ell(L) k_m(L) , \tag{9.67}$$

where

$$(Q_0, Q_1) = \begin{cases} (+1,+1) & \text{for type 1A} \\ (-1,-1) & \text{for type 1B} \\ (+1,-1) & \text{for type 2A} \\ (-1,+1) & \text{for type 2B} \end{cases} . \tag{9.68}$$

The anomaly coefficients are determined by the values of the wave functions $k_n(y)$ of the gauge fields at the UV and IR branes and the parity Q_j of the right-handed mode of the fermion field. The θ_H and z_L dependence of $a_{n\ell m}(\theta_H, z_L)$ is contained in $k_n(0)$ and $k_n(L)$.

The anomaly formula (9.67) is not only valid for $c = 0$, but also for other values of c. One can check that $a_{000}(\theta_H)$ displayed in Fig. 9.16 for $c = 0.25$ and 0.8 is reproduced by the formula (9.67). In Fig. 9.17(Left), the θ_H-dependence of various $a_{n\ell m}(\theta_H)$'s for $z_L = 10$ is displayed. In the right figure the z_L dependence of $a_{000}(\theta_H, z_L)$ is shown.

The anomaly coefficients depend solely on $k_n(0)$, $k_n(L)$, Q_0 and Q_1. They depend neither on the behavior of the wave functions of gauge fields and fermion field in the bulk region $0 < y < L$, nor on the bulk mass parameter c of the fermion field. The formula (9.67) represents holography in anomaly flow. Although each anomaly generated by specific fermion modes running along the internal triangular loop does depend on the detailed behavior of the wave functions of both gauge and fermion fields in the bulk, the total anomaly coefficients, after summing up all possible loop contributions, are determined by the information of the gauge fields at the UV and IR branes and of the parity conditions for the fermion field there. [97]

We comment that the formula (9.67) is valid in the flat space limit ($k \to 0$, $z_L \to 1$) as well. In $M^4 \times (S^1/Z_2)$, $a_{000} = 2$ for $0 \le \theta_H < \frac{1}{2}\pi$ and $\frac{3}{2}\pi < \theta_H \le 2\pi$, $a_{000} = 0$ for $\frac{1}{2}\pi < \theta_H < \frac{3}{2}\pi$, and $a_{000} = 2\sqrt{2}$ for $\theta_H = \frac{1}{2}\pi$ and $\frac{3}{2}\pi$. In the flat space a_{000} exhibits singular behavior at $\theta_H = \frac{1}{2}\pi$ and $\frac{3}{2}\pi$, where the level crossing of the mass spectrum of the gauge fields occurs.

The universality of the anomaly flow, expressed in the formula (9.67), has a profound implication in the GHU scenario. Chiral anomalies associated with gauge currents must be cancelled for the consistency of the theory in four dimensions. The fact that the anomaly coefficients are independent of the bulk mass parameters of fermions implies that anomaly cancellation can be achieved among various distinct fermions in the theory.

The conditions for the cancellation of the gauge anomalies are simple. Let the numbers of doublet fermions of type 1A, 1B, 2A and 2B be n_{1A}, n_{1B}, n_{2A} and n_{2B}, respectively. It follows from (9.67) that the anomalies

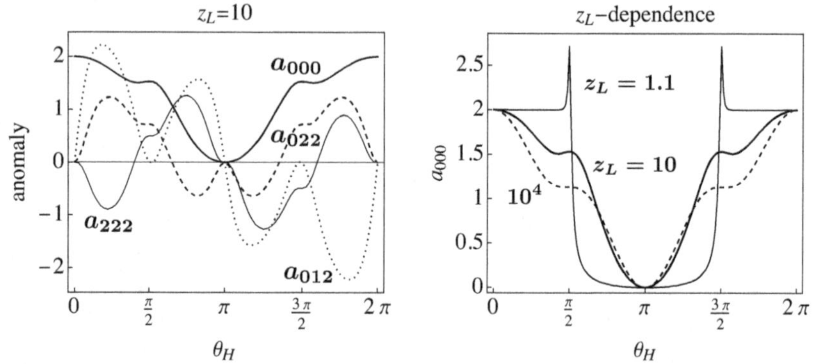

Fig. 9.17 In the left figure the θ_H-dependence of various $a_{n\ell m}(\theta_H)$'s for $z_L = 10$ is displayed. In the right figure $a_{000}(\theta_H, z_L)$ is shown for $z_L = 1.1$, 10 and 10^4.

are cancelled if

$$n_{1A} = n_{1B} \ , \quad n_{2A} = n_{2B} \ . \tag{9.69}$$

The conditions are generalized in the presence of brane fermions, namely fermions living only on the UV or IR brane. Suppose that there are $\hat{n}_R^{\rm UV}$ right-handed and $\hat{n}_L^{\rm UV}$ left-handed doublet brane fermions on the UV brane at $y = 0$, and $\hat{n}_R^{\rm IR}$ right-handed and $\hat{n}_L^{\rm IR}$ left-handed doublet brane fermions on the IR brane at $y = L$. As the $Z_\mu^{(n)}$ couplings of brane fermions are given by $(g_4/2)\, k_n(0)$ at the UV brane and by $(g_4/2)\, k_n(L)$ at the IR brane, the anomaly cancellation conditions become

$$\begin{aligned} &n_{1A} - n_{1B} + n_{2A} - n_{2B} + \hat{n}_R^{\rm UV} - \hat{n}_L^{\rm UV} = 0 \ , \\ &n_{1A} - n_{1B} - n_{2A} + n_{2B} + \hat{n}_R^{\rm IR} - \hat{n}_L^{\rm IR} = 0 \ . \end{aligned} \tag{9.70}$$

We stress that the conditions (9.69) and (9.70) do not depend on θ_H and z_L. Furthermore the conditions guarantee that not only the zero mode anomaly a_{000} but also all other anomalies $a_{n\ell m}$ are cancelled at once.

9.8 Gauge-Higgs grand unification

In the preceding chapters we have introduced and analyzed the GUT inspired $SO(5)\times U(1)_X \times SU(3)_C$ gauge-Higgs unification model which unifies the electromagnetic and weak interactions. It looks mandatory to incorporate strong interactions in the unification scheme as well. The fact that the electric charges of quarks are $+\frac{2}{3}e$ and $-\frac{1}{3}e$ compared to the electron charge $-e$, and the fact that the cancellation of gauge anomalies in the

SM is achieved with a miraculous combination of quarks and leptons are naturally explained in the framework of grand unification of strong and electroweak interactions.

In grand unification models in four dimensions $SU(5)$ and $SO(10)$ gauge groups have been frequently investigated, each of which contains the SM gauge group $\mathcal{G}_{\rm SM} = SU(3)_C \times SU(2)_L \times U(1)_Y$. In the $SU(5)$ grand unified theory (GUT), for instance, quarks and leptons are introduced in the $\overline{\mathbf{5}}$ and $\mathbf{10}$ representations of $SU(5)$. Further scalar fields in $\mathbf{24}$ and $\mathbf{5}$ representations are introduced. The $\mathbf{24}$ scalar field breaks $SU(5)$ to $\mathcal{G}_{\rm SM}$ by the Higgs mechanism, and the $\mathbf{5}$ scalar field breaks $SU(2)_L \times U(1)_Y$ to $U(1)_{\rm EM}$. The Higgs boson discovered in 2012 is a part of the $\mathbf{5}$ scalar field.

9.8.1 $SU(6)$ *unification*

How can one incorporate the 4D $\mathbf{5}$ scalar field as a part of the fifth-dimensional component of gauge fields in the context of gauge-Higgs unification? Lim and Maru first proposed an $SU(6)$ model in $M^4 \times (S^1/Z_2)$. [98] With orbifold boundary conditions $P_0 = P_1 = {\rm diag}\,(1,1,1,1,1,-1)$, $SU(6)$ symmetry is broken to $SU(5) \times U(1)$. The parity even $(+,+)$ component of $SU(6)$ gauge field A_y is $A_y^{j6} = (A_y^{6j})^*$ ($j = 1 \sim 5$), which contains, as a zero mode, the 4D $\mathbf{5}$ scalar field in $SU(5)$. However, there appears an extra $U(1)$ gauge field in the four-dimensional components, which contradicts the observation. Furthermore there is no mechanism to break $SU(5)$ to $\mathcal{G}_{\rm SM}$.

Alternatively one may impose orbifold boundary conditions $P_0 = {\rm diag}\,(1,1,1,1,1,-1)$ and $P_1 = {\rm diag}\,(1,1,-1,-1,-1,-1)$. Then $SU(6)$ symmetry breaks down to $SU(3)_C \times SU(2)_L \times U(1)_Y \times U(1)_X$ symmetry. The parity even $(+,+)$ component of $SU(6)$ gauge field A_y is $A_y^{a6} = A_y^{6a\,\dagger}$ ($a = 1,2,$), the zero mode of which serve as the 4D $SU(2)_L$ doublet field. There remains an extra $U(1)_X$ gauge field.

Fermion multiplets are introduced in the $\overline{\mathbf{6}}$, $\mathbf{20}$ and $\mathbf{35}$ representations of $SU(6)$. In the minimal model two $\overline{\mathbf{6}}$ fields and one $\mathbf{20}$ field are necessary in each generation to have quarks and leptons as zero modes. This fermion content, however, leads to gauge anomaly in $U(1)_X$. Improvement has been made both in $M^4 \times (S^1/Z_2)$ and in the RS warped space. [99–102] It is not easy to account for the quark-lepton mass spectrum in $SU(6)$ GHU models. As another approach $SU(5) \times SU(5) \times Z_2$ GHU has been proposed. [103,104]

9.8.2 $SO(11)$ *unification*

At this point it is instructive to recall the $SO(5) \times U(1)_X \times SU(3)_C$ gauge-Higgs unification model in the RS warped space we have examined in the previous chapters. In grand unified theory there are at least two energy scales. One is $M_{\rm GUT}$ characterizing the grand unification of all forces, and the other is $M_{\rm EW}$ characterizing the unification of electroweak interactions. In GHU there is an additional KK mass scale $m_{\rm KK}$. Typical values in the $SO(5) \times U(1)_X \times SU(3)_C$ GHU are $m_{\rm KK} \sim 13$ TeV and $M_{\rm EW} \sim m_Z \sim 90$ GeV.

We suppose that in the gauge-Higgs grand unification scenario the symmetry, gauge group $\mathcal{G}$, breaks down by orbifold boundary conditions to its subgroup $\mathcal{G}_{\rm BC}$ which contains $SO(4) \times U(1) \times SU(3)_C$, and then breaks down to $\mathcal{G}_{\rm SM}$ by a brane scalar field, and finally breaks down to $U(1)_{\rm EM} \times SU(3)_C$ by the Hosotani mechanism. At the $m_{\rm KK}$ scale a part of the A_y field should contain, as a zero mode, a 4D scalar field corresponding to the doublet Higgs field in the SM with the custodial symmetry, and the model should not contain extra 4D fermion fields other than quarks and leptons. To reproduce the observed phenomenology at low energies with high precision it is most natural to formulate a model in the RS warped space as in the $SO(5) \times U(1) \times SU(3)_C$ GHU.

The requirements above give severe constraints in the choice of the gauge group $\mathcal{G}$, orbifold boundary conditions (P_0, P_1), and fermion multiplets. Consider a $\mathcal{G} = SO(11)$ gauge theory in the RS space with the metric given by (6.23) or (6.24). In the $SO(5) \times U(1) \times SU(3)_C$ GHU the orbifold boundary conditions (7.10) reduce the symmetry to $SO(4) \times U(1) \times SU(3)_C$. In the $SO(11)$ theory we take

$$P_0^{\rm vec} = \begin{pmatrix} I_{10\times 10} & \mathbf{0}_{10\times 1} \\ \mathbf{0}_{1\times 10} & -1 \end{pmatrix},$$

$$P_1^{\rm vec} = \begin{pmatrix} I_{4\times 4} & \mathbf{0}_{4\times 7} \\ \mathbf{0}_{7\times 4} & -I_{7\times 7} \end{pmatrix}. \tag{9.71}$$

Here $I_{n\times n}$ is an n-by-n unit matrix, and $\mathbf{0}_{n\times m}$ is an n-by-m matrix with all elements being zero. With the boundary condition matrices (9.71), $SO(11)$ symmetry breaks down to $SO(10)$ symmetry on the UV brane at $y = 0$ and breaks down to $SO(4) \times SO(7)$ symmetry on the IR brane at $y = L$. With these two combined the $SO(11)$ symmetry is reduced to $\mathcal{G}_{\rm BC} = SO(4) \times SO(6)$ symmetry. Note that $SO(4) \simeq SU(2)_L \times SU(2)_R$ and $SO(6) \simeq SU(4) \supset SU(3) \times U(1)$. At this stage the $SO(4) \times SO(6)$

part of the four-dimensional components of the gauge potential $A_\mu(x,y)$ are parity even $(+,+)$ so that 4D gauge fields appear as zero modes. We decompose 11×11 matrices into $4 \oplus 6 \oplus 1$ blocks. Then the parity of anti-symmetric A_μ and A_y is given by

$$A_\mu : \begin{pmatrix} (++) & (+-) & (--) \\ (+-) & (++) & (-+) \\ (--) & (-+) & * \end{pmatrix} ,$$

$$A_y : \begin{pmatrix} (--) & (-+) & (++) \\ (-+) & (--) & (+-) \\ (++) & (+-) & * \end{pmatrix} . \tag{9.72}$$

The parity even part of $A_y(x,y)$ is $A_y^{a,11}$ $(a = 1 \sim 4)$, whose zero modes become an $SU(2)_L$ doublet in four dimensions. [59, 105]

9.8.3 *Fermions and a brane scalar*

We introduce fermion multiplets in the spinor representation $\Psi_{\mathbf{32}}(x,y)$ and vector representation $\Psi_{\mathbf{11}}(x,y)$ of $SO(11)$ in the bulk and singlet fields $\hat{\chi}_{\mathbf{1}}(x)$ on the UV brane. In addition we introduce a brane scalar field in the spinor representation $\hat{\Phi}_{\mathbf{16}}(x)$ of $SO(10)$ on the UV brane. $SO(10)$ gauge invariance is maintained on the UV brane.

As the brane scalar field $\hat{\Phi}_{(\mathbf{1},\mathbf{4})}$ in the $SO(5) \times U(1)_X \times SU(3)_C$ GHU spontaneously breaks $SO(4) \times U(1)_X$ to $SU(2)_L \times U(1)_Y$, the brane scalar field $\hat{\Phi}_{\mathbf{16}}$ with $\langle \hat{\Phi}_{\mathbf{16}} \rangle \neq 0$ spontaneously breaks $SO(10)$ gauge symmetry to $SU(5)$ gauge symmetry. Since $SU(5) \cap \mathcal{G}_{\rm BC} = SU(3)_C \times SU(2)_L \times U(1)_Y$, the symmetry is reduced to the SM gauge symmetry.

The spinor representation $\Psi_{\mathbf{32}}$ has 32 components, whose $SO(10)$ decomposition is given by $\mathbf{32} = \mathbf{16} + \overline{\mathbf{16}}$. The orbifold boundary condition matrices in the spinor representation corresponding to (9.71) are, in the matrix product representation described in Appendix A.2, given by

$$\begin{aligned} P_0^{\rm sp} &= \sigma^0 \otimes \sigma^0 \otimes \sigma^0 \otimes \sigma^0 \otimes \sigma^3 = I_{16\times 16} \otimes \sigma^3 , \\ P_1^{\rm sp} &= \sigma^0 \otimes \sigma^3 \otimes \sigma^0 \otimes \sigma^0 \otimes \sigma^0 = I_{2\times 2} \otimes \sigma^3 \otimes I_{8\times 8} . \end{aligned} \tag{9.73}$$

For each generation $(a = 1, 2, 3)$ we introduce $\Psi_{\mathbf{32}}^a$, $\Psi_{\mathbf{11}}^a$ and $\Psi_{\mathbf{11}}'^a$ with boundary conditions

$$\begin{aligned} \Psi_{\mathbf{32}}^a(x, y_j - y) &= -\gamma^5 P_j^{\rm sp} \Psi_{\mathbf{32}}^a(x, y_j + y) , \\ \Psi_{\mathbf{11}}^a(x, y_j - y) &= (-1)^j \gamma^5 P_j^{\rm vec} \Psi_{\mathbf{11}}^a(x, y_j + y) , \\ \Psi_{\mathbf{11}}'^a(x, y_j - y) &= (-1)^{j+1} \gamma^5 P_j^{\rm vec} \Psi_{\mathbf{11}}'^a(x, y_j + y) . \end{aligned} \tag{9.74}$$

Main parts of quarks and leptons are contained in $\Psi_{\mathbf{32}}^a$. Its content is summarized in Table 9.2. With the boundary conditions in Eq. (9.74) just quarks and leptons in the SM appear as zero modes. Each component of 5D Ψ has both left-handed and right-handed components so that $\Psi_{\mathbf{32}}^a$ has 64 chiral components. By the orbifold conditions at $y = 0$ and $y = L$, there appear zero modes in $64 \div 2 \div 2 = 16$ chiral components. In the $SU(5)$ GUT in four dimensions, as shown in (5.1), $\overline{\mathbf{5}}$ consists of ν_L, e_L and $(d_{jR})^c = (d_j^c)_L$, and $\mathbf{10}$ consists of u_{jL}, d_{jL} and $(u_{jR})^c = (u_j^c)_L$. In contrast, in the $SO(11)$ GHU the $\overline{\mathbf{5}}$ part in $\mathbf{16}$, for instance, consists of ν, e and $\hat{d}_j$.

Table 9.2 The content of $\Psi_{\mathbf{32}}^a$ in the first generation. In the third column $G_{\rm SM} = SU(3)_C \times SU(2)_L \times U(1)_Y$ and superscripts and subscripts represent the $U(1)_Y$ charge. In the last column parity (P_0, P_1) of the left-handed components is shown. The right-handed components have opposite parity. In the representation (A.29) 32 components of $\Psi_{\mathbf{32}}$ appears in the order of $\nu, e; \hat{d}_1, \hat{u}_1; u_3, d_3; \hat{d}_2, \hat{u}_2; u_1, d_1; \hat{e}, \hat{\nu}; u_2, d_2; \hat{d}_3, \hat{u}_3$ for the upper half ($\mathbf{16}$) components, and of $\hat{d}_3', \hat{u}_3'; u_2', d_2'; \hat{e}', \hat{\nu}'; u_1', d_1'; \hat{d}_2', \hat{u}_2'; u_3', d_3'; \hat{d}_1', \hat{u}_1'; \nu', e'$ for the lower half ($\overline{\mathbf{16}}$) components.

$\Psi_{\mathbf{32}}$						
$SO(10)$	$SU(5)$	G_{SM}	$Q_{\rm EM}$	Name	Zero mode	Parity of Ψ_L
$\mathbf{16}$	$\overline{\mathbf{5}}$ $\overline{\mathbf{5}}$	$(\mathbf{1}, \mathbf{2})_{-1/2}^{-1/2}$	0 -1	ν e	ν_L e_L	$(+,+)$
	$\mathbf{10}$ $\mathbf{10}$	$(\mathbf{3}, \mathbf{2})_{+1/6}^{+1/6}$	$+\frac{2}{3}$ $-\frac{1}{3}$	u_j d_j	u_{jL} d_{jL}	$(+,+)$
	$\mathbf{10}$ $\mathbf{1}$	$(\mathbf{1}, \mathbf{1})_0^{+1}$	$+1$ 0	$\hat{e}$ $\hat{\nu}$		$(+,-)$
	$\overline{\mathbf{5}}$ $\mathbf{10}$	$(\overline{\mathbf{3}}, \mathbf{1})_{-2/3}^{+1/3}$	$+\frac{1}{3}$ $-\frac{2}{3}$	$\hat{d}_j$ $\hat{u}_j$		$(+,-)$
$\overline{\mathbf{16}}$	$\overline{\mathbf{10}}$ $\overline{\mathbf{10}}$	$(\overline{\mathbf{3}}, \mathbf{2})_{-1/6}^{-1/6}$	$+\frac{1}{3}$ $-\frac{2}{3}$	$\hat{d}_j'$ $\hat{u}_j'$		$(-,+)$
	$\mathbf{5}$ $\mathbf{5}$	$(\mathbf{1}, \mathbf{2})_{+1/2}^{+1/2}$	$+1$ 0	$\hat{e}'$ $\hat{\nu}'$		$(-,+)$
	$\overline{\mathbf{10}}$ $\mathbf{5}$	$(\mathbf{3}, \mathbf{1})_{-1/3}^{+2/3}$	$+\frac{2}{3}$ $-\frac{1}{3}$	u_j' d_j'	u_{jR}' d_{jR}'	$(-,-)$
	$\mathbf{1}$ $\overline{\mathbf{10}}$	$(\mathbf{1}, \mathbf{1})_{-1}^{0}$	0 -1	ν' e'	ν_R' e_R'	$(-,-)$

The content of $\Psi_{\mathbf{11}}^a$ and $\Psi_{\mathbf{11}}'^a$ is summarized in Table 9.3. No component in $\Psi_{\mathbf{11}}$ has the quantum number of up (u) quarks. $(D_{jR}, \hat{D}_{jR})$ appear as zero modes from $\Psi_{\mathbf{11}}$, and $(D'_{jL}, \hat{D}'_{jL})$ appear as zero modes from $\Psi'_{\mathbf{11}}$. As $\Psi_{\mathbf{11}}$ and $\Psi'_{\mathbf{11}}$ have opposite parity, they can have a mass term in the bulk region $(0 \leq y \leq L)$ in the form of $\overline{\Psi}_{\mathbf{11}}\Psi'_{\mathbf{11}}$, with which the zero modes can have large masses.

Table 9.3 The content of $\Psi_{\mathbf{11}}^a$ and $\Psi_{\mathbf{11}}'^a$ in the first generation. In the last column parity (P_0, P_1) of the left-handed components of $\Psi_{\mathbf{11}}$ is shown. Parity of $\Psi'_{\mathbf{11}}$ is opposite to that of $\Psi_{\mathbf{11}}$.

$\Psi_{\mathbf{11}}$, $\Psi'_{\mathbf{11}}$						
$SO(10)$	$SU(5)$	G_{SM}	$Q_{\rm EM}$	Name $\Psi_{\mathbf{11}}$ $\Psi'_{\mathbf{11}}$	Zero mode	Parity of $\Psi_{\mathbf{11},L}$
$\mathbf{10}$	$\mathbf{5}$	$(\mathbf{1},\mathbf{2})_{+1/2}$	$+1$ 0	$\hat{E}$ $\hat{E}'$ $\hat{N}$ $\hat{N}'$		$(-,+)$
	$\overline{\mathbf{5}}$	$(\mathbf{1},\mathbf{2})_{-1/2}$	0 -1	N N' E E'		$(-,+)$
	$\mathbf{5}$	$(\mathbf{3},\mathbf{1})_{-1/3}$	$-\frac{1}{3}$	D_j D'_j	D_{jR} D'_{jL}	$(-,-)$
	$\overline{\mathbf{5}}$	$(\overline{\mathbf{3}},\mathbf{1})_{+1/3}$	$+\frac{1}{3}$	$\hat{D}_j$ $\hat{D}'_j$	$\hat{D}_{jR}$ $\hat{D}'_{jL}$	$(-,-)$
$\mathbf{1}$	$\mathbf{1}$	$(\mathbf{1},\mathbf{1})_0$	0	S S'		$(+,-)$

As $\Psi_{\mathbf{32}}$ contains all quarks and leptons as zero modes, it may look economical to build a model without $\Psi_{\mathbf{11}}$ and $\Psi'_{\mathbf{11}}$. However, with $\Psi_{\mathbf{32}}$ alone, quarks and leptons in each generation necessarily have degenerate masses. At this stage $\Psi_{\mathbf{11}}$, $\Psi'_{\mathbf{11}}$ and brane scalar $\hat{\Phi}_{\mathbf{16}}$ become important. There can be brane interactions on the UV brane. In the $SO(10)$ decomposition, $\Psi_{\mathbf{32}}$ is decomposed into $\Psi_{\mathbf{16}}$ and $\Psi_{\overline{\mathbf{16}}}$, and $\Psi_{\mathbf{11}}$ is decomposed into $\Psi_{\mathbf{10}}$ and $\Psi_{\mathbf{1}}$. Then one can form $SO(10)$ invariant interactions such as $\overline{\Psi}_{\mathbf{16}}\Gamma^a\hat{\Phi}_{\mathbf{16}}(\Psi_{\mathbf{10}})_a$ and $\overline{\Psi}_{\mathbf{16}}\hat{\Phi}_{\mathbf{16}}\Psi_{\mathbf{1}}$. Non-vanishing $\langle\hat{\Phi}_{\mathbf{16}}\rangle \neq 0$ generates mass terms connecting $\Psi_{\mathbf{32}}$ with $\Psi_{\mathbf{11}}$ and $\Psi'_{\mathbf{11}}$. With the masses generated by the Hosotani mechanism ($\theta_H \neq 0$) the realistic mass spectrum of quarks and leptons is reproduced.

Furthermore $SO(11)$-singlet Majorana fermion fields $\hat{\chi}_{\mathbf{1}}^a(x)$ are introduced on the UV brane. $\hat{\chi}_{\mathbf{1}} \equiv \chi$ satisfies the Majorana condition $\chi^c = \chi$. On the UV brane $\hat{\chi}_{\mathbf{1}}$ has, in addition to a Majorana mass term $M\bar{\chi}^c\chi$, a brane interaction of the form $\overline{\Psi}_{\mathbf{16}}\Phi_{\mathbf{16}}\chi$. With $\langle\hat{\Phi}_{\mathbf{16}}\rangle \neq 0$ it generates a mass term of the form $m_B\nu_R'^{\dagger}\chi_L$. The resultant mass matrix in the neutrino sector is given by Eq. (7.111), which leads to a tiny neutrino mass $m_{\nu_{eL}} \sim m_e^2 M/m_B^2$.

It is seen that gauge-Higgs unification is incorporated in the grand unification scheme in an interesting way, however, there remains a few problems to be solved. [105–109] The most serious one is about the mass spectrum of $(\hat{u}, \hat{u}')$ in $\Psi_{\mathbf{32}}$. As shown in Table 9.2, there are no zero modes in $(\hat{u}, \hat{u}')$. However, as shown in Chapter 6, Fig. 6.3, the mass of the lowest KK mode ($n = 1$ mode) of a fermion field Ψ with Ψ_L's parity $(+,-)$ $[(-,+)]$ and a bulk mass parameter $c > \frac{1}{2}$ $[c < -\frac{1}{2}]$ becomes very small for $z_L > 10^5$. For the first and second generations the bulk mass parameters of $\Psi_{\mathbf{32}}$ is $|c_{\mathbf{32}}| > \frac{1}{2}$. In GHU there are masses generated by the Hosotani mechanism, too. One can show that the mass of $\hat{u}$ or $\hat{u}'$ is about $m_u \cot \frac{1}{2}\theta_H$, which is much less than $m_{\rm KK}$. Since $\Psi_{\mathbf{11}}$ has no component with the quantum number of $\hat{u}$ or $\hat{u}'$, it is not possible to make $\hat{u}$ or $\hat{u}'$ heavier by brane interactions. One necessarily would have light, colored exotic particles of charge $\frac{2}{3}e$, which contradicts the observation. A satisfactory gauge-Higgs grand unification model has still been in quest.

Finally we would like to add that the fermion content of the GUT inspired $SO(5) \times U(1)_X \times SU(3)_C$ GHU introduced in Chapter 7 has been guided by the fermion content of the $SO(11)$ GHU. All fermion fields in Table 7.1 come from $\Psi_{\mathbf{32}}$ and $\Psi_{\mathbf{11}}$ in the $SO(11)$ theory.

Appendix A

Natural units, Dirac matrices and $SO(N)$

We summarize the notation used in this book.

A.1 Metric, natural units, Dirac matrices

Coordinates of four-dimensional Minkowski spacetime are $x^\mu = (x^0 = ct, x^1, x^2, x^3)$ where c is the light velocity. The world interval ds^2 is

$$\begin{aligned} ds^2 &= (dx^0)^2 - (dx^1)^2 - (dx^2)^2 - (dx^3)^2 \\ &= -\eta_{\mu\nu} dx^\mu dx^\nu \ , \\ \eta_{\mu\nu} &= \text{diag}\,(-1, 1, 1, 1) \ . \end{aligned} \tag{A.1}$$

$\eta_{\mu\nu}$ is the metric of the four-dimensional Minkowski spacetime. In the five-dimensional Minkowski spacetime we write $x^M = (x^\mu, x^5)$ and

$$-ds^2 = \eta_{MN} dx^M dx^N \ , \quad \eta_{MN} = \text{diag}\,(-1, 1, 1, 1, 1) \ . \tag{A.2}$$

In dealing with gravity and field theory in higher dimensions, it is convenient to take the spacial part of the metric $\eta_{jk} = \delta_{jk}$. In curved space

$$-ds^2 = g_{MN}(x)\, dx^M dx^N \tag{A.3}$$

and g_{MN} is called the metric of the spacetime.

The action of a point particle with mass m and charge e in electromagnetic fields is given by

$$\int \Big\{ -mc\, ds + \frac{e}{c} A_\mu dx^\mu \Big\} \tag{A.4}$$

Here $A_\mu dx^\mu = \eta_{\mu\nu} A^\mu dx^\nu$, and the four-potential $A^\mu = (\phi, \vec{A})$.

The Dirac equation is given by

$$\Big(\gamma^\mu D_\mu - \frac{mc}{\hbar}\Big) \psi = 0 \ , \quad D_\mu = \partial_\mu - i \frac{e}{\hbar c} A_\mu \ , \tag{A.5}$$

$$\{\gamma^\mu, \gamma^\nu\} = 2\eta^{\mu\nu} . \tag{A.6}$$

The Dirac matrices γ^μ satisfy the Dirac algebra (A.6). In the spinor representation, with the Pauli matrices $\vec{\sigma}$,

$$\gamma^\mu = \begin{pmatrix} & \sigma^\mu \\ \bar{\sigma}^\mu & \end{pmatrix} , \quad \sigma^\mu = (I_2, \vec{\sigma}) \ , \quad \bar{\sigma}^\mu = (-I_2, \vec{\sigma}) \ . \tag{A.7}$$

Note that

$$\sigma^\mu \partial_\mu \cdot \bar{\sigma}^\nu \partial_\nu = \bar{\sigma}^\mu \partial_\mu \cdot \sigma^\nu \partial_\nu = \eta^{\mu\nu} \partial_\mu \partial_\nu \equiv \Box_4 \ . \tag{A.8}$$

Furthermore, γ^5 is defined by

$$\gamma^5 = -i\gamma^0\gamma^1\gamma^2\gamma^3 = \begin{pmatrix} I_2 & \\ & -I_2 \end{pmatrix} ,$$

$$(\gamma^5)^2 = 1 \ , \quad \{\gamma^5, \gamma^\mu\} = 0 \tag{A.9}$$

States with $\gamma^5 = +1$ (-1) are right-handed (left-handed). The Dirac conjugate is defined by $\bar{\psi} = i\psi^\dagger \gamma^0$. $(\bar{\psi}\psi)^\dagger = \bar{\psi}\psi$ and $(\bar{\psi}\gamma^\mu\psi)^\dagger = -\bar{\psi}\gamma^\mu\psi$. The action for the Dirac field is

$$\int d^4x \, \bar{\psi} \Big(\gamma^\mu D_\mu - \frac{mc}{\hbar} \Big) \psi \ . \tag{A.10}$$

In terms of two-component spinors

$$\psi = \begin{pmatrix} \xi_R \\ \chi_L \end{pmatrix} ,$$

$$\bar{\psi}\psi = -i\chi_L^\dagger \xi_R + i\xi_R^\dagger \chi_L \ , \quad \bar{\psi}\gamma^5\psi = -i\chi_L^\dagger \xi_R - i\xi_R^\dagger \chi_L \ ,$$

$$\bar{\psi}\gamma^\mu\psi = i\xi_R^\dagger \bar{\sigma}^\mu \xi_R - i\chi_L^\dagger \sigma^\mu \chi_L \ . \tag{A.11}$$

In various manipulations, expressions become simpler in the natural unit in which $\hbar = c = 1$. This corresponds to identifying the units of time (T), length (L) and mass (M) by $T = L = M^{-1}$.

A.2 $SO(N)$ and Clifford algebra

$SO(N)$ is the rotation group in N dimensions. A set of $N \times N$ real orthogonal matrices O ($O^t = O^{-1}$, $\det O = 1$) is a representation of $SO(N)$. There are $\frac{1}{2}N(N-1)$ generators of the group $SO(N)$ which satisfy

$$T^{jk} = -T^{kj} \ , \quad 1 \le j, k \le N \ ,$$

$$[T^{jk}, T^{\ell m}] = i(\delta^{j\ell} T^{km} - \delta^{jm} T^{k\ell} - \delta^{k\ell} T^{jm} + \delta^{km} T^{j\ell}) \ . \tag{A.12}$$

In the vector representation they are given by

$$(T^{jk}_{\rm vec})_{pq} = -i(\delta^j{}_p\delta^k{}_q - \delta^j{}_q\delta^k{}_p) ,$$
$$\mathrm{Tr}\, T^{jk}_{\rm vec} T^{\ell m}_{\rm vec} = 2(\delta^{j\ell}\delta^{km} - \delta^{jm}\delta^{k\ell}) . \tag{A.13}$$

In this book $SO(3)$, $SO(4)$, $SO(5)$, $SO(10)$ and $SO(11)$ appear.

(i) $SO(3) \simeq SU(2)$
Defining $T^a = \frac{1}{2}\epsilon^{ajk}T^{jk}$ $(a,j,k = 1 \sim 3)$, we have

$$[T^a, T^b] = i\epsilon^{abc}T^c . \tag{A.14}$$

(ii) $SO(4) \simeq SU(2)_L \times SU(2)_R$
Let us define six generators by

$$T^a = \tfrac{1}{2}\epsilon^{ajk}T^{jk}, \quad K^a = T^{a4} ,$$
$$T^a_L = \tfrac{1}{2}(T^a + K^a), \quad T^a_R = \tfrac{1}{2}(T^a - K^a),$$
$$(a,j,k = 1 \sim 3) . \tag{A.15}$$

Then they satisfy

$$[T^a, T^b] = i\epsilon^{abc}T^c, \quad [K^a, K^b] = i\epsilon^{abc}T^c, \quad [T^a, K^b] = i\epsilon^{abc}K^c,$$
$$[T^a_L, T^b_L] = i\epsilon^{abc}T^c_L, \quad [T^a_R, T^b_R] = i\epsilon^{abc}T^c_R, \quad [T^a_L, T^b_R] = 0 . \tag{A.16}$$

In other words, $SU(2)_L$ with the generators $\{T^a_L\}$ and $SU(2)_R$ with the generators $\{T^a_R\}$ become independent of each other, implying $SO(4) \simeq SU(2)_L \times SU(2)_R$.

(iii) $SO(5)$
In addition to $\{T^a_L, T^a_R\}$ in (A.15) we define

$$\hat{T}^p = \frac{1}{\sqrt{2}}T^{p5} \quad (p = 1, \cdots, 4) . \tag{A.17}$$

Then the algebra of $SO(5)$ can be expressed in the form of

$$[T^a_L, T^b_L] = i\epsilon^{abc}T^c_L \;, \; [T^a_R, T^b_R] = i\epsilon^{abc}T^c_R \;, \; [T^a_L, T^b_R] = 0 \;,$$
$$[T^a_L, \hat{T}^b] = \tfrac{i}{2}\delta^{ab}\hat{T}^4 + \tfrac{i}{2}\epsilon^{abc}\hat{T}^c \;, \; [T^a_R, \hat{T}^b] = -\tfrac{i}{2}\delta^{ab}\hat{T}^4 + \tfrac{i}{2}\epsilon^{abc}\hat{T}^c \;,$$
$$[\hat{T}^a, \hat{T}^b] = \tfrac{i}{2}\epsilon^{abc}(T^c_L + T^c_R) \;,$$
$$[T^a_L, \hat{T}^4] = -\tfrac{i}{2}\hat{T}^a \;, \; [T^a_R, \hat{T}^4] = \tfrac{i}{2}\hat{T}^a \;, \; [\hat{T}^a, \hat{T}^4] = \tfrac{i}{2}(T^a_L - T^a_R) \;,$$

$$[T_L^a + T_R^a, \hat{T}^b] = i\epsilon^{abc}\hat{T}^c \ , \ [T_L^a + T_R^a, \hat{T}^4] = 0 \ ,$$

$$[T_L^a - T_R^a, \hat{T}^b] = i\delta^{ab}\hat{T}^4 \ , \ [T_L^a - T_R^a, \hat{T}^4] = -i\hat{T}^a \ ,$$

$$(1 \le a, b, c \le 3). \tag{A.18}$$

10 generators of $SO(10)$ $\{T^\alpha\} = \{T_L^a, T_R^a, \hat{T}^p\}$ satisfy $\mathrm{Tr}\, T_{\mathrm{vec}}^\alpha T_{\mathrm{vec}}^\beta = \delta^{\alpha\beta}$ in the vector representation.

The spinor representation in the group $SO(N)$ is important. The Clifford algebra is given by

$$\{\Gamma^j \, , \, \Gamma^k\} = 2\delta^{jk} \quad (j,k = 1, \cdots, N) \ . \tag{A.19}$$

With δ^{jk} replaced by η^{jk}, it becomes the Dirac algebra (A.6). In terms of Γ^j we define

$$T_{\mathrm{sp}}^{jk} = -\frac{i}{4}\,[\Gamma^j, \Gamma^k] \ . \tag{A.20}$$

The set $\{T_{\mathrm{sp}}^{jk}\}$ satisfies the $SO(N)$ algebra (A.12).

Under $SO(N)$ transformations an N component vector $\vec{V}$ transforms as

$$\vec{V} \to \vec{V}' = \exp\Big\{\frac{i}{2}\sum_{j\neq k}\omega^{jk}T_{\mathrm{vec}}^{jk}\Big\}\,\vec{V} \ . \tag{A.21}$$

A spinor ψ transforms as

$$\psi \to \psi' = \exp\Big\{\frac{i}{2}\sum_{j\neq k}\omega^{jk}T_{\mathrm{sp}}^{jk}\Big\}\,\psi \ . \tag{A.22}$$

$\{T_{\mathrm{sp}}^{jk}\}$ are called the generators in the spinor representation.

We list below matrix representations of Γ^j.

(a) $SO(3)$

In terms of Pauli matrices $\vec{\sigma}$

$$\Gamma^j = \sigma^j \quad (j = 1, 2, 3) \ . \tag{A.23}$$

The generators T^{jk} are given by

$$T_{\mathrm{vec}}^{12} = \begin{pmatrix} 0 & -i & 0 \\ i & 0 & 0 \\ 0 & 0 & 0 \end{pmatrix}, \ T_{\mathrm{vec}}^{23} = \begin{pmatrix} 0 & 0 & 0 \\ 0 & 0 & -i \\ 0 & i & 0 \end{pmatrix}, \ T_{\mathrm{vec}}^{31} = \begin{pmatrix} 0 & 0 & i \\ 0 & 0 & 0 \\ -i & 0 & 0 \end{pmatrix},$$

$$T_{\mathrm{sp}}^{12} = \tfrac{1}{2}\sigma^3 \ , \ T_{\mathrm{sp}}^{23} = \tfrac{1}{2}\sigma^1 \ , \ T_{\mathrm{sp}}^{31} = \tfrac{1}{2}\sigma^2 \tag{A.24}$$

Note the factor $\frac{1}{2}$ in $T_{\rm sp}^{jk}$ in (A.24). A vector returns to the original value under 2π rotation, whereas a two-component spinor needs 4π rotation to return to the original one.

(b) $SO(4)$, $SO(5)$

In both $SO(4)$ and $SO(5)$, Γ^j are given by 4×4 matrices. As an example one can take

$$\begin{aligned}
\Gamma^1 &= \sigma^1 \otimes \sigma^1 \\
\Gamma^2 &= \sigma^2 \otimes \sigma^1 \\
\Gamma^3 &= \sigma^3 \otimes \sigma^1 \\
\Gamma^4 &= \sigma^0 \otimes \sigma^2 \\
\Gamma^5 &= \sigma^0 \otimes \sigma^3 = -\Gamma^1\Gamma^2\Gamma^3\Gamma^4 \,.
\end{aligned} \tag{A.25}$$

Here $\sigma^0 = I_2$ and in the direct product $\otimes$ of matrices, as in

$$\sigma^j \otimes \sigma^1 = \begin{pmatrix} 0 & \sigma^j \\ \sigma^j & 0 \end{pmatrix}, \tag{A.26}$$

we place the first matrix in the position of the elements of the second matrix. The relation $(A_1 \otimes B_1)(A_2 \otimes B_2) = A_1A_2 \otimes B_1B_2$ holds. Γ^5 is expressed as an product of $\Gamma^1, \cdots, \Gamma^4$.

In the spinor representation T_L^a and T_R^a defined in (A.15) are

$$\begin{aligned}
T_L^a &= \tfrac{1}{2}\sigma^a \otimes \begin{pmatrix} 1 & 0 \\ 0 & 0 \end{pmatrix} , \\
T_R^a &= \tfrac{1}{2}\sigma^a \otimes \begin{pmatrix} 0 & 0 \\ 0 & 1 \end{pmatrix} .
\end{aligned} \tag{A.27}$$

In other words, the upper two component of a four component spinor ψ of $SO(4), SO(5)$ transforms as a two component spinor of $SU(2)_L$, whereas the lower two component of ψ transforms as a two component spinor of $SU(2)_R$. On the other hand $\hat{T}^p$ in (A.17) of $SO(5)$ are given by

$$\begin{aligned}
\hat{T}^a &= -\frac{1}{2\sqrt{2}}\,\sigma^a \otimes \sigma^2 \quad (a = 1, 2, 3) \,, \\
\hat{T}^4 &= \frac{1}{2\sqrt{2}}\,\sigma^0 \otimes \sigma^1 \,,
\end{aligned} \tag{A.28}$$

and therefore they mix the upper and lower components of ψ.

(c) $SO(10)$, $SO(11)$

The above prescription is easily generalized to $SO(N)$. In the case of $SO(10)$ and $SO(11)$, which become important in grand unification of forces,

Γ^j are given by $2^5 \times 2^5$ matrices. Explicitly they are given by

$$\begin{aligned}
\Gamma^1 &= \sigma^1 \otimes \sigma^1 \otimes \sigma^1 \otimes \sigma^1 \otimes \sigma^1 \\
\Gamma^2 &= \sigma^2 \otimes \sigma^1 \otimes \sigma^1 \otimes \sigma^1 \otimes \sigma^1 \\
\Gamma^3 &= \sigma^3 \otimes \sigma^1 \otimes \sigma^1 \otimes \sigma^1 \otimes \sigma^1 \\
\Gamma^4 &= \sigma^0 \otimes \sigma^2 \otimes \sigma^1 \otimes \sigma^1 \otimes \sigma^1 \\
\Gamma^5 &= \sigma^0 \otimes \sigma^3 \otimes \sigma^1 \otimes \sigma^1 \otimes \sigma^1 \\
\Gamma^6 &= \sigma^0 \otimes \sigma^0 \otimes \sigma^2 \otimes \sigma^1 \otimes \sigma^1 \\
\Gamma^7 &= \sigma^0 \otimes \sigma^0 \otimes \sigma^3 \otimes \sigma^1 \otimes \sigma^1 \\
\Gamma^8 &= \sigma^0 \otimes \sigma^0 \otimes \sigma^0 \otimes \sigma^2 \otimes \sigma^1 \\
\Gamma^9 &= \sigma^0 \otimes \sigma^0 \otimes \sigma^0 \otimes \sigma^3 \otimes \sigma^1 \\
\Gamma^{10} &= \sigma^0 \otimes \sigma^0 \otimes \sigma^0 \otimes \sigma^0 \otimes \sigma^2 \\
\Gamma^{11} &= \sigma^0 \otimes \sigma^0 \otimes \sigma^0 \otimes \sigma^0 \otimes \sigma^3 = -i\,\Gamma^1 \cdots \Gamma^{10} \,.
\end{aligned} \tag{A.29}$$

In the case of $SO(N = 2n)$ the generators $T_{\rm sp}^{jk}$ $(j, k = 1, \cdots, 2n)$ satisfy $[T_{\rm sp}^{jk}, \Gamma^{2n+1}] = 0$ so that the upper 2^{n-1} components and the lower 2^{n-1} components of a spinor independently transform under $SO(2n)$ transformations. In the case of $SO(4)$, by identifying $\gamma^0 = -i\Gamma^4$ and $\gamma^j = \Gamma^j$ $(j = 1, 2, 3)$ one obtains the Dirac matrices γ^μ in Section (A.1). $\Gamma^5 = \gamma^5$, and the upper (lower) two-components of a four-component spinor become a right-handed (left-handed) Weyl spinor.

Bispinor representation of $SO(4)$ vectors

As seen in (A.15) and (A.16), $SO(4) \simeq SU(2)_L \times SU(2)_R$. A four-component $SO(4)$ vector is a $(\mathbf{2}, \mathbf{2})$ representation of $SU(2)_L \times SU(2)_R$, or a bispinor. This property is utilized quite often in gauge-Higgs unification.

Consider an $SO(4)$ four-vector $\vec{\phi} = (\phi_1, \phi_2, \phi_3, \phi_4)^t$. Under an infinitesimal $SU(2)_L$, $SU(2)_R$ transformation

$$\Omega = 1 + i(\vec{\omega}_L \vec{T}_L + \vec{\omega}_R \vec{T}_R) \tag{A.30}$$

$\vec{\phi}$ is transformed as $\vec{\phi} \to \vec{\phi}' = \Omega\,\vec{\phi} \equiv \vec{\phi} + \delta\vec{\phi}$ where

$$\begin{pmatrix} \delta\phi_1 \\ \delta\phi_2 \\ \delta\phi_3 \\ \delta\phi_4 \end{pmatrix} = \frac{1}{2} \begin{pmatrix} (\omega_L^1 - \omega_R^1)\phi_4 - (\omega_L^2 + \omega_R^2)\phi_3 + (\omega_L^3 + \omega_R^3)\phi_2 \\ (\omega_L^1 + \omega_R^1)\phi_3 + (\omega_L^2 - \omega_R^2)\phi_4 - (\omega_L^3 + \omega_R^3)\phi_1 \\ -(\omega_L^1 + \omega_R^1)\phi_2 + (\omega_L^2 + \omega_R^2)\phi_1 + (\omega_L^3 - \omega_R^3)\phi_4 \\ -(\omega_L^1 - \omega_R^1)\phi_1 - (\omega_L^2 - \omega_R^2)\phi_2 - (\omega_L^3 - \omega_R^3)\phi_3 \end{pmatrix} . \tag{A.31}$$

Here the representation $T_{\rm vec}^{jk}$ in (A.13) has been used.

It follows from (A.31) that under an $SU(2)_L$ transformation $\Omega_L = 1 + i\vec{\omega}_L \vec{T}_L$

$$\delta \begin{pmatrix} \phi_2 + i\phi_1 \\ \phi_4 - i\phi_3 \end{pmatrix} = \frac{i}{2} \vec{\omega}_L \vec{\sigma} \begin{pmatrix} \phi_2 + i\phi_1 \\ \phi_4 - i\phi_3 \end{pmatrix},$$

$$\delta \begin{pmatrix} \phi_4 + i\phi_3 \\ -\phi_2 + i\phi_1 \end{pmatrix} = \frac{i}{2} \vec{\omega}_L \vec{\sigma} \begin{pmatrix} \phi_4 + i\phi_3 \\ -\phi_2 + i\phi_1 \end{pmatrix}. \tag{A.32}$$

As $\vec{T}_L = \frac{1}{2}\vec{\sigma}$ for two-component spinors in $SU(2)_L$, we observe that

$$\Phi_{L1} = \begin{pmatrix} \phi_2 + i\phi_1 \\ \phi_4 - i\phi_3 \end{pmatrix}, \quad \Phi_{L2} = \begin{pmatrix} \phi_4 + i\phi_3 \\ -\phi_2 + i\phi_1 \end{pmatrix} \tag{A.33}$$

transform, under Ω_L, as

$$\Phi_{Lj} \to \Phi'_{Lj} = \Omega_L \, \Phi_{Lj} \ . \tag{A.34}$$

That is, Φ_{Lj} is an $SU(2)_L$ spinor.

Similarly under an $SU(2)_R$ transformation $\Omega_R = 1 + i\vec{\omega}_R \vec{T}_R$

$$\Phi_{R1} = \begin{pmatrix} -\phi_2 - i\phi_1 \\ \phi_4 + i\phi_3 \end{pmatrix}, \quad \Phi_{R2} = \begin{pmatrix} -\phi_4 + i\phi_3 \\ -\phi_2 + i\phi_1 \end{pmatrix} \tag{A.35}$$

transform as

$$\Phi_{Rj} \to \Phi'_{Rj} = \Omega_R \, \Phi_{Rj} \ , \tag{A.36}$$

i.e. Φ_{Rj} is an $SU(2)_R$ spinor. As $\sigma_2 \vec{\sigma}^{\,t} \sigma_2 = -\vec{\sigma}$, $\sigma_2 (\Omega_R)^t \sigma_2 = \Omega_R^{-1}$ so that

$$\Phi^t_{Rj}(-i\sigma_2) \to \Phi^t_{Rj}(-i\sigma_2)\Omega_R^{-1} \ . \tag{A.37}$$

Note that

$$\Phi^t_{R1}(-i\sigma_2) = (\phi_4 + i\phi_3, \phi_2 + i\phi_1) \ ,$$
$$\Phi^t_{R2}(-i\sigma_2) = (-\phi_2 + i\phi_1, \phi_4 - i\phi_3) \ . \tag{A.38}$$

Now we define a 2×2 matrix $\hat{\phi}$ by

$$\hat{\phi} = \begin{pmatrix} \phi_4 + i\phi_3 & \phi_2 + i\phi_1 \\ -\phi_2 + i\phi_1 & \phi_4 - i\phi_3 \end{pmatrix} = \phi_4 + i\vec{\phi}\,\vec{\sigma} \ . \tag{A.39}$$

As $[T^a_L, T^b_R] = 0$, it follows from (A.33), (A.34), (A.37), (A.38) that under a general $SU(2)_L \times SU(2)_R$ transformation

$$\hat{\phi} \to \hat{\phi}' = \Omega_L \hat{\phi}\, \Omega_R^{-1} \ . \tag{A.40}$$

$\hat{\phi}$ is the bispinor representation of a vector $\vec{\phi}$.

The above argument is valid irrespective of $\vec{\phi}$ being real or complex. When $\vec{\phi}$ is real,

$$\Phi_{L2} = i\sigma_2 \Phi^*_{L1} \equiv \tilde{\Phi}_{L1} \ , \quad \Phi_{R1} = i\sigma_2 \Phi^*_{R2} = \tilde{\Phi}_{R2} \ . \tag{A.41}$$

The 4D Higgs field in the $SO(5) \times U(1)$ gauge-Higgs unification, in (7.19), corresponds to Φ_{L1} with real $\vec{\phi}$.

A.3 $SU(N)$

$SU(2)$, $SU(3)$, $SU(5)$, $SU(6)$ etc. often appear in unified theories in particle physics. A set of $N \times N$ special unitary complex matrices ($U^\dagger = U^{-1}$, $\det U = 1$) is one representation of the group $SU(N)$. There are $N^2 - 1$ generators of $SU(N)$, which can be expressed as $N \times N$ hermitian matrices with vanishing trace.

$$U = \exp\left\{ i \sum_{a=1}^{N^2-1} \omega_a T_a \right\} ,$$

$$T_a^\dagger = T_a \ , \quad \mathrm{Tr}\, T_a = 0 \ . \tag{A.42}$$

For $SU(2)$ the generators are given by

$$T_a = \tfrac{1}{2}\sigma^a \quad (a = 1, 2, 3) \ . \tag{A.43}$$

For $SU(3)$ one can take

$$\begin{aligned}
&T_a = \tfrac{1}{2}\lambda_a \quad (a = 1, \cdots, 8) \ , \\
&\lambda_1 = \begin{pmatrix} 0&1&0 \\ 1&0&0 \\ 0&0&0 \end{pmatrix} , \ \lambda_2 = \begin{pmatrix} 0&-i&0 \\ i&0&0 \\ 0&0&0 \end{pmatrix} , \ \lambda_3 = \begin{pmatrix} 1&0&0 \\ 0&-1&0 \\ 0&0&0 \end{pmatrix} , \\
&\lambda_4 = \begin{pmatrix} 0&0&1 \\ 0&0&0 \\ 1&0&0 \end{pmatrix} , \ \lambda_5 = \begin{pmatrix} 0&0&-i \\ 0&0&0 \\ i&0&0 \end{pmatrix} , \ \lambda_6 = \begin{pmatrix} 0&0&0 \\ 0&0&1 \\ 0&1&0 \end{pmatrix} , \\
&\lambda_7 = \begin{pmatrix} 0&0&0 \\ 0&0&-i \\ 0&i&0 \end{pmatrix} , \ \lambda_8 = \frac{1}{\sqrt{3}} \begin{pmatrix} 1&0&0 \\ 0&1&0 \\ 0&0&-2 \end{pmatrix} , \\
&\mathrm{Tr}\, \lambda_a \lambda_b = 2\delta_{ab} \ .
\end{aligned} \tag{A.44}$$

Generalization to $SU(N)$ is straightforward.

Appendix B

Effective Potential

In this appendix we describe how to evaluate the effective potential $V_{\text{eff}}(\theta_H)$. In field theory there are infinite degrees of freedom such that naive evaluation of V_{eff} gives rise to a divergent result. What is of interest is the θ_H-dependent part of $V_{\text{eff}}(\theta_H)$, which turns out finite and can be evaluated in perturbation theory even in the RS warped space.

B.1 Regularization

The KK mass spectrum of each field in five-dimensional spacetime is given in the form $\{m_n\}$. In gauge-Higgs unification the KK mass spectrum depends on the AB phase θ_H ($m_n = m_n(\theta_H)$). As seen in Sections 4.2 and 4.3, $V_{\text{eff}}(\theta_H)$ in the one loop approximation is given by

$$V_{\text{eff}}(\theta_H) = \sum_n \mp\frac{i}{2}\int\frac{d^4p}{(2\pi)^4}\ln\left\{p^2 + m_n(\theta_H)^2 - i\epsilon\right\} . \tag{B.1}$$

Here the sign $-$ $(+)$ corresponds to bosons (fermions). By making a Wick rotation in the p_0 integral ($p_0 = ip_4$) one finds

$$V_{\text{eff}}(\theta_H) = \sum_n \pm\frac{1}{2}\int\frac{d^4p_E}{(2\pi)^4}\ln\left\{p_E^2 + m_n(\theta_H)^2\right\} . \tag{B.2}$$

Though the momentum integration is divergent, the θ_H-dependent part is finite.

To evaluate the integral, we consider

$$V(m^2; d) = \frac{1}{2}\int\frac{d^dp}{(2\pi)^d}\ln(p^2 + m^2) \tag{B.3}$$

where $p^2 = p_1^2 + \cdots + p_d^2$. The generalized zeta($\zeta$)-function is defined by

$$\zeta(s) = \zeta(s; m^2, d) = \int\frac{d^dp}{(2\pi)^d}\frac{1}{(p^2 + m^2)^s} . \tag{B.4}$$

For $\Re s > \frac{1}{2}d$ the integral in (B.4) is finite and $\zeta(s)$ is defined as an analytic function of a complex variable s. For $\Re s \leq \frac{1}{2}d$, $\zeta(s)$ is defined by analytic continuation. In terms of $\zeta(s)$ we define

$$V(m^2;d) = -\frac{1}{2}\zeta'(0) \;, \tag{B.5}$$

which is called the regularization of $V(m^2;d)$. The "regularization" implies to make a quantity under consideration finite by some means. Furthermore with the use of identities

$$\begin{aligned}\int d^d p &= \frac{2\pi^{d/2}}{\Gamma(\frac{1}{2}d)}\int_0^\infty dp\, p^{d-1}, \\ \frac{1}{A^s} &= \frac{1}{\Gamma(s)}\int_0^\infty du\, u^{s-1}e^{-Au}\end{aligned} \tag{B.6}$$

one can derive

$$\zeta(s;m^2,d) = \frac{\Gamma(s-\frac{1}{2}d)}{2^d\pi^{d/2}\Gamma(s)}\, m^{d-2s} \;. \tag{B.7}$$

It is seen that $\zeta(s)$ has a pole at $s = \frac{1}{2}d$ and diverges there.

With (B.7) $\zeta(s;m^2,d)$ can be viewed as an analytic function of d as well. We take the limit $d \to 4$ at the end. For the time being we consider complex $d \neq 0, 2, 4, \cdots$. Since $\Gamma(s) \sim 1/s$ near $s = 0$, one finds

$$V(m^2;d) = -\frac{\Gamma(-\frac{1}{2}d)}{2(4\pi)^{d/2}}\, m^d \;. \tag{B.8}$$

Reflecting the fact that the expression (B.3) is divergent at $d = 4$, $V(m^2;d)$ in (B.8) has a pole at $d = 4$ and diverges there.

$V_{\rm eff}(\theta_H)$ in (B.1) in d dimensions is expressed as

$$V_{\rm eff}(\theta_H) = \mp\frac{\Gamma(-\frac{1}{2}d)}{2(4\pi)^{d/2}}\sum_n m_n(\theta_H)^d \quad \text{for} \left\{ \begin{array}{l} \text{bosons (B)} \\ \text{fermions (F)} \end{array} \right. . \tag{B.9}$$

In gauge-Higgs unification θ_H is an AB phase as a result of the gauge invariance, and the KK mass spectrum has a property $\{m_n(\theta_H + 2\pi)\} = \{m_n(\theta_H)\}$. After carrying out the infinite sum $\sum_n$, one finds $V_{\rm eff}(\theta_H + 2\pi) = V_{\rm eff}(\theta_H)$. It is shown below that the θ_H-dependent part of $V_{\rm eff}(\theta_H)$ becomes finite even at $d = 4$.

B.2 Infinite sum in flat space

In gauge-Higgs unification in flat spacetime such as $M^4 \times S^1$ or $M^4 \times (S^1/Z_2)$ the KK mass spectrum is easily found to take the form of

$$m_n^2 = \frac{(n+a)^2}{R^2} + \Lambda^2 \ , \quad a = \frac{\ell\theta_H}{2\pi} + c \ . \tag{B.10}$$

Here $\ell = 0, 1, 2, \cdots$ and $0 \le c < 1$. Λ corresponds to the original mass in the Lagrangian, and vanishes for gauge fields. In field theory on orbifolds the range of n can be taken as $-\infty < n < \infty$ by considering a pair of KK towers of fields together.

The function necessary to evaluate $V_{\rm eff}$ in (B.9) is

$$f(a\,;d) = \Gamma(-\tfrac{1}{2}d) \sum_{n=-\infty}^{\infty} \left\{ \frac{(n+a)^2}{R^2} + \Lambda^2 \right\}^{d/2} . \tag{B.11}$$

The infinite sum converges for $\Re\, d < -1$. $f(a\,;d)$ for $\Re\, d \ge -1$ is defined by analytic continuation. With the use of (B.6) it can be written as

$$f = R^{-d} \int_0^\infty du\, u^{-(d/2)-1} \sum_{n=-\infty}^{\infty} e^{-[(n+a)^2 + (\Lambda R)^2]u} \ . \tag{B.12}$$

Further, by employing the Poisson formula

$$\sum_{n=-\infty}^{\infty} e^{-b(n+a)^2} = \sqrt{\frac{\pi}{b}} \sum_{n=-\infty}^{\infty} e^{-(\pi^2 n^2/b) + 2\pi i n a} \ , \tag{B.13}$$

which can be derived from its Fourier expansion, the expression is transformed to

$$\begin{aligned} f = \frac{\sqrt{\pi}}{R^d} \bigg\{ & (\Lambda R)^{d+1} \Gamma\big[-\tfrac{1}{2}(d+1) \big] \\ & + \sum_{n=1}^{\infty} \frac{2\cos 2\pi a n}{(\pi n)^{d+1}} \int_0^\infty dt\, t^{(d-1)/2} e^{-t-(\pi n \Lambda R)^2/t} \bigg\} \ . \end{aligned} \tag{B.14}$$

The first term on the right-hand side of (B.14) is independent of a, and therefore is θ_H-independent, and may be dropped.

The integral representation of the modified Bessel function $K_\nu(z) = K_{-\nu}(z)$ is given by

$$K_\nu(z) = \frac{1}{2} \Big(\frac{z}{2}\Big)^{-\nu} \int_0^\infty dt\, t^{\nu-1} \exp\Big(-t - \frac{z^2}{4t} \Big) \ , \qquad (z \ne 0,\ |\arg z| < \tfrac{1}{4}\pi) \ , \tag{B.15}$$

so that f in (B.14) is expressed, up to an a-independent constant term, as

$$f(a;d) = \frac{2\Gamma\left[\frac{1}{2}(d+1)\right]}{\pi^{d+(1/2)} R^d} \sum_{n=1}^{\infty} \frac{\cos 2\pi n a}{n^{d+1}} B_{\frac{d+1}{2}}(2\pi n \Lambda R) ,$$

$$B_\nu(z) = \frac{2}{\Gamma(\nu)} \left(\frac{z}{2}\right)^\nu K_\nu(z) , \quad B_\nu(0) = 1 . \tag{B.16}$$

$V_{\rm eff}$ in (B.9) is expressed as

$$V_{\rm eff} = \mp \frac{\Gamma\left[\frac{1}{2}(d+1)\right]}{\pi^{(d+1)/2}(2\pi R)^d} \sum_{n=1}^{\infty} \frac{\cos 2\pi n a}{n^{d+1}} B_{\frac{d+1}{2}}(2\pi n \Lambda R) \quad \text{for} \left\{ \begin{matrix} \mathrm{B} \\ \mathrm{F} \end{matrix} \right. . \tag{B.17}$$

In particular, for $d = 4$

$$V_{\rm eff} = \mp \frac{3}{64\pi^6 R^4} \sum_{n=1}^{\infty} \frac{\cos 2\pi n a}{n^5} B_{\frac{5}{2}}(2\pi n \Lambda R) \quad \text{for} \left\{ \begin{matrix} \mathrm{B} \\ \mathrm{F} \end{matrix} \right. . \tag{B.18}$$

It is seen that the θ_H-dependent part of $V_{\rm eff}(\theta_H)$ is finite.

B.3 Integral representation of $V_{\rm eff}(\theta_H)$

The KK mass spectrum $\{m_n = k\lambda_n\}$ in gauge-Higgs unification models formulated in the Randall-Sundrum (RS) warped space cannot be explicitly expressed. The spectrum $\{\lambda_n\}$ is determined, typically as zeros of an equation like (7.62)

$$2S(1;\lambda_n)C'(1;\lambda_n) + \lambda_n \sin^2\theta_H = 0 . \tag{B.19}$$

We are going to show that information like (B.19) is sufficient to find $V_{\rm eff}(\theta_H)$ at the one loop level. [47]

Suppose that the mass spectrum $\{m_n\}$ is given by zeros of a certain function $\rho(z)$;

$$m_n = z_n, \quad \rho(z_n) = 0 \quad (n = 1, 2, \cdots) . \tag{B.20}$$

For the moment we assume that all $z_n = m_n$ are real and positive, that the spectrum is not degenerate, and that $\rho(z)$ is analytic for $\Re z > 0$. (The argument can be easily generalized in the presence of a zero mode $m_0 = 0$. See below.) $\rho(z)$ is expanded in a Taylor series $\rho(z) = a_n(z - z_n) + \cdots$ around z_n so that $\rho'(z)/\rho(z) = (z - z_n)^{-1} + \cdots$. With the use of the residue theorem one finds

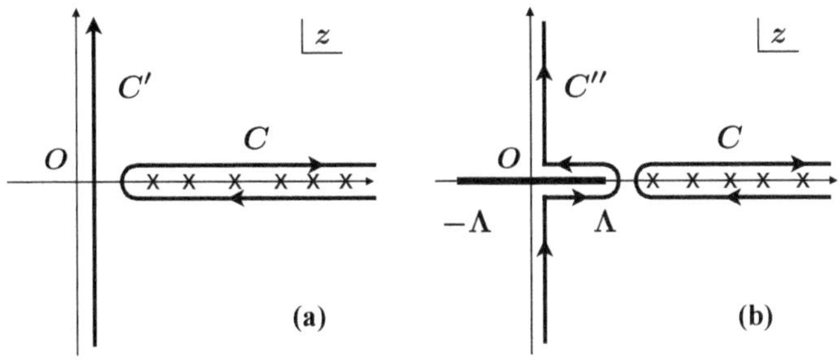

Fig. B.1 Integration paths: Cross symbols × represent the location of zeros of the function $\rho(z)$. (a) The path C is deformed to C'. (b) When $\rho(z)$ has a branch cut on the real axis $-\Lambda \le z \le \Lambda$, the path C is deformed to C''.

$$\begin{aligned}\sum_{n=1}^{\infty} m_n^d &= -\frac{1}{2\pi i}\int_C dz\, z^d \frac{d}{dz}\ln\rho(z)\\ &= \frac{d}{2\pi i}\int_C dz\, z^{d-1}\ln\rho(z)\ . \end{aligned} \tag{B.21}$$

Here the integration path C is given by Fig. B.1(a). Furthermore, assuming that $|z^{d-1}\ln\rho| < R^{-\alpha}$, $\alpha > 1$ for $|z| = R \gg 1$, and that $\ln\rho(z)$ is analytic for $\Re z \ge 0$ except for the poles on the real axis, one can deform the path C to C' in Fig. B.1(a). We take C' to be the imaginary axis $z = iy$. When $\rho(iy) = \rho(-iy)$, we obtain

$$\sum_{n=1}^{\infty} m_n^d = \frac{d}{\pi}\sin\frac{\pi d}{2}\int_0^{\infty} dy\, y^{d-1}\ln\rho(iy)\ . \tag{B.22}$$

Inserting (B.22) into (B.9) and recalling $\Gamma(z)\Gamma(1-z) = \pi/\sin\pi z$, we obtain

$$V_{\text{eff}}(\theta_H) = \pm\frac{1}{(4\pi)^{d/2}\Gamma\left(\frac{1}{2}d\right)}\int_0^{\infty} dy\, y^{d-1}\ln\rho(iy;\theta_H) \quad \text{for } \begin{cases}\text{B}\\ \text{F}\end{cases}. \tag{B.23}$$

As one example let us consider the KK mass spectrum

$$m_n(\theta)^2 = \frac{1}{R^2}\left(n + \frac{\theta}{2\pi}\right)^2 \quad (n = 0, \pm1, \pm2, \cdots) \tag{B.24}$$

for a boson field on $M^4 \times S^1$. This spectrum is obtained as the solutions of $\sin^2\pi R z = \sin^2\frac{1}{2}\theta$ ($\Re z \ge 0$). Hence one can take

$$\rho(z;\theta) = 1 - \frac{\sin^2\frac{1}{2}\theta}{\sin^2\pi R z}\ . \tag{B.25}$$

For $0 < \theta < 2\pi$ there is no zero mode. Applying (B.23), one finds for $d = 4$

$$V_{\rm eff}(\theta) = \frac{1}{16\pi^6 R^4} \int_0^\infty dq\, q^3 \ln\left\{ 1 + \frac{\sin^2 \frac{1}{2}\theta}{\sinh^2 q} \right\} . \tag{B.26}$$

The resulting formula (B.26) is valid even for $\theta = 0$ when there appears a zero mode.

As another example let us consider the KK mass spectrum (B.19) of the W boson tower in the $SO(5) \times U(1)$ gauge-Higgs unification in the RS space. As shown in Eq. (7.133), in this case with $m_n = kz_n$

$$\rho(z; \theta_H) = 1 - \frac{\cos 2\theta_H}{1 + (4/z) S(1; z) C'(1; z)} . \tag{B.27}$$

Here the functions S and C' are defined by (C.14). The contribution of the W boson tower to $V_{\rm eff}(\theta_H)$ (per one degree of freedom) is given by

$$\begin{aligned} &V_{\rm eff}(\theta_H) \\ &= \frac{(kz_L^{-1})^4}{16\pi^2} \int_0^\infty dq\, q^3 \ln\left\{ 1 - \frac{\cos 2\theta_H}{1 + 4q^2 z_L^{-1} \hat{F}_{1,1}(qz_L^{-1}, q) \hat{F}_{0,0}(qz_L^{-1}, q)} \right\} , \\ &\hat{F}_{\alpha\alpha}(u, v) = I_\alpha(u) K_\alpha(v) - K_\alpha(u) I_\alpha(v) , \end{aligned} \tag{B.28}$$

where $I_\alpha(u)$ and $K_\alpha(u)$ are the modified Bessel functions in Appendix C.3. The q integral in (B.28) rapidly converges. The contribution from $q > 10$ can be numerically ignored. The KK mass scale is $m_{\rm KK} = 1/R$ in flat spacetime, whereas $m_{\rm KK} \sim \pi k z_L^{-1}$ in the RS space. It is seen that (B.26) and (B.28) have a similar structure.

The case in which the KK mass spectrum is given, instead of (B.20), by

$$m_n = \sqrt{z_n^2 + \Lambda^2} , \quad \rho(z_n) = 0 \quad (n = 1, 2, \cdots) \tag{B.29}$$

can be treated in a similar manner. [110] In this case we consider $\bar{\rho}(z) = \rho(\sqrt{z^2 - \Lambda^2})$ in place of $\rho(z)$. The function $\bar{\rho}(z)$ has branch points at $z = \pm\Lambda$, and has a branch cut on the real axis $-\Lambda \le z \le \Lambda$. In deforming the integration path, one needs to use the path C'' in Fig. B.1(b) by avoiding the branch cut. In the cases under consideration in this book, the contributions coming from the paths above and below the cut cancel each other. Therefore we find

$$V_{\rm eff}(\theta_H) = \pm \frac{1}{(4\pi)^{d/2} \Gamma(\frac{1}{2}d)} \int_0^\infty dy\, y^{d-1} \ln \rho(i\sqrt{y^2 + \Lambda^2}; \theta_H) \quad \text{for } \begin{cases} \text{B} \\ \text{F} \end{cases} . \tag{B.30}$$

B.4 Finiteness of $V_{\rm eff}^{\rm 1\,loop}(\theta_H)$

The results obtained above, (B.18), (B.26) and (B.28), show that the θ_H-dependent part of the effective potential at the one loop level, $V_{\rm eff}^{\rm 1\,loop}(\theta_H)$, is finite. Gauge theory defined in five-dimensional spacetime is not renormalizable so the finiteness of $V_{\rm eff}^{\rm 1\,loop}(\theta_H)$ is not obvious. The finiteness is a special property of gauge-Higgs unification. The finite Higgs boson mass is generated at the quantum level. It is independent of the cutoff scale at high energies. The gauge hierarchy problem is solved, irrespective of physics at short distance and ultra high energy scale.

In Section 4.7 we have observed in gauge theory on $M^4 \times S^1$ that the finiteness of $V_{\rm eff}^{\rm 1\,loop}(\theta_H)$ is guaranteed by the gauge invariance. Under large gauge transformations which retain the boundary conditions θ_H is transformed to $\theta_H + 2\pi$. As the theory is gauge invariant, the KK mass spectrum $\{m_n(\theta_H)\}$ also remains invariant. It follows that $V_{\rm eff}(\theta_H + 2\pi) = V_{\rm eff}(\theta_H)$ and $V_{\rm eff}(\theta_H)$ is expanded in a Fourier series; $V_{\rm eff}(\theta_H) = \sum_n v_n e^{in\theta_H}$. The proposition we claim is that except for v_0, $V_{\rm eff}(\theta_H) - v_0$ is finite at the one loop level.

$V_{\rm eff}^{\rm 1\,loop}(\theta_H)$ is given in the form of (B.2). In the cases of $M^4 \times S^1$ and $M^4 \times (S^1/Z_2)$ the spectrum is given by (B.10). When $\Lambda = 0$, the argument in Section 4.7 applies as it is. By differentiating (B.2) with respect to θ_H sufficiently many times, $d^q V_{\rm eff}^{\rm 1\,loop}/d\theta_H^q$ becomes finite and the proposition is proved. For $\Lambda > 0$, it is enough to consider the difference from the $\Lambda = 0$ case. For $|n| \gg \Lambda R$ the difference in the mass becomes small so that $V_{\rm eff}^{\rm 1\,loop}\big|_\Lambda - V_{\rm eff}^{\rm 1\,loop}\big|_0$ becomes finite. In the RS space $m_n(\theta_H)$ has more complex θ_H dependence. However the spectrum $m_n(\theta_H)/m_{\rm KK}$ for $n \gg 1$ approaches to that in the flat space. Consequently the θ_H-dependent part of $V_{\rm eff}^{\rm 1\,loop}(\theta_H)$ becomes finite in the RS space as well.

Appendix C

Bessel Functions

In field theory in Randall-Sundrum (RS) warped space, Bessel functions play an important role in place of trigonometric functions in flat spacetime. In this appendix we summarize the properties of the Bessel functions which have been frequently used in this book.

C.1 Properties of Bessel functions

$J_\nu(z)$ and $Y_\nu(z)$, Bessel functions of the first and second kind, are two linearly independent solutions to the Bessel equation

$$\left\{ \frac{d^2}{dz^2} + \frac{1}{z}\frac{d}{dz} + 1 - \frac{\nu^2}{z^2} \right\} u(z) = 0 \ . \tag{C.1}$$

They are given by

$$\begin{aligned} J_\nu(z) &= \Big(\frac{z}{2}\Big)^\nu \sum_{n=0}^\infty \frac{(-z^2/4)^n}{n!\,\Gamma(\nu+n+1)} \ , \\ Y_\nu(z) &= \begin{cases} \dfrac{1}{\sin\pi\nu} \big\{ \cos\pi\nu \cdot J_\nu(z) - J_{-\nu}(z) \big\} & \text{for } \nu \neq \text{an integer}, \\ \dfrac{1}{\pi}\left[\dfrac{\partial J_\nu(z)}{\partial\nu} - (-1)^n \dfrac{\partial J_{-\nu}(z)}{\partial\nu} \right]_{\nu=n} & \text{for } \nu = n. \end{cases} \end{aligned} \tag{C.2}$$

For $|z| \ll 1$

$$\begin{aligned} Y_0(z) &\sim \frac{2}{\pi}\Big(\ln\frac{z}{2} + \gamma \Big)\Big(1 - \frac{z^2}{4} \Big), \\ Y_1(z) &\sim -\frac{2}{\pi z} + \frac{z}{\pi}\Big(\ln\frac{z}{2} + \gamma - \frac{1}{2} \Big). \end{aligned} \tag{C.3}$$

Let $Z_\nu(z)$ collectively express $J_\nu(z)$ and $Y_\nu(z)$. $Z_\nu(z)$ satisfies the recurrence relation

$$\frac{d}{dz}\, z^{\pm\nu} Z_\nu(\lambda z) = \pm\lambda z^{\pm\nu} Z_{\nu\mp 1}(\lambda z) \ . \tag{C.4}$$

For $|z| \gg 1$ ($z \neq$ a negative integer)

$$\begin{aligned} J_\nu(z) &\sim \sqrt{\frac{2}{\pi z}} \cos\left(z - \frac{(2\nu+1)\pi}{4}\right) , \\ Y_\nu(z) &\sim \sqrt{\frac{2}{\pi z}} \sin\left(z - \frac{(2\nu+1)\pi}{4}\right) . \end{aligned} \tag{C.5}$$

Furthermore the relations

$$\begin{aligned} & Z_{\nu-1}(z) + Z_{\nu+1}(z) = \frac{2\nu}{z}\, Z_\nu(z) , \\ & \frac{dZ_\nu(z)}{dz} = \frac{\nu}{z}\, Z_\nu(z) - Z_{\nu+1}(z) = Z_{\nu-1}(z) - \frac{\nu}{z}\, Z_\nu(z) , \\ & J_\nu(z) Y_{\nu-1}(z) - Y_\nu(z) J_{\nu-1}(z) = \frac{2}{\pi z} \end{aligned} \tag{C.6}$$

hold. $Z_{-\nu}(z)$ is related to $Z_\nu(z)$ by

$$\begin{pmatrix} J_{-\nu}(z) \\ Y_{-\nu}(z) \end{pmatrix} = \begin{pmatrix} \cos\nu\pi & -\sin\nu\pi \\ \sin\nu\pi & \cos\nu\pi \end{pmatrix} \begin{pmatrix} J_\nu(z) \\ Y_\nu(z) \end{pmatrix} . \tag{C.7}$$

In field theory in the RS space the combination of J_ν and Y_ν

$$F_{\alpha,\beta}(u,v) = J_\alpha(u) Y_\beta(v) - Y_\alpha(u) J_\beta(v) \tag{C.8}$$

becomes important. $F_{\alpha,\beta}(u,v)$ satisfies

$$\begin{aligned} & F_{\alpha,\alpha-1}(u,u) = -F_{\alpha,\alpha+1}(u,u) = \frac{2}{\pi u} , \\ & F_{\alpha-1,\alpha}(u,v) F_{\alpha,\alpha-1}(u,v) = F_{\alpha-1,\alpha-1}(u,v) F_{\alpha,\alpha}(u,v) - \frac{4}{\pi^2 uv} , \\ & F_{\alpha+1,\beta}(u,v) + F_{\alpha-1,\beta}(u,v) = \frac{2\alpha}{u}\, F_{\alpha,\beta}(u,v) . \end{aligned} \tag{C.9}$$

For $\alpha \neq$ an integer,

$$\begin{aligned} F_{\alpha,\alpha}(u,v) &= \frac{1}{\sin\pi\alpha} \{ J_{-\alpha}(u) J_\alpha(v) - J_\alpha(u) J_{-\alpha}(v) \} , \\ F_{\alpha,\alpha-1}(u,v) &= \frac{1}{\sin\pi\alpha} \{ J_\alpha(u) J_{1-\alpha}(v) + J_{-\alpha}(u) J_{\alpha-1}(v) \} , \\ F_{\alpha-1,\alpha}(u,v) &= \frac{-1}{\sin\pi\alpha} \{ J_{\alpha-1}(u) J_{-\alpha}(v) + J_{1-\alpha}(u) J_\alpha(v) \} . \end{aligned} \tag{C.10}$$

Also

$$F_{-\alpha,-\beta}(u,v) = \cos(\alpha-\beta)\pi\, F_{\alpha,\beta}(u,v) - \sin(\alpha-\beta)\pi\left\{J_\alpha(u)J_\beta(v) + Y_\alpha(u)Y_\beta(v)\right\} \tag{C.11}$$

holds.

For $u, v \ll 1$ and $\alpha \neq$ an integer,

$$\begin{aligned}
F_{\alpha,\alpha}(u,v) &\sim \frac{1}{\pi\alpha}\left\{\left(\frac{v}{u}\right)^\alpha - \left(\frac{u}{v}\right)^\alpha - \frac{uv}{4(1-\alpha)}\left[\left(\frac{v}{u}\right)^{\alpha-1} - \left(\frac{u}{v}\right)^{\alpha-1}\right]\right. \\
&\qquad \left. - \frac{uv}{4(1+\alpha)}\left[\left(\frac{v}{u}\right)^{\alpha+1} - \left(\frac{u}{v}\right)^{\alpha+1}\right]\right\}, \\
F_{0,0}(u,v) &\sim \frac{2}{\pi}\ln\frac{v}{u}, \\
F_{1,1}(u,v) &\sim \frac{1}{\pi}\left(\frac{v}{u} - \frac{u}{v}\right) + \frac{uv}{2\pi}\ln\frac{v}{u}, \\
F_{\alpha-1,\alpha}(u,v) &= -F_{\alpha,\alpha-1}(v,u) \\
&\sim -\frac{2}{\pi u}\left\{\left(\frac{u}{v}\right)^\alpha\left[1 - \frac{v^2}{4(1-\alpha)} - \frac{u^2}{4\alpha}\right] + \frac{u^2}{4\alpha(1-\alpha)}\left(\frac{v}{u}\right)^\alpha\right\}, \\
F_{0,1}(u,v) &= -F_{1,0}(v,u) \sim -\frac{2}{\pi v} + \frac{v}{\pi}\left(\ln\frac{v}{u} - \frac{1}{2} + \frac{u^2}{2v^2}\right).
\end{aligned} \tag{C.12}$$

In particular, for $\lambda \ll 1$ and $\lambda z \ll 1$, one finds

$$\begin{aligned}
F_{\alpha,\alpha}(\lambda,\lambda z) &\sim \frac{z^\alpha - z^{-\alpha}}{\pi\alpha} - \frac{\lambda^2 z}{4\pi\alpha}\left\{\frac{z^{\alpha-1} - z^{1-\alpha}}{1-\alpha} + \frac{z^{\alpha+1} - z^{-\alpha-1}}{1+\alpha}\right\}, \\
F_{0,0}(\lambda,\lambda z) &\sim \frac{2}{\pi}\ln z, \\
F_{1,1}(\lambda,\lambda z) &\sim \frac{z - z^{-1}}{\pi} + \frac{\lambda^2 z}{2\pi}\ln z, \\
F_{\alpha-1,\alpha}(\lambda,\lambda z) &\sim -\frac{2z^{-\alpha}}{\pi\lambda}\left\{1 - \frac{\lambda^2 z^2}{4(1-\alpha)} - \frac{\lambda^2}{4\alpha} + \frac{\lambda^2 z^{2\alpha}}{4\alpha(1-\alpha)}\right\}, \\
F_{\alpha,\alpha-1}(\lambda,\lambda z) &\sim \frac{2z^{\alpha-1}}{\pi\lambda}\left\{1 - \frac{\lambda^2}{4(1-\alpha)} - \frac{\lambda^2 z^2}{4\alpha} + \frac{\lambda^2 z^{2-2\alpha}}{4\alpha(1-\alpha)}\right\}, \\
F_{0,1}(\lambda,\lambda z) &\sim -\frac{2}{\pi\lambda z} + \frac{\lambda z}{\pi}\left(\ln z - \frac{1}{2} + \frac{1}{2z^2}\right), \\
F_{1,0}(\lambda,\lambda z) &\sim \frac{2}{\pi\lambda} + \frac{\lambda}{\pi}\left(\ln z + \frac{1}{2} - \frac{z^2}{2}\right).
\end{aligned} \tag{C.13}$$

C.2 Basis functions in the RS space

The RS space is a curved space such that the form of field equations depends on the spin of a field. The metric of the RS space is given by (6.11), (6.20) and (6.24). We assume that the warp factor $z_L = e^{kL} \gg 1$.

For a vector field of spin 1, particularly for gauge fields, it is convenient to introduce a set of basis functions

$$\begin{aligned} C(z;\lambda) &= \frac{\pi}{2}\lambda z z_L F_{1,0}(\lambda z, \lambda z_L) \ , \\ S(z;\lambda) &= -\frac{\pi}{2}\lambda z F_{1,1}(\lambda z, \lambda z_L) \ , \\ C'(z;\lambda) &= \frac{\pi}{2}\lambda^2 z z_L F_{0,0}(\lambda z, \lambda z_L) \ , \\ S'(z;\lambda) &= -\frac{\pi}{2}\lambda^2 z F_{0,1}(\lambda z, \lambda z_L) \ . \end{aligned} \tag{C.14}$$

These functions satisfy

$$\begin{aligned} & -z\frac{d}{dz}\frac{1}{z}\frac{d}{dz}\begin{pmatrix} C(z;\lambda) \\ S(z;\lambda) \end{pmatrix} = \lambda^2 \begin{pmatrix} C(z;\lambda) \\ S(z;\lambda) \end{pmatrix}, \\ & -\frac{d}{dz}z\frac{d}{dz}\frac{1}{z}\begin{pmatrix} C'(z;\lambda) \\ S'(z;\lambda) \end{pmatrix} = \lambda^2 \begin{pmatrix} C'(z;\lambda) \\ S'(z;\lambda) \end{pmatrix}, \\ & C(z_L;\lambda) = z_L \ , \quad S(z_L;\lambda) = 0 \ , \\ & C'(z_L;\lambda) = 0 \ , \quad S'(z_L;\lambda) = \lambda \ , \\ & CS' - SC' = \lambda z \ . \end{aligned} \tag{C.15}$$

Mode functions of the four-dimensional components A_μ are expressed in terms of C and S, whereas those of the fifth dimensional component A_z are expressed in terms of C' and S'.

For the left-handed and right-handed components of fermion fields of spin $\frac{1}{2}$ we define

$$\begin{aligned} \begin{pmatrix} C_L \\ S_L \end{pmatrix}(z;\lambda,c) &= \pm\frac{\pi}{2}\lambda\sqrt{z z_L} F_{c+\frac{1}{2},c\mp\frac{1}{2}}(\lambda z, \lambda z_L) \ , \\ \begin{pmatrix} C_R \\ S_R \end{pmatrix}(z;\lambda,c) &= \mp\frac{\pi}{2}\lambda\sqrt{z z_L} F_{c-\frac{1}{2},c\pm\frac{1}{2}}(\lambda z, \lambda z_L) \ . \end{aligned} \tag{C.16}$$

These functions satisfy

$$D_+(c)\begin{pmatrix}C_L\\S_L\end{pmatrix}=\lambda\begin{pmatrix}S_R\\C_R\end{pmatrix},$$

$$D_-(c)\begin{pmatrix}C_R\\S_R\end{pmatrix}=\lambda\begin{pmatrix}S_L\\C_L\end{pmatrix}, \quad D_\pm(c)=\pm\frac{d}{dz}+\frac{c}{z},$$

$$C_R=C_L=1, \quad S_R=S_L=0, \quad \text{at } z=z_L,$$

$$C_LC_R-S_LS_R=1. \tag{C.17}$$

Consequently we have

$$D_-D_+\begin{pmatrix}C_L\\S_L\end{pmatrix}=\lambda^2\begin{pmatrix}C_L\\S_L\end{pmatrix},$$

$$D_+D_-\begin{pmatrix}C_R\\S_R\end{pmatrix}=\lambda^2\begin{pmatrix}C_R\\S_R\end{pmatrix}. \tag{C.18}$$

The parameter c is the bulk mass parameter appearing in the Dirac equation in the RS space. The change of the sign of c corresponds to interchanging the left-handed and right-handed components. As $F_{-c+\frac{1}{2},-c-\frac{1}{2}}(u,v)=-F_{c+\frac{1}{2},c-\frac{1}{2}}(u,v)$ and $F_{-c-\frac{1}{2},-c-\frac{1}{2}}(u,v)=F_{c+\frac{1}{2},c+\frac{1}{2}}(u,v)$, the relations

$$\begin{aligned}C_L(z;\lambda,-c)&=C_R(z;\lambda,c),\\S_L(z;\lambda,-c)&=-S_R(z;\lambda,c),\end{aligned} \tag{C.19}$$

hold. For zero modes $\lambda=0$, one sees

$$S_{L/R}(z;0,c)=0, \quad C_{L/R}(z;0,c)\neq 0. \tag{C.20}$$

We also note that gauge field mode functions are related to fermion mode functions by

$$\begin{aligned}C(z;\lambda)&=\sqrt{zz_L}\,C_L(z;\lambda,\tfrac{1}{2}),\\S(z;\lambda)&=\sqrt{\frac{z}{z_L}}\,S_L(z;\lambda,\tfrac{1}{2}),\\C'(z;\lambda)&=\lambda\sqrt{zz_L}\,S_R(z;\lambda,\tfrac{1}{2}),\\S'(z;\lambda)&=\lambda\sqrt{\frac{z}{z_L}}\,C_R(z;\lambda,\tfrac{1}{2}).\end{aligned} \tag{C.21}$$

It is seen that gauge fields correspond to fermions with $c=\frac{1}{2}$.

In writing down wave functions of various fields, the usage of the following rescaled functions becomes useful.

$$\begin{aligned}
\check{S}(z;\lambda) &= \frac{C(1;\lambda)}{S(1;\lambda)}\, S(z;\lambda)\ , \\
\begin{pmatrix} \check{S}_L \\ \check{C}_R \end{pmatrix}(z;\lambda,c) &= \frac{C_L(1;\lambda,c)}{S_L(1;\lambda,c)} \begin{pmatrix} S_L \\ C_R \end{pmatrix}(z;\lambda,c)\ , \\
\begin{pmatrix} \check{S}_R \\ \check{C}_L \end{pmatrix}(z;\lambda,c) &= \frac{C_R(1;\lambda,c)}{S_R(1;\lambda,c)} \begin{pmatrix} S_R \\ C_L \end{pmatrix}(z;\lambda,c)\ .
\end{aligned} \tag{C.22}$$

For $\lambda z, \lambda z_L \ll 1$, the basis functions for gauge fields, (C.14), are approximated by

$$\begin{aligned}
C(z;\lambda) &\sim z_L \Big\{ 1 + \frac{\lambda^2 z^2}{2} \Big(\ln\frac{z_L}{z} + \frac{1}{2} \Big) \Big\}, \\
S(z;\lambda) &\sim -\frac{\lambda z_L}{2} \Big\{ 1 - \frac{z^2}{z_L^2} + \frac{\lambda^2 z^2}{2} \ln\frac{z_L}{z} \Big\}, \\
C'(z;\lambda) &\sim \lambda^2 z z_L \ln\frac{z_L}{z}\ , \\
S'(z;\lambda) &\sim \frac{\lambda z}{z_L} \Big\{ 1 - \frac{\lambda^2 z_L^2}{2} \Big(\ln\frac{z_L}{z} - \frac{1}{2} + \frac{z^2}{2z_L^2} \Big) \Big\},
\end{aligned} \tag{C.23}$$

whereas the basis functions for fermions, (C.16), are approximated by

$$\begin{aligned}
C_L(z;\lambda,c) &\sim \Big(\frac{z}{z_L}\Big)^{-c} \Big\{ 1 - \frac{\lambda^2 z^2}{2(1-2c)} - \frac{\lambda^2 z_L^2}{2(1+2c)} + \frac{\lambda^2 z z_L}{1-4c^2} \Big(\frac{z}{z_L}\Big)^{2c} \Big\}, \\
C_R(z;\lambda,c) &\sim \Big(\frac{z}{z_L}\Big)^{c} \Big\{ 1 - \frac{\lambda^2 z_L^2}{2(1-2c)} - \frac{\lambda^2 z^2}{2(1+2c)} + \frac{\lambda^2 z z_L}{1-4c^2} \Big(\frac{z}{z_L}\Big)^{-2c} \Big\}, \\
S_L(z;\lambda,c) &\sim -\frac{\lambda z_L}{1+2c} \Big(\frac{z}{z_L}\Big)^{-c} + \frac{\lambda z}{1+2c} \Big(\frac{z}{z_L}\Big)^{c}, \\
S_R(z;\lambda,c) &\sim \frac{\lambda z_L}{1-2c} \Big(\frac{z}{z_L}\Big)^{c} - \frac{\lambda z}{1-2c} \Big(\frac{z}{z_L}\Big)^{-c}.
\end{aligned} \tag{C.24}$$

Note that the subleading terms in (C.24) are important to satisfy the differential equations relating (C_L, S_L) and (C_R, S_R) in Eq. (C.17). For $z = 1$, $z_L \gg 1$ and $\lambda z_L \ll 1$

$$\begin{aligned}
C(1;\lambda) &\sim z_L\ , \\
S(1;\lambda) &\sim -\tfrac{1}{2}\lambda z_L\ , \\
C'(1;\lambda) &\sim \lambda^2 z_L \ln z_L\ , \\
S'(1;\lambda) &\sim \lambda z_L^{-1} \Big(1 - \frac{\lambda^2 z_L^2}{2} \ln z_L \Big)\ ,
\end{aligned} \tag{C.25}$$

and

$$
\begin{aligned}
&C_L(1;\lambda,c) \sim z_L^c \quad \text{for } \lambda^2 z_L^{1-2c} \ll 1 \ , \\
&C_R(1;\lambda,c) \sim z_L^{-c} \quad \text{for } \lambda^2 z_L^{1+2c} \ll 1 \ , \\
&S_L(1;\lambda,c) \sim \begin{cases} -\dfrac{\lambda z_L^{c+1}}{2(c+\frac{1}{2})} & \text{for } c > -\frac{1}{2} \\ -\lambda z_L^{1/2} \ln z_L & \text{for } c = -\frac{1}{2} \\ \dfrac{\lambda z_L^{-c}}{2(c+\frac{1}{2})} & \text{for } c < -\frac{1}{2} \end{cases} \ , \\
&S_R(1;\lambda,c) \sim \begin{cases} \dfrac{\lambda z_L^{c}}{2(c-\frac{1}{2})} & \text{for } c > \frac{1}{2} \\ \lambda z_L^{1/2} \ln z_L & \text{for } c = \frac{1}{2} \\ \dfrac{\lambda z_L^{1-c}}{2(\frac{1}{2}-c)} & \text{for } c < \frac{1}{2} \end{cases} \ .
\end{aligned}
\tag{C.26}
$$

In particular, the product $S_L S_R$ behaves as

$$
S_L(1;\lambda,c) S_R(1;\lambda,c) \sim \begin{cases} -\dfrac{\lambda^2 z_L^{2|c|+1}}{4c^2-1} & \text{for } |c| > \frac{1}{2} \\ -\frac{1}{2}\lambda^2 z_L^2 \ln z_L & \text{for } c = \pm\frac{1}{2} \\ -\dfrac{\lambda^2 z_L^2}{1-4c^2} & \text{for } |c| < \frac{1}{2} \end{cases} \ . \tag{C.27}
$$

For fermion fields with a vector-like mass $m = k\tilde{m}$, mode functions are expressed in terms of

$$
\begin{aligned}
\mathcal{C}_{L1}(z;\lambda,c,\tilde{m}) &= C_L(z;\lambda,c+\tilde{m}) + C_L(z;\lambda,c-\tilde{m}) \ , \\
\mathcal{C}_{L2}(z;\lambda,c,\tilde{m}) &= S_L(z;\lambda,c+\tilde{m}) - S_L(z;\lambda,c-\tilde{m}) \ , \\
\mathcal{S}_{L1}(z;\lambda,c,\tilde{m}) &= S_L(z;\lambda,c+\tilde{m}) + S_L(z;\lambda,c-\tilde{m}) \ , \\
\mathcal{S}_{L2}(z;\lambda,c,\tilde{m}) &= C_L(z;\lambda,c+\tilde{m}) - C_L(z;\lambda,c-\tilde{m}) \ , \\
\mathcal{C}_{R1}(z;\lambda,c,\tilde{m}) &= C_R(z;\lambda,c+\tilde{m}) + C_R(z;\lambda,c-\tilde{m}) \ , \\
\mathcal{C}_{R2}(z;\lambda,c,\tilde{m}) &= S_R(z;\lambda,c+\tilde{m}) - S_R(z;\lambda,c-\tilde{m}) \ , \\
\mathcal{S}_{R1}(z;\lambda,c,\tilde{m}) &= S_R(z;\lambda,c+\tilde{m}) + S_R(z;\lambda,c-\tilde{m}) \ , \\
\mathcal{S}_{R2}(z;\lambda,c,\tilde{m}) &= C_R(z;\lambda,c+\tilde{m}) - C_R(z;\lambda,c-\tilde{m}) \ .
\end{aligned}
\tag{C.28}
$$

These functions satisfy

$$
\begin{aligned}
&D_+(c)\begin{pmatrix}\mathcal{C}_{L1}\\ \mathcal{C}_{L2}\end{pmatrix} = \lambda\begin{pmatrix}\mathcal{S}_{R1}\\ \mathcal{S}_{R2}\end{pmatrix} - \frac{\tilde{m}}{z}\begin{pmatrix}\mathcal{S}_{L2}\\ \mathcal{S}_{L1}\end{pmatrix},\\
&D_+(c)\begin{pmatrix}\mathcal{S}_{L1}\\ \mathcal{S}_{L2}\end{pmatrix} = \lambda\begin{pmatrix}\mathcal{C}_{R1}\\ \mathcal{C}_{R2}\end{pmatrix} - \frac{\tilde{m}}{z}\begin{pmatrix}\mathcal{C}_{L2}\\ \mathcal{C}_{L1}\end{pmatrix},\\
&D_-(c)\begin{pmatrix}\mathcal{C}_{R1}\\ \mathcal{C}_{R2}\end{pmatrix} = \lambda\begin{pmatrix}\mathcal{S}_{L1}\\ \mathcal{S}_{L2}\end{pmatrix} - \frac{\tilde{m}}{z}\begin{pmatrix}\mathcal{S}_{R2}\\ \mathcal{S}_{R1}\end{pmatrix},\\
&D_-(c)\begin{pmatrix}\mathcal{S}_{R1}\\ \mathcal{S}_{R2}\end{pmatrix} = \lambda\begin{pmatrix}\mathcal{C}_{L1}\\ \mathcal{C}_{L2}\end{pmatrix} - \frac{\tilde{m}}{z}\begin{pmatrix}\mathcal{C}_{R2}\\ \mathcal{C}_{R1}\end{pmatrix},\\
&\mathcal{S}_{Rj} = \mathcal{S}_{Lj} = D_-(c)\,\mathcal{C}_{Rj} = D_+(c)\,\mathcal{C}_{Lj} = 0 \quad \text{at } z = z_L \ ,
\end{aligned} \tag{C.29}
$$

and

$$
\begin{aligned}
&\mathcal{C}_{Rj}(z;\lambda,-c,\tilde{m}) = \mathcal{C}_{Lj}(z;\lambda,c,\tilde{m}) \ ,\\
&\mathcal{S}_{Rj}(z;\lambda,-c,\tilde{m}) = -\mathcal{S}_{Lj}(z;\lambda,c,\tilde{m}) \ ,\\
&\mathcal{C}_{R/Lj}(z;\lambda,c,-\tilde{m}) = (-1)^{j-1}\,\mathcal{C}_{R/Lj}(z;\lambda,c,\tilde{m}) \ ,\\
&\mathcal{S}_{R/Lj}(z;\lambda,c,-\tilde{m}) = (-1)^{j-1}\,\mathcal{S}_{R/Lj}(z;\lambda,c,\tilde{m}) \ .
\end{aligned} \tag{C.30}
$$

The following relations are useful.

$$
\begin{aligned}
&\mathcal{S}_{L1}\mathcal{S}_{R1} - \mathcal{S}_{L2}\mathcal{S}_{R2} + 2 = \mathcal{C}_{L1}\mathcal{C}_{R1} - \mathcal{C}_{L2}\mathcal{C}_{R2} - 2\\
&\qquad = S_L^+S_R^- + S_L^-S_R^+ + C_L^+C_R^- + C_L^-C_R^+ \equiv \mathcal{B}_0(z;\lambda,c,\tilde{m}) \ ,\\
&\mathcal{S}_{L1}\mathcal{C}_{L1} - \mathcal{S}_{L2}\mathcal{C}_{L2} = 2(S_L^+C_L^- + S_L^-C_L^+) \equiv \mathcal{B}_1(z;\lambda,c,\tilde{m}) \ ,\\
&\mathcal{S}_{R1}\mathcal{C}_{R1} - \mathcal{S}_{R2}\mathcal{C}_{R2} = 2(S_R^+C_R^- + S_R^-C_R^+) \ ,\\
&\mathcal{S}_{R1}\mathcal{C}_{L1} - \mathcal{S}_{L2}\mathcal{C}_{R2} = 2(S_R^+C_L^- + S_R^-C_L^+) \ ,\\
&\mathcal{S}_{L1}\mathcal{C}_{R1} - \mathcal{S}_{R2}\mathcal{C}_{L2} = 2(S_L^+C_R^- + S_L^-C_R^+) \ ,\\
&\mathcal{S}_{R1}\mathcal{C}_{L2} - \mathcal{S}_{L2}\mathcal{C}_{R1} = -\mathcal{S}_{L1}\mathcal{C}_{R2} + \mathcal{S}_{R2}\mathcal{C}_{L1}\\
&\qquad = S_L^+S_R^- - S_L^-S_R^+ - C_L^+C_R^- + C_L^-C_R^+ \ ,\\
&S_{L/R}^{\pm} = S_{L/R}(z;\lambda,c\pm\tilde{m}) \ ,\\
&C_{L/R}^{\pm} = C_{L/R}(z;\lambda,c\pm\tilde{m}) \ .
\end{aligned} \tag{C.31}
$$

C.3 Modified Bessel functions

In evaluating the effective potential $V_{\rm eff}(\theta_H)$ one needs the modified Bessel functions $I_\nu(z)$ and $K_\nu(z)$ obtained by analytic continuation of $J_\nu(z)$ and $Y_\nu(z)$;

$$\begin{aligned} I_\nu(z) &= e^{-i\nu\pi/2}\, J_\nu(e^{i\pi/2}z) \quad (-\pi < \arg z < \tfrac{1}{2}\pi) \\ &= \Big(\frac{z}{2}\Big)^\nu \sum_{n=0}^\infty \frac{(z^2/4)^n}{n!\,\Gamma(\nu+n+1)} \quad (\arg z \neq \pi), \\ K_\nu(z) &= \frac{\pi}{2}\frac{I_{-\nu}(z)-I_\nu(z)}{\sin\nu\pi}\ . \end{aligned} \tag{C.32}$$

For sufficiently large $|z|$

$$\begin{aligned} I_\nu(z) &\sim \frac{1}{\sqrt{2\pi z}}\, e^z\ , \\ K_\nu(z) &\sim \sqrt{\frac{\pi}{2z}}\, e^{-z}\ . \end{aligned} \tag{C.33}$$

The analytic continuation of the function $F_{\alpha,\beta}(u,v)$ in (C.8) is

$$\begin{aligned} F_{\alpha,\beta}(iu,iv) &= -\frac{2}{\pi}\, e^{i(\alpha-\beta)\pi/2}\, \hat{F}_{\alpha,\beta}(u,v)\ , \\ \hat{F}_{\alpha,\beta}(u,v) &= I_\alpha(u)K_\beta(v) - e^{-i(\alpha-\beta)\pi}\, K_\alpha(u)I_\beta(v)\ . \end{aligned} \tag{C.34}$$

In terms of $\hat{F}_{\alpha,\beta}(u,v)$ we define

$$\begin{aligned} \hat{C}(q) &= q\hat{F}_{1,0}(qz_L^{-1},q)\ , \\ \hat{S}(q) &= iqz_L^{-1}\hat{F}_{1,1}(qz_L^{-1},q)\ , \\ \hat{C}'(q) &= q^2z_L^{-1}\hat{F}_{0,0}(qz_L^{-1},q)\ , \\ \hat{S}'(q) &= -iq^2z_L^{-2}\hat{F}_{0,1}(qz_L^{-1},q) \end{aligned} \tag{C.35}$$

for gauge fields. For fermion fields with $c>0$ we define

$$\begin{aligned} \hat{C}_L(q;c) &= qz_L^{-1/2}\hat{F}_{c+\frac{1}{2},c-\frac{1}{2}}(qz_L^{-1},q)\ , \\ \hat{S}_L(q;c) &= iqz_L^{-1/2}\hat{F}_{c+\frac{1}{2},c+\frac{1}{2}}(qz_L^{-1},q)\ , \\ \hat{C}_R(q;c) &= qz_L^{-1/2}\hat{F}_{c-\frac{1}{2},c+\frac{1}{2}}(qz_L^{-1},q)\ , \\ \hat{S}_R(q;c) &= -iqz_L^{-1/2}\hat{F}_{c-\frac{1}{2},c-\frac{1}{2}}(qz_L^{-1},q)\ . \end{aligned} \tag{C.36}$$

For $c<0$, we use the relations

$$\hat{C}_L(q;-c) = \hat{C}_R(q;c)\ , \quad \hat{S}_L(q;-c) = -\hat{S}_R(q;c)\ . \tag{C.37}$$

It is easy to confirm that

$$\begin{aligned}
&S_L(1;iqz_L^{-1},c)S_R(1;iqz_L^{-1},c) = \hat{S}_L(q;c)\hat{S}_R(q;c) \ , \\
&C_R(1;iqz_L^{-1},c)S_R(1;iqz_L^{-1},c) = \hat{C}_R(q;c)\hat{S}_R(q;c) \ , \\
&\mathcal{B}_0(1;iqz_L^{-1},c,\tilde{m}) = \hat{\mathcal{B}}_0(q,c,\tilde{m}) \ , \\
&\mathcal{B}_1(1;iqz_L^{-1},c,\tilde{m}) = \hat{\mathcal{B}}_1(q,c,\tilde{m}) \ ,
\end{aligned} \tag{C.38}$$

where

$$\begin{aligned}
&\hat{\mathcal{B}}_0(q;c,\tilde{m}) = (\hat{S}_L^+\hat{S}_R^- + \hat{S}_L^-\hat{S}_R^+ + \hat{C}_L^+\hat{C}_R^- + \hat{C}_L^-\hat{C}_R^+)(q;c,\tilde{m}) \ , \\
&\hat{\mathcal{B}}_1(q;c,\tilde{m}) = 2(\hat{S}_L^+\hat{C}_L^- + \hat{S}_L^-\hat{C}_L^+)(q;c,\tilde{m}) \ , \\
&\begin{pmatrix} \hat{S}_{L/R}^{\pm} \\ \hat{C}_{L/R}^{\pm} \end{pmatrix}(q;c,\tilde{m}) = \begin{pmatrix} \hat{S}_{L/R} \\ \hat{C}_{L/R} \end{pmatrix}(q;c\pm\tilde{m}) \ .
\end{aligned} \tag{C.39}$$

We also note

$$\frac{1}{\lambda}S(1;\lambda)C'(1;\lambda)\Big|_{\lambda=iqz_L^{-1}} = \hat{S}_L(q;\tfrac{1}{2})\hat{S}_R(q;\tfrac{1}{2}) \ . \tag{C.40}$$

Bibliography

[1] C.N. Yang and R.L. Mills, Conservation of Isotopic Spin and Isotopic Gauge Invariance, *Phys. Rev.* **96**, 191 (1954).

[2] Y. Nambu, Quasiparticles and Gauge Invariance in the Theory of Superconductivity, *Phys. Rev.* **117**, 648 (1960).

[3] Y. Nambu and G. Jona-Lasinio, Dynamical Model of Elementary Particles Based on an Analogy with Superconductivity. I, *Phys. Rev.* **122**, 345 (1961).

[4] J. Goldstone, Field Theories with ⟨⟨ Superconductor ⟩⟩ Solutions, *Nuovo Cimento* **19**, 154 (1961).

[5] P.W. Higgs, Broken Symmetries, Massless Particles and Gauge Fields, *Phys. Lett.* **12**, 132 (1964); Broken Symmetries and the Masses of Gauge Bosons, *Phys. Rev. Lett.* **13**, 508 (1964).

[6] F. Englert and R. Brout, Broken Symmetry and the Mass of Gauge Vector Mesons, *Phys. Rev. Lett.* **13**, 321 (1964).

[7] S.L. Glashow, Partial-Symmetries of Weak Interactions, *Nucl. Phys.* **22**, 579 (1961).

[8] S. Weinberg, A Model of Leptons, *Phys. Rev. Lett.* **19**, 1264 (1967).

[9] A. Salam, Weak and Electromagnetic Interactions, in *Elementary Particle Physics*, Nobel Symp No. 8. (1969).

[10] ATLAS Collaboration, Observation of a New Particle in Search for the Standard Model Higgs Boson with the ATLAS Detector at the LHC, *Phys. Lett.* **B716**, 1 (2012).

[11] CMS Collaboration, Observation of a New Boson at a Mass of 125 GeV with the CMS Experiment at the LHC, *Phys. Lett.* **B716**, 30 (2012).

[12] Y. Hosotani, Dynamical Mass Generation by Compact Extra Dimensions, *Phys. Lett.* **B126**, 309 (1983).

[13] Y. Aharonov and D. Bohm, Significance of Electromagnetic Potentials in the Quantum Theory, *Phys. Rev.* **115**, 485 (1959).

[14] Y. Hosotani, Dynamics of Nonintegrable Phases and Gauge Symmetry Breaking, *Ann. Phys. (N.Y.)* **190**, 233 (1989).

[15] S. Coleman, Secret Symmetry: An Introduction to Spontaneous Symmetry Breakdown and Gauge Fields, in S. Coleman, *Aspects of Symmetry*, Cambridge University Press, (1985).

[16] S. Coleman and E. Weinberg, Radiative Corrections as the Origin of Spontaneous Symmetry Breaking, *Phys. Rev.* **D7**, 1888 (1973).
[17] R. Jackiw, Functional Evaluation of the Effective Potential, *Phys. Rev.* **D9**, 1686 (1974).
[18] K. Shiraishi, Finite Temperature and Density Effect on Symmetry Breaking by Wilson Loops, *Z. Phys.* **C35**, 37 (1987).
[19] A. T. Davies and A. McLachlan, Gauge Group Breaking by Wilson Loops, *Phys. Lett.* **B200**, 305 (1988); Congruency Class Effects in the Hosotani Model, *Nucl. Phys.* **B317**, 237 (1989).
[20] J.E. Hetrick and C-L. Ho, Dynamical Symmetry Breaking from Toroidal Compactification, *Phys. Rev.* **D40**, 4085 (1989).
[21] A. McLachlan, Flux-Breaking in Space-times with Toroidal Compactification, *Nucl. Phys.* **B338**, 188 (1990).
[22] C-L. Ho and Y. Hosotani, Symmetry Breaking by Wilson Lines and Finite-Temperature Effects, *Nucl. Phys.* **B345**, 445 (1990).
[23] H. Hatanaka, Matter Representations and Gauge Symmetry Breaking via Compactified Space, *Prog. Theor. Phys.* **102**, 407 (1999).
[24] K. Takenaga, Gauge Symmetry Breaking through the Hosotani Mechanism in Softly Broken Supersymmetric QCD, *Phys. Rev.* **D66**, 085009 (2002).
[25] Y. Hosotani, Dynamical Gauge Symmetry Breaking by Wilson Lines in the Electroweak Theory, *Proc. 2004 Int. Workshop, on Dynamical Symmetry Breaking*, page 17, Nagoya University, (2004). (arXiv:hep-ph/0504272)
[26] Y. Hosotani, N. Maru, K. Takenaga and T. Yamashita, Two Loop Finiteness of Higgs Mass and Potential in the Gauge-Higgs Unification, *Prog. Theor. Phys.* **118**, 1053 (2007).
[27] A. Yamada, The UV Sensitivity of the Higgs Potential in Gauge-Higgs Unification, *Prog. Theor. Exp. Phys.* **2021**, 093B01 (2021).
[28] G. Cossu, H. Hatanaka, Y. Hosotani and J. Noaki, Polyakov Loops and the Hosotani Mechanism on the Lattice, *Phys. Rev.* **D89**, 094509 (2014).
[29] G. Cossu and M. D'Elia, Finite Size Phase Transitions in QCD with Adjoint Fermions, *J. High Energy Phys.* **07**, 048 (2009).
[30] A. Pastor-Gutierrez and M. Yamada, Phase Structure of Extra-dimensional Gauge theories with Fermions, *Phys. Rev.* **D109**, 076018 (2024).
[31] S. Elitzur, Impossibility of Spontaneously Breaking Local Symmetries, *Phys. Rev.* **D12**, 3978 (1975).
[32] H. Georgi and S.L. Glashow, Unity of All Elementary-Particle Forces, *Phys. Rev. Lett.* **32**, 438 (1974).
[33] A. Pomarol and M. Quiros, The Standard Model from Extra Dimensions, *Phys. Lett.* **B438**, 255 (1998).
[34] N. Haba, M. Harada, Y. Hosotani and Y. Kawamura, Dynamical Rearrangement of Gauge Symmetry on the Orbifold S^1/Z_2, *Nucl. Phys.* **B657**, 169 (2003); *Erratum, ibid.* **669**, 381 (2003).
[35] N. Haba, Y. Hosotani and Y. Kawamura, Classification and Dynamics of Equivalence Classes in SU(N) Gauge Theory on the Orbifold S^1/Z_2, *Prog. Theoret. Phys.* **111**, 265 (2004).
[36] K. Takeuchi and T. Inagaki, New Classification Method for Equivalence

Classes on S^1/Z_2 and T^2/Z_3 Orbifolds, *Prog. Theoret. Exp. Phys.* **2024**, 033B03 (2024).

[37] T. Appelquist, H.C. Cheng and B.A. Dobrescu, Bounds on Universal Extra Dimensions, *Phys. Rev.* **D64**, 035002 (2001).

[38] Y. Kawamura, Gauge Symmetry Reduction from the Extra Space S^1/Z_2, *Prog. Theoret. Phys.* **103**, 613 (2000); Triplet-Doublet Splitting, Proton Stability and an Extra Dimension, *Prog. Theoret. Phys.* **105**, 999 (2001).

[39] M. Kubo, C.S. Lim and H. Yamashita, The Hosotani Mechanism in Bulk Gauge Theories with an Orbifold Extra Space S^1/Z_2, *Mod. Phys. Lett.* **A17**, 2249 (2002).

[40] I. Antoniadis, K. Benakli and M. Quiros, Finite Higgs Mass without Supersymmetry, *New J. Phys.* **3**, 20 (2001).

[41] G. Burdman and Y. Nomura, Unification of Higgs and Gauge Fields in Five Dimensions, *Nucl. Phys.* **B656**, 3 (2003).

[42] C. Csaki, C. Grojean and H. Murayama, Standard Model Higgs from Higher Dimensional Gauge Fields, *Phys. Rev.* **D67**, 085012 (2003).

[43] C.A. Scrucca, M. Serone and L. Silvestrini, Electroweak Symmetry Breaking and Fermion Masses from Extra Dimensions, *Nucl. Phys.* **B669**, 128 (2003).

[44] Y. Hosotani, S. Noda and K. Takenaga, Dynamical Gauge-Higgs Unification in the Electroweak Theory, *Phys. Lett.* **B607**, 276 (2005).

[45] G. Cacciapaglia, C. Csaki and S.C. Park, Fully Radiative Electroweak Symmetry Breaking, *J. High Energy Phys.* **0603**, 099 (2006).

[46] M. Serone, Holographic Methods and Gauge-Higgs Unification in Flat Extra Dimensions, *New J. Phys.* **12**, 075013 (2010).

[47] A. Falkowski, Holographic Pseudo-Goldstone Boson, *Phys. Rev.* **D75**, 025017 (2007).

[48] Y. Hosotani and Y. Sakamura, Anomalous Higgs Couplings in the $SO(5) \times U(1)$ Gauge-Higgs Unification in Warped Spacetime, *Prog. Theoret. Phys.* **118**, 935 (2007).

[49] L. Randall and R. Sundrum, A Large Mass Hierarchy from a Small Extra Dmension, *Phys. Rev. Lett.* **83**, 3370 (1999).

[50] T. Gherghetta and A. Pomarol, Bulk Fields and Supersymmetry in a Slice of AdS, *Nucl. Phys.* **B586**, 141 (2000).

[51] K. Agashe, R. Contino and A. Pomarol, The Minimal Composite Higgs Model, *Nucl. Phys.* **B719**, 165 (2005).

[52] A. D. Medina, N. R. Shah and C. E. M. Wagner, Gauge-Higgs Unification and Radiative Electroweak Symmetry Breaking in Warped Extra Dimensions, *Phys. Rev.* **D76**, 095010 (2007).

[53] Y. Hosotani, K. Oda, T. Ohnuma and Y. Sakamura, Dynamical Electroweak SymmetryBreaking in $SO(5) \times U(1)$ Gauge-Higgs Unification with Top and Bottom Quarks, *Phys. Rev.* **D78**, 096002 (2008); Erratum *ibid.* **D79**, 079902 (2009).

[54] S. Funatsu, H. Hatanaka, Y. Hosotani, Y. Orikasa and T. Shimotani, Novel Universality and Higgs Decay $H \to \gamma\gamma, gg$ in the $SO(5) \times U(1)$ Gauge-Higgs Unification, *Phys. Lett.* **B722**, 94 (2013).

[55] J. Yoon and M. E. Peskin, *"Dissection of an $SO(5) \times U(1)$ Gauge-Higgs Unification Model"*, *Phys. Rev.* **D100**, 015001 (2019).

[56] S. Funatsu, H. Hatanaka, Y. Hosotani, Y. Orikasa and N. Yamatsu, GUT Inspired $SO(5) \times U(1) \times SU(3)$ Gauge-Higgs Unification, *Phys. Rev.* **D99**, 095010 (2019).

[57] S. Funatsu, H. Hatanaka, Y. Hosotani, Y. Orikasa and N. Yamatsu, CKM Matrix and FCNC Suppression in $SO(5) \times U(1) \times SU(3)$ Gauge-Higgs Unification, *Phys. Rev.* **D101**, 055016 (2020).

[58] S. Funatsu, H. Hatanaka, Y. Hosotani, Y. Orikasa and N. Yamatsu, Effective Potential and Universality in GUT-inspired Gauge-Higgs Unification, *Phys. Rev.* **D102**, 015005 (2020).

[59] Y. Hosotani and N. Yamatsu, Gauge-Higgs Grand Unification, *Prog. Theoret. Exp. Phys.* **2015**, 111B01 (2015).

[60] Y. Hosotani and N. Yamatsu, Gauge-Higgs Seesaw Mechanism in 6-Dimensional Grand Unification, *Prog. Theoret. Exp. Phys.* **2017**, 091B01 (2017).

[61] P. Minkowski, $\mu \to e\gamma$ at a Rate of One out of 10^9 Muon Decays?, *Phys. Lett.* **B67**, 421 (1977).

[62] T. Yanagida, Horizontal Gauge Symmetry and Masses of Neutrinos, in *Proceedings of Workshop on Unified Theory and Baryon Number of the Universe*, edited by O. Sawada and A. Sugamoto (KEK, Japan, 1979).

[63] M. Gell-Mann, P. Ramond and R. Slansky, Complex Spinors and Unified Theories, in *Supergravity*, edited by P. van Nieuwenhuizen and D. Freedman (North-Holland, Amsterdam, 1980) p. 317.

[64] R.N. Mohapatra, Mechanism for Understanding Small Neutrino Mass in Superstring Theories, *Phys. Rev. Lett.* **56**, 561 (1986).

[65] R.N. Mohapatra and J.W.F. Valle, Neutrino Mass and Baryon-number Nonconservation in Superstring Models, *Phys. Rev.* **D34**, 1642 (1986).

[66] K. Oda and A. Weiler, Wilson Lines in Warped Space: Dynamical Symmetry Breaking and Restoration, *Phys. Lett.* **B606**, 408 (2005).

[67] R.L. Workman et al. (Particle Data Group), Reviews, Tables & Plots, *Prog. Theoret. Exp. Phys.* **2022**, 083C01 (2022).

[68] J. Haller, A. Hoecker, R. Kogler, K. Mönig and J. Stelzer, Status of the Global Electroweak fit with Gfitter in the Light of New Precision Measurements, *Proc. Sci. ICHEP2022* **2022**, 897 (2022).

[69] M. Awramik, M. Czakon, A. Freitas and G. Weiglein, Precise Prediction for the W-boson Mass in the Standard Model, *Phys. Rev.* **D69**, 053006 (2004); *arXiv* hep-ph/0311148v3 (8 Nov 2021).

[70] J. de Blas, M. Pierini, L. Reina and L. Silvestrini, Impact of the Recent Measurements of the Top-quark and W-boson Masses on electroweak precision fits, *Phys. Rev. Lett.* **129**, 271801 (2022).

[71] Y. Hosotani, S. Funatsu, H. Hatanaka, Y. Orikasa and N. Yamatsu, W Boson Mass in Gauge-Higgs Unification, *Phys. Rev.* **D108**, 115036 (2023).

[72] T. Aaltonen et al. (CDF Collaboration), High-precision Measurement of the W Boson Mass with the CDF II Detector, *Science* **376**, 170 (2022).

[73] The ATLAS Collaboration, Improved W Boson Mass Measurement Using

$\sqrt{s} = 7\,\text{TeV}$ Proton-proton Collisions with the ATLAS Detector, Report No. ATLAS-CONF-2023-004.

[74] S. Funatsu, H. Hatanaka, Y. Hosotani, Y. Orikasa and N. Yamatsu, Signals of W' and Z' Bosons at the LHC in the $SU(3)\times SO(5)\times U(1)$ Gauge-Higgs Unification, *Phys. Rev.* **D105**, 055015 (2022).

[75] J. F. Gunion, H. E. Haber, G. L. Kane and S. Dawson, The Higgs Hunter's Guide, *Front. Phys.* **80**, 1 (2000).

[76] S. Funatsu, H. Hatanaka, Y. Hosotani and Y. Orikasa, Distinct Signals of the Gauge-Higgs Unification in e^+e^- Collider Experiments, *Phys. Lett.* **B775**, 297 (2017).

[77] J. Yoon and M. E. Peskin, Fermion Pair Production in $SO(5)\times U(1)$ Gauge-Higgs Unification Models, arXiv:1811.07877 [hep-ph].

[78] S. Funatsu, Forward-backward Asymmetry in the Gauge-Higgs Unification at the International Linear Collider, *Eur. Phys. J.* **C79**, 854 (2019).

[79] S. Funatsu, H. Hatanaka, Y. Hosotani, Y. Orikasa and N. Yamatsu, Fermion Pair Production at e^-e^+ Linear Collider Experiments in GUT Inspired Gauge-Higgs Unification, *Phys. Rev.* **D102**, 015029 (2020).

[80] A. Irles, J.P. Márquez, R. Pöschl, F. Richard, A. Saibel, H. Yamamoto and N. Yamatsu, Probing Gauge-Higgs Unification Models at the ILC with Quark-antiquark Forward-backward Asymmetry at Center-of-mass Energies above the Z Mass, *Eur. Phys. J.* **C84**, 537 (2024).

[81] J. M. Cornwall, D. N. Levin and G. Tiktopoulos, Derivation of Gauge Invariance from High-Energy Unitarity Bounds on the S Matrix, *Phys. Rev.* **D10**, 1145 (1974); Erratum, *Phys. Rev.* **D11**, 972 (1975).

[82] M. S. Chanowitz and M. K. Gaillard, The TeV Physics of Strongly Interacting W's and Z's, *Nucl. Phys.* **B261**, 379 (1985).

[83] M. E. Peskin and D. V. Schroeder, *An Introduction to Quantum Field Theory*, Chapter 21, ISBN 0-201-50397-2, Addison-Wesley Publishing Company (1995).

[84] N. Yamatsu, S. Funatsu, H. Hatanaka, Y. Hosotani and Y. Orikasa, W and Z Boson Pair Production at Electron-positron Colliders in Gauge-Higgs Unification, *Phys. Rev.* **D108**, 115014 (2023).

[85] S. Funatsu, H. Hatanaka, Y. Hosotani and Y. Orikasa, Collider Signals of W' and Z' Bosons in the Gauge-Higgs Unification, *Phys. Rev.* **D95**, 035032 (2017).

[86] M.E. Peskin and T. Takeuchi, New Constraint on a Strongly Interacting Higgs Sector, *Phys. Rev. Lett.* **65**, 964 (1990).

[87] M.E. Peskin and T. Takeuchi, Estimation of Oblique Electroweak Corrections, *Phys. Rev.* **D46**, 381 (1992).

[88] L. Lavoura and J.P. Silva, Oblique Corrections from Vectorlike Singlet and Doublet Quarks, *Phys. Rev.* **D47**, 2046 (1993).

[89] Y. Hosotani, S. Funatsu, H. Hatanaka, Y. Orikasa and N. Yamatsu, Coupling Sum Rules and Oblique Corrections in Gauge-Higgs Unification, *Prog. Theoret. Exp. Phys.* **2023**, 063B01 (2023).

[90] S.L. Adler, Axial Vector Vertex in Spinor Electrodynamics, *Phys. Rev.* **177**, 2426 (1969).

[91] J.S. Bell and R. Jackiw, A PCAC Puzzle: $\pi^0 \to \gamma\gamma$ in the σ Model, *Nuovo Cim.* **A60**, 47 (1969).

[92] K. Fujikawa, Path-integral Measure for Gauge-invariant Fermion Theories, *Phys. Rev. Lett.* **42**, 1195 (1979).

[93] K. Fujikawa, Path Integral for Gauge Theories with Fermions, *Phys. Rev.* **D21**, 2848 (1980).

[94] C. Bouchiat, J. Iliopoulos and Ph. Meyer, An Anomaly-free Version of Weinberg's Model, *Phys. Lett.* **B38**, 519 (1972).

[95] D.J. Gross and R. Jackiw, Effects of Anomalies on Quasi-renormalizable Theories, *Phys. Rev.* **D6**, 477 (1972).

[96] S. Funatsu, H. Hatanaka, Y. Hosotani, Y. Orikasa and N. Yamatsu, Anomaly Flow by an Aharonov-Bohm Phase, *Prog. Theoret. Exp. Phys.* **2022**, 043B04 (2022).

[97] Y. Hosotani, Universality in Anomaly Flow, *Prog. Theoret. Exp. Phys.* **2022**, 073B01 (2022).

[98] C.S. Lim and N. Maru, Towards a Realistic Grand Gauge-Higgs Unification, *Phys. Lett.* **B653**, 320 (2007).

[99] N. Maru, H. Takahashi and Y. Yatagai, Gauge Coupling Unification in Simplified Grand Gauge-HIggs Unification, *Phys. Rev.* **D106**, 055033 (2022).

[100] N. Maru and R. Nago, New Models of $SU(6)$ Grand Gauge-Higgs Unification, arXiv:2405.07463.

[101] A. Angelescu, A. Bally, S. Blasi and F. Goertz, Minimal $SU(6)$ Gauge-Higgs Grand Unification, *Phys. Rev.* **D105**, 035026 (2022).

[102] A. Angelescu, A. Bally, F. Goertz and S. Weber, $SU(6)$ Gauge-Higgs Grand Unification: Minimal Viable Models and Flavor, *J. High Energy Phys.* **04**, 012 (2023).

[103] K. Kojima, K. Takenaga and T. Yamashita, Grand Gauge-Higgs Unification, *Phys. Rev.* **D84**, 051701(R) (2011).

[104] K. Kojima, K. Takenaga and T. Yamashita, Gauge Symmetry Breaking Patterns in an $SU(5)$ Grand Gauge-Higgs Unification Model, *Phys. Rev.* **D95**, 015021 (2017).

[105] A. Furui, Y. Hosotani and N. Yamatsu, Toward Realistic Gauge-Higgs Grand Unification, *Prog. Theoret. Exp. Phys.* **2016**, 093B01 (2016).

[106] N. Yamatsu, Gauge Coupling Unification in Gauge-Higgs Grand Unification, *Prog. Theoret. Exp. Phys.* **2016**, 043B02 (2016).

[107] Y. Hosotani and N. Yamatsu, Electroweak Symmetry Breaking and Mass Spectra in Six-dimensional Gauge-Higgs Grand Unification, *Prog. Theoret. Exp. Phys.* **2018**, 023B05 (2018).

[108] C. Englert, D.J. Miller and D.D. Smaranda, Phenomenology of GUT-inspired Gauge-Higgs Unification, *Phys. Lett.* **B802**, 135261 (2020).

[109] C. Englert, D.J. Miller and D.D. Smaranda, The Weinberg Angle and 5D RGE Effects in a $SO(11)$ GUT Theory, *Phys. Lett.* **B807**, 135548 (2020).

[110] H. Hatanaka and Y. Hosotani, SUSY Breaking Scales in the Gauge-Higgs Unification, *Phys. Lett.* **B713**, 481 (2012).

Index

$A^{\hat{4}}$ tower, 163
$M^n \times S^1$, 25
R_ξ gauge, 29
S, T, U parameters, 211
$SO(11)$ unification, 224
$SO(N)$, 230
$SU(3)$ gauge-Higgs unification, 75
$SU(6)$ unification, 223
$SU(N)$, 236
W boson, 4, 17
W boson mass, 191
W couplings, 178
W tower, 129, 158
W' bosons, 199
W_R couplings, 181
W_R tower, 129, 158
Z boson, 4, 17
Z couplings, 182
Z tower, 130, 160
Z' bosons, 199
Z' interference effect, 201
Z_R couplings, 184
Z_R tower, 130, 160
γ couplings, 181
γ tower, 130, 160
μ decay, 192
θ_H universality, 153, 154, 156, 158
1-loop effective potential, 36

A model, 117
AB phase, 68, 115, 214
Aharonov-Bohm (AB) phase, 25, 28, 51
Aharonov-Bohm effect, 4, 23, 24
anomaly coefficient, 216, 217
anomaly flow, 213, 218, 219
anomaly formula, 219, 221

B model, 118
bare weak mixing angle, 193
bare weak mixing angle θ_W^0, 130
basis functions in the RS space, 248
Bessel equation, 245
Bessel functions, 245
bispinor representation, 234, 235
brane interactions, 121
brane scalar, 121
brane scalar field, 225
bulk mass parameter, 101, 249

Cabbibo-Kobayashi-Maskawa matrix, 21
cancellation of gauge anomalies, 213, 221
charged lepton, 138
charged lepton tower, 166
chiral anomalies, 216
chiral fermion, 63
chiral fermion problem, 60
chiral structure, 170, 171
Clifford algebra, 230, 232
conformal coordinate, 89

cosmological constant, 86
coupling sum rule, 207, 208
covariant derivative, 9, 19, 100
cubic and quartic self-couplings, 199
cubic couplings, 174
curvature 2-form, 86, 100

dark fermion F tower, 168
dark fermion V tower, 168
dark fermions, 124, 141
Debye screening, 48
Dirac algebra, 230
Dirac equation, 59, 229
Dirac matrices, 230
Dirichlet condition, 62, 102
distribution density of AB phases, 50
distribution density of gauge invariant AB phases, 56
doublet-triplet splitting problem, 73, 74
down-type quark, 135
down-type quark tower, 165
dynamical gauge symmetry breaking, 38

effective action, 35
effective Dirichlet condition, 123
effective interactions of the Higgs field, 157
effective potential, 34, 35, 146, 237
effective potential for AB phases, 37
effective potential in field theory, 34
effective potential in quantum mechanics, 33
Einstein equations, 86
electroweak gauge theory, 3
electroweak symmetry breaking, 146
Elitzur's theorem, 55
equivalence classes of boundary conditions, 44, 69

Fadeev-Popov quantization, 29
Fermi constant, 191
Fermion fields in the RS space, 101
fermion zero modes, 170
field strength, 12
field theory at finite temperature, 47
fifth dimension, 4
finiteness of $V_{\rm eff}(\theta_H)$, 45
finiteness of $V_{\rm eff}^{\rm 1\,loop}(\theta_H)$, 243
fixed points, 61
flavor changing neutral currents (FCNC), 21
forward-backward asymmetry, 196, 205

gauge anomalies, 213
gauge eigenstates, 20
gauge fields in the RS space, 93
gauge hierarchy problem, 21, 47, 110
gauge interactions, 16
gauge invariance, 9
gauge symmetry breaking, 28
gauge theory, 1
gauge-fixing function, 29
gauge-Higgs couplings, 186
gauge-Higgs grand unification, 222
gauge-Higgs unification (GHU), 5, 110
gauge-invariant gauge symmetry breaking, 52
generalized zeta(ζ)-function, 237
ghost term, 29
Glashow-Salam-Weinberg model, 17
grand unified theory, 57
GUT inspired GHU, 118, 153

Haar measure, 52
Higgs boson, 4, 17, 115, 133, 196
Higgs field, 17, 110, 116, 148
Higgs mechanism, 3, 7, 10, 14, 65
Higgs modes, 42
Higgs tower, 163
holography in anomaly flow, 221
Hosotani mechanism, 4, 23, 28, 31, 33, 38, 41, 56, 110

infinite sum in flat space, 239
integral representation of $V_{\rm eff}(\theta_H)$, 240
interference term, 203
IR brane, 88

Kaluza-Klein (KK) expansion, 29
Kawamura model, 73
KK expansion, 59, 62, 64, 71, 75, 80, 95, 104, 105
KK mass scale, 92
KK modes, 29
KK parity, 72

Landau, Feynman and unitary gauges, 30
large gauge transformation, 44, 68
Large Hadron Collider (LHC), 4
lattice gauge theory, 49
left-right (LR) asymmetry, 205
leptons, 15, 120
loop expansion ($\hbar$ expansion), 35

Majorana condition, 120
Majorana masses, 120
mass eigenstates, 20
mass hierarchy, 134
mass spectrum of gauge fields, 127
modified Bessel functions, 253

Nambu's mechanism, 3, 14
Nambu-Goldstone (NG) boson, 8, 14
Nambu-Goldstone theorem, 8, 14
natural unit, 230
Neumann condition, 62, 96, 102
neutrino, 139
neutrino tower, 167
non-Abelian gauge theory, 2, 11
normalized distribution density, 52

oblique corrections, 209
oblique parameters, 212
orbifold boundary condition, 66
orbifolds S^1/Z_2, 61

parity, 62, 63
Peskin-Takeuchi, 211
physical gauge symmetry, 70
physical symmetry, 44
Polyakov loop, 50
Pontecorvo-Maki-Nakagawa-Sakata matrix, 21

QCD at finite temperature, 48
quantum chromodynamics (QCD), 2
quantum electrodynamics (QED), 2
quarks, 15, 117
quartic couplings, 176

Randall-Sundrum solution, 87
Randall-Sundrum warped space (RS space), 85, 88, 111
residual gauge invariance, 28
residual gauge symmetry, 68
Ricci tensor, 86

scalar curvature, 86
scalar fields in the RS space, 89
Schrödinger equation, 23
seesaw mechanism, 141
selection rules, 72
self-couplings of gauge fields, 173
self-couplings of the Higgs boson, 199
spin connection, 86, 100
spontaneous gauge symmetry breaking, 4
spontaneous symmetry breaking, 4, 7, 8
standard model (SM), 4, 15, 109
symmetry of boundary condition, 69

tetrad, 85
twisted gauge, 82, 125
twisted gauge in flat space, 81
twisted gauge in the RS space, 125

UED (Universal Extra Dimensions) model, 70
unification of forces, 2
unitarity at the tree level, 207, 209
unitary gauge, 10
universality of the anomaly flow, 221
up-type quark, 133
up-type quark tower, 164
UV brane, 88

warp factor, 89
weak mixing angle, 18

weak mixing angle in the on-shell scheme, 193

Weinberg angle, 18

Wilson line integral, 27

Yang-Mills theory, 2, 11

Yukawa couplings, 20, 188

Yukawa interactions, 17

zero mode, 63, 65, 66, 103, 104

About the Author

Yutaka Hosotani is an Emeritus Professor of Osaka University. He received his Ph.D. (Doctor of Science) from University of Tokyo in 1979, and was awarded the 2007 Nishina Memorial Prize for the discovery of the Hosotani mechanism.

www.ingramcontent.com/pod-product-compliance
Lightning Source LLC
LaVergne TN
LVHW011811200926
848484LV00005B/1143
9789819800971